The Art of Proving Binomial Identities

Discrete Mathematics and Its Applications

Series Editors
Miklos Bona
Donald L. Kreher
Douglas West
Patrice Ossona de Mendez

https://www.crcpress.com/Discrete-Mathematics-and-Its-Applications/book-series/
CHDISMTHAPP?page=1&order=dtitle&size=12&view=list&status=published,forthco
ming

The Art of Proving Binomial Identities

Michael Z. Spivey

CRC Press
Taylor & Francis Group
Boca Raton London New York

CRC Press is an imprint of the
Taylor & Francis Group, an **Informa** business

CRC Press
Taylor & Francis Group
6000 Broken Sound Parkway NW, Suite 300
Boca Raton, FL 33487-2742

First issued in paperback 2022

Version Date: 20190313

ISBN 13: 978-1-03-247558-5 (pbk)
ISBN 13: 978-0-8153-7942-3 (hbk)

DOI: 10.1201/9781351215824

Visit the Taylor & Francis Web site at
http://www.taylorandfrancis.com

and the CRC Press Web site at
http://www.crcpress.com

Contents

Preface

I first ran across the binomial coefficients when studying for a junior high mathematics competition many years ago. At that time it was never clearly explained to me why the number in, say, entry 5 of row 7 of Pascal's triangle was the coefficient of x^5 in the expansion of $(x+1)^7$, nor why the same number told you how many committees of size 5 you could make from 7 people. Even more fundamentally, how in the world could it be that the cute pattern that generates Pascal's triangle has any use at all, let alone two such apparently disparate applications?

As I learned more mathematics, though, I kept seeing the binomial coefficients show up again and again—most often in a combinatorial setting, but also in contexts like probability, infinite series, special functions, and even Leibniz's formula for derivatives. I also started seeing identities satisfied by the binomial coefficients. These were nearly always proved by some algebraic manipulations, but sometimes I saw a combinatorial, probabilistic, or generating function argument. Very occasionally there was something else involved, such as calculus. I found the variety of methods one could use to prove these identities to be fascinating. However, the methods generally showed up in different places. There are a lot of good texts out there that list binomial identities or that discuss some methods for proving them, but I did not see a single text focused on the binomial coefficients that tried to bring together all of these methods that I had come across. So I decided to write one.

This book does not try to list every possible binomial identity in existence. Such a task would be practically impossible. More importantly, the book's focus is not on lists but on techniques. In fact, the book takes the point of view that it's often not enough to prove a binomial identity once; instead, we will prove many identities multiple times, from several different perspectives. An identity represents an idea that is true, and like most truths, you understand them better as you look at them from more angles. Besides the introduction, this book has chapters on basic algebraic manipulations, combinatorial methods, calculus, probability, generating functions, recurrence relations and finite differences, special numbers (such as Fibonacci, Stirling, Bell, Lah, Catalan, and Bernoulli numbers), complex numbers and linear algebra, and a final one on mechanical summation procedures. All of this is what I am calling the *art* of proving binomial identities.

The text does have an index of identities and theorems in the back, however. This is there partly so that the reader who finds a reference to a particular identity can easily find that identity. However, this index serves another important purpose: It makes it easy to find alternative proofs or other uses of a particular identity. For example, the index will point you to seven different ways to prove Identity 21, $\sum_{k=0}^{n} \binom{n}{k} k = n2^{n-1}$. These proofs use algebra, combinatorics, calculus, probability, generating functions, finite differences, and mechanical summation.

The book is aimed at advanced undergraduates or beginning graduate students. I hope that professional mathematicians can find something useful, new, or interesting in here as well.

To the instructor

One of my favorite features of this book is that it brings many different areas of mathematics to bear on a single topic. I used an early draft of this text to teach a special topics class in the spring of 2017, and in a lot of ways it served as a kind of capstone course for the students. They got to see basic algebra, calculus, probability, complex numbers, linear algebra, infinite series, combinatorics, and more all sewn together using the single thread of the binomial coefficients. Any course based on this book could use this breadth as one of its overarching themes.

Many of the exercises in the text ask students to prove identities. There are a lot of these exercises. They're there not because we're interested in giving a comprehensive list of binomial identities but to provide practice—lots of practice—in using the techniques we discuss. There are also many exercises in the book that help to build skills and don't ask for the proof of a binomial identitiy.

In terms of preparation, the students in my course were junior and senior mathematics majors, plus a couple of very strong sophomores. They had all taken the calculus sequence and linear algebra, but beyond that they had chosen a variety of other mathematics courses before signing up for the special topics course.

In general, I would say that the main prerequisite for a course based on this book is the level of mathematical maturity expected of a junior or senior majoring in mathematics. However, some background in thinking probabilistically is very helpful for Chapter 5, and a course in linear algebra is helpful for half of Chapter 9. If the instructor cannot count on students having experience with these topics, this material may be skipped.

Here are some thoughts on specific chapters for a course using this book as its text.

- Chapter 1, the introduction, can be covered quickly. Not all of the different methods for proving the row sum formula (Identity 4) need to be discussed in detail; Proofs 4 and 5 in particular are there to give a taste of material in later chapters.

- The generalization of the binomial coefficient in Chapter 2 is crucial. Binomial inversion keeps showing up, so it should be covered as well. The other algebraic techniques serve as a nice way to help students become comfortable thinking about and proving binomial identities and so should be covered, too.

- Chapter 3, on combinatorial methods, is a lot of fun for the students. The material on inclusion-exclusion and sign-reversing involutions can be challenging for some and so should be gone over carefully. Based on my experience with the special topics class, students do not need to have taken a course in combinatorics to understand the material in this chapter.

- The first two sections of Chapter 4 are easy for the students, since they mostly just involve basic differentiation and integration. The third section on the beta integral and the gamma function is a bit more challenging. The algebra is more involved, and more complicated integration techniques like partial fractions decomposition and trigonometric substitution appear.

- The material in Chapter 5 mostly features discrete probability and is largely self-contained, but students who have taken a course in probability will be used to thinking probabilistically and so will get more out of this chapter. The last two sections, on the probabilistic version of inclusion-exclusion and a probabilistic view of the beta integral, are more challenging and may be skipped. In particular, the last section is the one place in the chapter that requires an understanding of probability on a continuous space.

- The generating function material in Chapter 6 is important and useful, even though the algebra will sometimes be more complicated than anything the students would have seen up to that point. All of it should be covered.

- The recurrence relation material in the first section of Chapter 7 should be covered. (It also requires an understanding of generating functions gained from the previous chapter.) The finite difference material in the second section of Chapter 7 is less important. It should be fine to cover the basic idea that higher-order finite differences involve the binomial coefficients and then pick and choose a couple of the major identities to focus on. The last section is interesting, in that it shows how you could derive the factorial definition of the binomial coefficients without knowing it in advance, but it may be skipped in the interest of time.

- Much of the material in Chapter 8 is combinatorial in flavor, and so in some respects it serves as a more advanced chapter on combinatorial methods. The Fibonacci, Stirling, and Catalan numbers are the most important numbers discussed here; the other sections may be skipped in the interest of time. The section on Catalan numbers contains more advanced lattice path arguments than are presented in Chapter 3.

- The sections on complex numbers and linear algebra in Chapter 9 are independent and so may be covered in either order—or even skipped, depending on time. The material on complex numbers is self-contained, but the material on linear algebra does require some basic familiarity with matrix operations. Determinants and linear independence appear in a few places, but they are not crucial to understanding how to use the primary techniques discussed.

- Chapter 10 contains the powerful Gosper-Zeilberger algorithm (both in Gosper's original form as well as Zeilberger's extension). This algorithm is best automated, but we describe it and give several examples. Hypergeometric series are covered as well. This chapter contains very heavy algebra and can easily be skipped.

Suggested outline for a four-semester-hour course:

- Chapters 1–4

- Sections 5.1–5.3

- Chapter 6

- Section 7.1 and selections (instructor's choice) from 7.2

- Chapters 8–9

Depending on students' probability backgrounds, Chapter 5 may be skipped or gone over more slowly, skipping Section 5.3. Depending on students' linear algebra backgrounds, Section 9.2 may be skipped or gone over more slowly.

Suggested outline for a three-semester-hour course:

- Chapters 1–2

- Sections 3.1–3.4

- Sections 4.1–4.2

- Sections 5.1–5.2

- Chapter 6

- Section 7.1

- Sections 8.1, 8.2, and 8.5

- Section 9.1 or 9.2

Again, depending on students' probability backgrounds, Chapter 5 may be skipped or gone over more slowly. Depending on students' linear algebra backgrounds, Section 9.2 may be skipped or gone over more slowly.

To the student

When I was a student I usually treated my math texts as reference books. I'd look up the details of a theorem to see whether it applied in the proof I was working on, or I would find an example similar to the problem I was struggling with to determine whether it could give insight into where to go next. Some of that was because I was fortunate to have several instructors whose lectures were clear enough that I never had to master the skill of truly reading a math text.

I think I did myself a disservice with that practice, though. Like a novel, a good mathematics text has a narrative. Understanding that narrative gives you the overarching picture of what the text is all about. Learning to read a mathematics text and pick out its narrative is more difficult than learning to read a novel, but it can be done, and it is very much worth it when you're trying to learn mathematics.

This textbook is called *The Art of Proving Binomial Identities*, and that tells you a lot about its narrative. It is about the binomial coefficients, also known as the numbers in Pascal's triangle. These somewhat innocuous-looking numbers satisfy very, very many identities. But why? What makes them so special that this would be the case? Well, the answer partly depends on the angle from which you choose to view the binomial coefficients. So, we're going to look at the binomial coefficients from lots of different angles. We'll look at them algebraically, and combinatorially, and from a calculus standpoint. We'll look at them probabilistically, and tackle them with generating functions, and see how they arise when taking finite differences. We'll look at some relatives of theirs—certain other famous numbers. We'll peer at them through the lenses of complex numbers and linear algebra. Finally, we'll see how you can automate the proofs of a lot of binomial identities. Hopefully, after all of this, you will understand the binomial coefficients in a way you never did before—just like watching a person in a variety of different contexts helps you understand more deeply who that person is.

One of the things that makes reading a mathematics text different from reading many other kinds of books is that if you truly want to understand the ideas involved you have to practice them on your own. So, at the end of each chapter there are a large number of exercises that you may use to practice what you've been reading. Near the back of the book are hints—or in some cases, outright solutions—for every problem in this book. They are there to help you if you get stuck. However, I strongly, strongly urge you to give each problem that you attempt your best try before flipping to the back. I know it's frustrating to be stuck on a problem. (It happens to all of us, even professional mathematicians.) But some of the best learning experiences (and the ones you remember the longest) come from getting yourself unstuck. Figuring something out on your own also gives you a very nice sense of accomplishment, as well as more self-confidence.

But the hints and solutions are there because, again, we all do get stuck. When (if) you do turn to the back, I recommend reading just enough to see where you should go next and ignoring the rest. Then flip back to the page where the problem is, and continue working through it on your own, as much as you can. If you get stuck again, and you're stuck for a while, feel free to look at the hints once more.

At the end of each chapter (except the first) are notes that give additional context to the topics in that chapter or that provide recommended resources for further reading. If some technique particularly "clicks" with you, this is where you could go to learn more.

I hope you enjoy *The Art of Proving Binomial Identities*.

Some words of appreciation

This book would not have been possible without my participation in the mathematics question-and-answer site, Mathematics Stack Exchange, for much of the first year or two of its existence. Many of the proofs in this book come directly or indirectly from ideas I first saw there. Any attempt to thank specific people from Math.SE will leave some out, but I would at least like to thank the following folks: Robin Chapman, Michael Hardy, Will Jagy, Rob Johnson, Jan Mangaldan, Eric Naslund, and Qiaochu Yuan. I have yet to meet any of them in person, and some of them may be unaware that I found inspiration from their contributions to the site, but they have my thanks. Thanks also to my students at the University of Puget Sound who have taken a course or done an independent study based on material in this book: Sam Burdick, Rachel Chaiser, Xeno Fish, Thomas Gagne, Rachel Gilligan, Jiawen (Christy) Li, Eli Moore, Johnny Moore, Maria Ross, Ramsey Rossmann, Chris Spalding, and Cameron Verinsky. I appreciate their feedback and the multiple mistakes they spotted.

Mike Spivey
Tacoma, Washington

1

Introducing the Binomial Coefficients

The infinite collection of integers known as *Pascal's triangle* contains an enormous number of patterns. Let's construct a few rows of this triangle and take a look at some of the patterns we see. First, place a 1 at the top to start the triangle. Then let every other number in the triangle be the sum of the two numbers diagonally above it to the left and right, with the understanding that blank spaces are counted as zeroes. If we do this for several rows, we get the triangle in Figure 1.1.

					1					
				1		1				
			1		2		1			
		1		3		3		1		
	1		4		6		4		1	
1		5		10		10		5		1
1	6		15		20		15		6	1

FIGURE 1.1
Rows 0 through 6 of Pascal's triangle

A quick look at the triangle shows that the two outermost diagonals consist entirely of 1's. Another pattern is that the next-to-outermost diagonals contain the counting numbers $1, 2, 3, 4$, and so forth.

A somewhat more complicated pattern comes from adding up the numbers in each row. Row 0 consists of just the number 1. The sum of row 1 is $1+1 = 2$. The sum of row 2 is $1 + 2 + 1 = 4$, and the sums of rows 3, 4, 5, and 6 are 8, 16, 32, and 64, respectively. If this pattern continues the sum of row 7 would be 128. In general, it looks like the sum of row n is 2^n. How would we know that for certain, though? In other words, how could we prove that it is true?

This book is about just that: proving the truth of patterns in Pascal's triangle. Many of these patterns are beautiful and surprising. Sometimes the proofs are as well—from simple one-line proofs to proofs that look like stories to proofs that make unexpected connections to other parts of mathematics. In fact, as we work our way through the book we'll use tools from areas like algebra, calculus, infinite series, combinatorics, probability, special functions, complex numbers, and linear algebra. The *binomial coefficients*—as the numbers in Pascal's triangle are normally called—will be our common thread winding through and tying together these different fields of mathematics.

The patterns we will be proving will be formalized as *identities*. We'll see a variety of proofs of the same identity, too. Each new proof of an identity (using, say, algebra, or calculus, or a counting argument, or some other technique) will deepen our understanding of why the underlying pattern is true. In this chapter we'll illustrate this idea by giving five different proofs of the pattern we conjectured before: The sum of the numbers in row n of Pascal's triangle really is 2^n.

Before we dive into these proofs, though, we need a symbol for the binomial coefficients. The binomial coefficient in entry k of row n will be denoted $\binom{n}{k}$. (In older texts you may also see $C(n, k)$, but we will consistently use $\binom{n}{k}$ throughout this book.) Also, one important thing to remember when working with Pascal's triangle is that you start counting both entries and rows with 0, not 1.

We also need to give a careful definition of the binomial coefficients. This is a bit trickier than it might seem, even though we have already implicitly defined them through the rule for generating Pascal's triangle. Part of the reason is that there are actually several different definitions of the binomial coefficients. Each of these definitions has its own advantages—both for thinking about problems that can be solved using the binomial coefficients and for proving identities involving the binomial coefficients. In this chapter we will consider four definitions of the binomial coefficients, and we will prove that they are equivalent.

The other reason that defining the binomial coefficients is a bit tricky is that there are different restrictions one can put on the values of n and k in $\binom{n}{k}$. In this chapter, we will start with the strongest set of restrictions normally imposed; namely, that n and k are nonnegative integers with $n \geq k$. In Chapter 2 we will see what happens when we remove some of these restrictions.

French mathematician **Blaise Pascal** (1623–1662) showed an aptitude for mathematics at a young age, publishing a treatise on conic sections while he was still a teenager. Also during his teenage years he constructed one of the first mechanical calculators in order to assist his father, who was tax commissioner in the city of Rouen. As an adult, Pascal contributed to the study of fluids, and (with Pierre de Fermat) he helped lay the foundations for probability theory. Pascal also published on theology and philosophy, of which his argument known as *Pascal's wager* is probably the most famous. The Pascal, a unit of pressure, is named for him. Pascal was in poor health for much of his life and died at the age of 39.

With these caveats out of the way, here are our four definitions of the binomial coefficients when n and k are integers with $n \geq k \geq 0$.

1. *The combinatorial definition:* $\binom{n}{k}$ is the number of subsets of size k that can be formed from a set of n elements. For example, here are the subsets of $\{1, 2, 3, 4\}$.

Subsets of size zero:	$\{\}$
Subsets of size one:	$\{1\}, \{2\}, \{3\}, \{4\}$
Subsets of size two:	$\{1, 2\}, \{1, 3\}, \{1, 4\}, \{2, 3\}, \{2, 4\}, \{3, 4\}$
Subsets of size three:	$\{1, 2, 3\}, \{1, 2, 4\}, \{1, 3, 4\}, \{2, 3, 4\}$
Subsets of size four:	$\{1, 2, 3, 4\}$

   There are 1, 4, 6, 4, and 1 of these, respectively.

2. *The recursive definition:* $\binom{n}{k}$ is the solution in nonnegative integers n and k to the two-variable recurrence

$$\left|\begin{matrix} n \\ k \end{matrix}\right| = \left|\begin{matrix} n-1 \\ k \end{matrix}\right| + \left|\begin{matrix} n-1 \\ k-1 \end{matrix}\right|,$$

   valid for $n - 1 \geq k \geq 1$, with boundary conditions $\left|\begin{smallmatrix} n \\ 0 \end{smallmatrix}\right| = \left|\begin{smallmatrix} n \\ n \end{smallmatrix}\right| = 1$ for all nonnegative integers n. This definition is the one that gives us Pascal's triangle: $\left|\begin{smallmatrix} n-1 \\ k \end{smallmatrix}\right|$ and $\left|\begin{smallmatrix} n-1 \\ k-1 \end{smallmatrix}\right|$ are the numbers diagonally above to the left and right of $\left|\begin{smallmatrix} n \\ k \end{smallmatrix}\right|$.

   For future reference, we'll call this

   Identity 1 (Pascal's Recurrence). *For integers n and k with $n \geq k \geq 0$,*

$$\binom{n}{k} = \binom{n-1}{k} + \binom{n-1}{k-1},$$

   valid for integers $n - 1 \geq k \geq 1$, with boundary conditions $\binom{n}{0} = \binom{n}{n} = 1$ for all nonnegative integers n.

3. *The binomial theorem definition:* $\binom{n}{k}$ is the coefficient of x^k in the expansion of $(x+1)^n$ in powers of x. (This is where $\binom{n}{k}$ gets the name "binomial coefficient.") For example, if we expand $(x+1)^4$ we get $1 + 4x + 6x^2 + 4x^3 + x^4$.

4. *The factorial definition:* $\binom{n}{k} = \frac{n!}{k!(n-k)!}$. For example,

$$\binom{4}{0} = \frac{4!}{0!4!} = 1 \qquad \binom{4}{1} = \frac{4!}{1!3!} = 4 \qquad \binom{4}{2} = \frac{4!}{2!2!} = 6$$

$$\binom{4}{3} = \frac{4!}{1!3!} = 4 \qquad \binom{4}{4} = \frac{4!}{4!0!} = 1.$$

   Also for future reference, we'll call this

Identity 2. *For integers n and k with $n \geq k \geq 0$,*

$$\binom{n}{k} = \frac{n!}{k!(n-k)!}.$$

Why do these four very different definitions give exactly the same set of numbers? We will answer that question by proving that they do. Before we dive into the proofs, though, it is probably worth spending some time discussing the value of having multiple definitions of the binomial coefficients. Each of these definitions, as we alluded to before, has its advantages. For example, the combinatorial definition allows one to think about what quantities are counted by the binomial coefficients, as well as to construct combinatorial arguments explaining binomial identities. The recursive definition lets one generate the binomial coefficients easily, in the form of Pascal's triangle. The binomial theorem definition can be differentiated and integrated, as we shall see in Chapter 4, leading to many other binomial identities. The factorial definition is particularly useful for algebraic manipulation, since factorials obey some nice algebra rules. There are other uses of these definitions, but these examples give some flavor. In mathematics as in life in general, truly coming to understand something often requires viewing it from multiple perspectives. The various definitions help us do that with the binomial coefficients.

On to the proofs of equivalence.

Proof that (1) $\implies$ (2). Suppose we take $\binom{n}{k}$ to be the number of subsets of size k that can be formed from n elements. These can be partitioned into two groups: Those that don't contain the element n, and those that do. The number of subsets that don't contain n is the number of subsets of size k that can be formed from $n-1$ elements, or $\binom{n-1}{k}$. The number that do contain n is the number of subsets of size $k-1$ that can be formed from $n-1$ elements, as you're essentially choosing the other $k-1$ elements to join n out of the other $n-1$ elements from the original set. This can be done in $\binom{n-1}{k-1}$ ways. For the boundary conditions, there is only one set of size zero that can be formed from n elements: the empty set. Similarly, there is only one set of size n that can be formed from n elements: the complete set of all n elements. $\quad\square$

Proof that (2) $\implies$ (3). Next, let's take $\binom{n}{k}$ to be the solution to the recurrence $\left|\begin{smallmatrix}n\\k\end{smallmatrix}\right| = \left|\begin{smallmatrix}n-1\\k\end{smallmatrix}\right| + \left|\begin{smallmatrix}n-1\\k-1\end{smallmatrix}\right|$, with boundary conditions $\left|\begin{smallmatrix}n\\0\end{smallmatrix}\right| = \left|\begin{smallmatrix}n\\n\end{smallmatrix}\right| = 1$ for all nonnegative integers n. Multiply both sides of the recurrence by x^k and sum as k goes from 1 to $n-1$ to get

$$\sum_{k=1}^{n-1} \binom{n}{k} x^k = \sum_{k=1}^{n-1} \binom{n-1}{k} x^k + \sum_{k=1}^{n-1} \binom{n-1}{k-1} x^k. \tag{1.1}$$

Here, we must sum as k goes from 1 to $n-1$ to make sure all the binomial coefficient expressions are defined. (In Chapter 2 we will drop some of these restrictions.)

Now, let's define

$$G(n) = \sum_{k=0}^{n} \binom{n}{k} x^k.$$

Our goal in the rest of this proof is to show that $G(n) = (x+1)^n$.

Reindex the second expression on the right side of Equation (1.1) so that the binomial coefficients match. This gives

$$\sum_{k=1}^{n-1} \binom{n}{k} x^k = \sum_{k=1}^{n-1} \binom{n-1}{k} x^k + \sum_{k=0}^{n-2} \binom{n-1}{k} x^{k+1}.$$

Adding terms to the sums so that we can combine and simplify yields

$$\sum_{k=0}^{n} \binom{n}{k} x^k - \binom{n}{0} x^0 - \binom{n}{n} x^n$$

$$= \sum_{k=0}^{n-1} \binom{n-1}{k} x^k - \binom{n-1}{0} x^0 + x \sum_{k=0}^{n-1} \binom{n-1}{k} x^k - \binom{n-1}{n-1} x^n$$

$$\implies G(n) - 1 - x^n = (1+x) \sum_{k=0}^{n-1} \binom{n-1}{k} x^k - 1 - x^n$$

$$\implies G(n) = (1+x)G(n-1).$$

As recurrences go, $G(n) = (1+x)G(n-1)$ is fairly easy to solve. (We consider more complicated recurrences in Chapter 7.) Substituting $n-1$ for n, we have $G(n-1) = (1+x)G(n-2)$. This means that $G(n) = (1+x)^2 G(n-2)$. Applying this idea further, we obtain $G(n) = (1+x)^3 G(n-3) = (1+x)^4 G(n-4) = \cdots = (1+x)^n G(0)$. Since $G(0) = \binom{0}{0} = 1$, we finally obtain

$$\sum_{k=0}^{n} \binom{n}{k} x^k = G(n) = (x+1)^n.$$

$\square$

In both the previous proof and the next one $(x+1)^n$ plays the role of a *generating function* for the binomial coefficients in row n of Pascal's triangle. By this we mean simply that when $(x+1)^n$ is expanded in powers of x, the coefficient of x^k is $\binom{n}{k}$. Thus the binomial theorem definition of $\binom{n}{k}$ can be also be thought of as the generating function definition of $\binom{n}{k}$. A manipulation on a sequence can often be modeled by a simpler, related manipulation on the corresponding generating function, and this makes generating functions a powerful tool for proving identities (binomial and otherwise). Generating functions form the subject of Chapter 6.

Proof that (3) $\implies$ (4). Now, let $\binom{n}{k}$ be the coefficient of x^k in the expansion of $(x+1)^n$ in powers of x, so that

$$(x+1)^n = \sum_{k=0}^{n} \binom{n}{k} x^k.$$

Let's construct the Maclaurin series for $f(x) = (x+1)^n$. Remember that the Maclaurin series for a function f is given by

$$\sum_{k=0}^{\infty} \frac{f^{(k)}(0)x^k}{k!}.$$

We have $f'(x) = n(x+1)^{n-1}$, $f''(x) = n(n-1)(x+1)^{n-2}$, and in general,

$$f^{(k)}(x) = n(n-1)\cdots(n-k+1)(x+1)^{n-k} = \begin{cases} \dfrac{n!}{(n-k)!}(x+1)^{n-k}, & k \le n; \\ 0, & k > n. \end{cases}$$

This means that the Maclaurin series for $(x+1)^n$ is finite, which means it actually equals $(x+1)^n$. Thus we have

$$(x+1)^n = \sum_{k=0}^{n} \frac{n!}{(n-k)!k!} x^k.$$

This means that

$$\sum_{k=0}^{n} \binom{n}{k} x^k = \sum_{k=0}^{n} \frac{n!}{(n-k)!k!} x^k.$$

However, the only way two polynomials can be equal is for their coefficients to be equal. Therefore,

$$\binom{n}{k} = \frac{n!}{k!(n-k)!}.$$

$\square$

For the last of our four proofs we should distinguish between two important concepts. A *permutation* of a set of n objects is an ordering of those n objects. For example, $4, 1, 3, 2$ and $2, 3, 1, 4$ are different permutations of the set $\{1, 2, 3, 4\}$. A *combination* of k out of n objects is a selection of k of the n objects without regard to order. Thus $4, 1, 3, 2$ and $2, 3, 1, 4$ are considered the same combination of the four numbers 1, 2, 3, and 4. Selecting k out of n objects without regard to order is the same process as selecting a subset of size k from n elements, though. As this is the combinatorial definition of $\binom{n}{k}$, an equivalent way to view $\binom{n}{k}$ is as the number of combinations of k objects that can be formed from n total objects. (This is the reason for the older notation $C(n, k)$ for the binomial coefficient $\binom{n}{k}$.)

Proof that (4) $\implies$ (1). Let $\binom{n}{k} = \frac{n!}{k!(n-k)!}$, for $n \geq k \geq 0$. The expression $n!$ counts the number of permutations of n objects, as there are n choices for the object that appears first, $n-1$ choices for the object that appears second given that the first has already been placed, $n-2$ choices for the object that appears third given that the first two have been placed, and so forth. The same logic shows that $n(n-1)(n-2)\cdots(n-k+1) = \frac{n!}{(n-k)!}$ counts the number of ways to create a permutation of size k from n objects. However, forming such a permutation is equivalent to first choosing k objects from n and then ordering those k objects. Since there are $k!$ ways to order the k chosen objects, $\frac{n!}{k!(n-k)!}$ must count the number of ways to choose k objects from n *without* ordering them; i.e., the number of combinations of k objects that can be formed from n objects. This is the same as the number of subsets of size k that can be formed from n elements. $\qquad\square$

This proves that the four definitions of $\binom{n}{k}$ are equivalent. For fun (and a useful generalization), let's look at (1) $\implies$ (3) as well. If we write out the expansion of $(x+1)^n$ we get

$$\overbrace{(x+1)(x+1)\cdots(x+1)}^{n \text{ factors}}.$$

This product is obtained by considering all possible ways to choose x or 1 from the first factor, x or 1 from the second factor, and so forth. Any term containing x^k in the resulting sum must come from choosing x from k of the factors and 1 from the remaining $n-k$ factors. Since there are $\binom{n}{k}$ ways to choose x from k of the factors, the coefficient of x^k in the final sum must be $\binom{n}{k}1^{n-k} = \binom{n}{k}$. In fact, a very slight tweak to this argument gives us a combinatorial proof of the more general binomial theorem for nonnegative integers n.

Identity 3 (Binomial Theorem). *For n a nonnegative integer,*

$$(x+y)^n = \sum_{k=0}^{n} \binom{n}{k} x^k y^{n-k}.$$

Proof. Expanding $(x+y)^n$ will result in a large number of terms, each of which contains, for some k between 0 and n, exactly k instances of x and $n-k$ instances of y. In other words, each term will be of the form $x^k y^{n-k}$ for some k. There are $\binom{n}{k}$ ways to choose from which factors the k instances of x come, leaving the remaining $n-k$ factors to contribute the instances of y. Summing over all possible values of k gives the general theorem. $\qquad\square$

In Chapter 2 we will discuss how Identity 3 can be extended to real values of n. Also, we don't have the tools to do it right now, but once we take a closer look at recurrence relations in Chapter 7 we'll give a direct proof that (2) $\implies$ (4).

So, what have we done here? The proofs (1) $\implies$ (2), (4) $\implies$ (1), and (1) $\implies$ (3) all involve some kind of counting argument; this makes them *combinatorial proofs*. The proof (2) $\implies$ (3) uses a recurrence relation argument, or, from another point of view, a generating function. The remaining proof, (3) $\implies$ (4), uses calculus. We will see all of these techniques in greater depth later in this book: combinatorial methods in Chapter 3 and some in Chapter 8, calculus in Chapter 4, generating functions in Chapter 6, and recurrence relations in Chapter 7.

Let's now turn to the example we started with and give five methods for proving that the sum of the elements in row n of Pascal's triangle is 2^n.

Identity 4.

$$\sum_{k=0}^{n} \binom{n}{k} = 2^n.$$

Proof 1. The simplest proof, now that we know it, is to use the binomial theorem. Since $\sum_{k=0}^{n} \binom{n}{k} x^k y^{n-k} = (x + y)^n$, letting $x = 1$ and $y = 0$ in Identity 3 immediately yields

$$\sum_{k=0}^{n} \binom{n}{k} = 2^n.$$

$\square$

(Substituting values into the binomial theorem and its variants can be a powerful tool for producing binomial identities. We will see more of this in Chapter 4.)

Proof 2. For a combinatorial approach, how many subsets of any size can be formed from a set of n elements? If we condition on the number of subsets of size k and add up, we see there are $\sum_{k=0}^{n} \binom{n}{k}$ such subsets. Alternatively, all the subsets can be generated by deciding whether each of the n elements will be placed in a subset or not. The first element can be in a subset or not, the second element can be in a subset or not, and so forth, with two choices for each of the n elements. Multiplying these together we get that there are 2^n total subsets. Thus

$$\sum_{k=0}^{n} \binom{n}{k} = 2^n.$$

$\square$

Proof 3. For a probabilistic approach, imagine flipping a fair coin n times. What is the probability that exactly k of the flips come up heads? The probability that we get k heads followed by $n - k$ tails is $\left(\frac{1}{2}\right)^k \left(\frac{1}{2}\right)^{n-k}$. However, the same probability must hold if we have $n - k$ tails followed by k heads, or for any other way of interspersing k heads and $n - k$ tails among n coin flips. Since there are $\binom{n}{k}$ ways of selecting which of the k flips are to be heads, the probability of obtaining exactly k flips must be $\binom{n}{k}\left(\frac{1}{2}\right)^{k}\left(\frac{1}{2}\right)^{n-k}$. We know, though, that the sum of the probabilities of all the possible outcomes must be 1, so we have

$$1 = \sum_{k=0}^{n} \binom{n}{k}\left(\frac{1}{2}\right)^{k}\left(\frac{1}{2}\right)^{n-k} = \frac{1}{2^n}\sum_{k=0}^{n}\binom{n}{k}.$$

In other words,

$$\sum_{k=0}^{n}\binom{n}{k} = 2^n.$$

$\square$

Proof 4. In Chapter 7 we prove the following formula (Identity 188): If $f(n) = g(n+1) - g(n)$, $F(n) = \sum_{k=0}^{n}\binom{n}{k}f(k)$, and $G(n) = \sum_{k=0}^{n}\binom{n}{k}g(k)$, then $G(n)$ and $F(n)$ are related by the formula

$$G(n) = 2^n\left(g(0) + \sum_{k=1}^{n}\frac{F(k-1)}{2^k}\right).$$

If $g(n) = 1$, then $f(n) = g(n+1) - g(n) = 0$. Thus $F(n) = 0$. The formula then yields $G(n) = 2^n(1+0) = 2^n$. (If $f(n) = g(n+1) - g(n)$, then we say $f(n)$ is the *finite difference* of $g(n)$, and we write $f(n) = \Delta g(n)$. This can be thought of as a discrete version of the derivative.) $\square$

Proof 5. Pascal's recurrence (Identity 1), $\binom{n}{k} = \binom{n-1}{k} + \binom{n-1}{k-1}$, is the special case $\alpha = \alpha' = \beta = \beta' = 0$, $\gamma = \gamma' = 1$, of the more general two-term recurrence

$$\left|{n\atop k}\right| = (\alpha(n-1) + \beta k + \gamma)\left|{n-1\atop k}\right| + (\alpha'(n-1) + \beta'(k-1) + \gamma')\left|{n-1\atop k-1}\right|. \quad (1.2)$$

We discuss this recurrence in Chapter 7. We also prove the following result, Identity 201: If $\left|{n\atop k}\right|$ satisfies a recurrence of the form of Equation (1.2), with $\beta + \beta' = 0$ and boundary conditions $\left|{0\atop 0}\right| = 1$, $\left|{0\atop k}\right| = 0$ for $k \neq 0$, then the nth row sum of $\left|{n\atop k}\right|$ is given by

$$\sum_{k=0}^{n}\left|{n\atop k}\right| = \prod_{i=0}^{n-1}((\alpha+\alpha')i + \gamma + \gamma').$$

In Chapter 2 we show that extending any of our four definitions of $\binom{n}{k}$ to allow for $k > n$ leads to $\binom{0}{0} = 1$ and $\binom{0}{k} = 0$ for $k \neq 0$. By Identity 201, then, we have.

$$\sum_{k=0}^{n}\binom{n}{k} = \prod_{i=1}^{n} 2 = 2^n.$$

$\square$

As with the proofs that our four definitions of the binomial coefficients are equivalent, this gives us a variety of approaches to proving that the sum of the elements in row n of Pascal's triangle is 2^n. Proof 1 uses a known binomial coefficient identity, Proof 2 uses combinatorics, Proof 3 uses probability, Proof 4 uses a formula involving finite differences, and Proof 5 uses a result on two-term triangular recurrence relations. In Chapter 6 we will see a sixth proof of this formula using generating functions. All of these methods are part of the art of proving binomial coefficient identities.

Let's take a moment now to preview what we will see throughout the rest of the book.

1. Chapter 1 has introduced the binomial coefficients and given a taste of some different techniques for proving binomial identities.

2. Chapter 2 works through some basic ideas we will need before going further. It generalizes $\binom{n}{k}$ by removing some of the restrictions on the values of n and k, with emphasis on the special case of $\binom{n}{k}$ when n is negative. It also discusses algebraic manipulation as a tool for proving binomial identities, with emphasis on the absorption identity and binomial inversion. (Exercise 1 in this chapter gets us started with proving binomial identities via algebraic manipulation.)

3. Chapter 3 is about combinatorial methods for proving binomial identities. We'll think more in depth about how to view the two sides of a binomial identity as counting something. The chapter begins with some basic combinatorial ideas using mostly interpretations involving committees and coin flips. It then moves on to lattice paths and selection in which an item may be chosen more than once. The chapter ends with two more advanced techniques: sign-reversing involutions and the principle of inclusion-exclusion.

4. Chapter 4 discusses calculus as a tool for proving binomial identities, with emphasis on differentiating and integrating the binomial theorem. The last section introduces the beta integral and the gamma function; these special functions will give us the most general expressions for $\binom{n}{k}$ we will see in this text.

5. Chapter 5 is about probabilistic methods for proving binomial identities. We'll discuss three important distributions whose probability mass functions all involve the binomial coefficients: the binomial distribution, the negative binomial distribution, and the hypergeometric distribution. We'll also see how the use of expected values and indicator variables can be used to prove binomial identities, look at the probabilistic variant of the principle of inclusion-exclusion, and give a probabilistic proof of the beta integral representation of $\binom{n}{k}$.

6. Chapter 6 introduces the important concept of generating functions. These allow one to represent a sequence of numbers using a single function. Since

manipulations on the sequence can often be modeled using simpler manipulations on the function, generating functions are an extremely powerful tool for proving identities. We will look at both ordinary and exponential generating functions.

7. Chapter 7 discusses recurrence relations and finite differences as tools for proving binomial identities. Binomial coefficients turn out to be instrumental in representing higher-order finite differences, and we exploit this relationship in multiple ways to prove binomial identities. At the end of the chapter we give a direct proof that the factorial definition of the binomial coefficients follows from Pascal's recurrence.

8. Chapter 8 is on special numbers related to the binomial coefficients: Fibonacci numbers, both kinds of Stirling numbers, Bell numbers, Lah numbers, Catalan numbers, and Bernoulli numbers. All of these numbers have representations in terms of, or satisfy interesting identities involving, binomial coefficients. This chapter's text and exercises feature combinatorial arguments, recurrence relations, generating functions, calculus, and probability; thus Chapter 8 represents something of a culmination of the ideas in the book up to that point.

9. Chapter 9 features two techniques that do not fit anywhere else but are not quite substantial enough to form independent chapters: complex numbers and linear algebra. Complex numbers are particularly useful for proving otherwise difficult-to-evaluate alternating binomial sums featuring $\binom{n}{2k}$, $\binom{n}{3k}$, etc. Finally, concepts from linear algebra—particularly matrices—give a whole new perspective on binomial identities proved earlier in the text. The important technique of binomial inversion that we will see in Chapter 2 also turns out to be matrix inversion in disguise.

10. In Chapter 10 we discuss Gosper's algorithm and its extension by Zeilberger. The resulting Gosper-Zeilberger algorithm can be used to automate the proofs of many binomial identities through the use of computers. (In a sense, then, this chapter is more about the *science* of proving binomial identities than the art thereof, but it's such a powerful tool that it deserves mention.) We give several examples, and we discuss the more general class of hypergeometric terms to which the Gosper-Zeilberger algorithm applies.

Finally, a few comments on definitions and notation.

- Throughout this text we take 0^0 to be 1. This is the right definition of 0^0 when working with binomial coefficients, in the sense that the formulas continue to hold when 0^0 is defined thusly.

- The variables x, y, and z are real numbers unless stated otherwise. All other variables are nonnegative integers unless stated otherwise.

- The sequence $(a_0, a_1, a_2, \ldots) = (a_n)_{n=0}^{\infty}$ will frequently be denoted (a_n).

- The *Iverson bracket* $[P]$ for the statement P is defined to be 0 if P is true and 1 if P is false. It allows expressions that involve conditional statements to be expressed more simply and manipulated more easily and thus will come in handy throughout the book. For example, Identity 12 for the alternating row sum of the binomial coefficients looks like

$$\sum_{k=0}^{n} \binom{n}{k}(-1)^k = \begin{cases} 1, & n = 0; \\ 0, & n \geq 1; \end{cases}$$

in conditional notation but with Iverson notation becomes the more compact

$$\sum_{k=0}^{n} \binom{n}{k}(-1)^k = [n = 0].$$

1.1 Exercises

1. Use algebraic manipulation and the factorial definition of the binomial coefficients (Identity 2) to prove the following identities, where k, m, and n are assumed to be nonnegative integers. (We include restrictions on the variables to ensure that the binomial coefficient expressions are always defined when using factorials. After we introduce the generalized binomial coefficient in Chapter 2 many of these restrictions may be dropped.)

Identity 1 (Pascal's Recurrence).

$$\binom{n}{k} = \binom{n-1}{k} + \binom{n-1}{k-1}, \quad n-1 \geq k \geq 1.$$

Identity 5 (Symmetry).

$$\binom{n}{k} = \binom{n}{n-k}, \quad n \geq k \geq 0.$$

Identity 6 (Absorption).

$$k\binom{n}{k} = n\binom{n-1}{k-1}, \quad n \geq k \geq 1.$$

Identity 7 (Trinomial Revision).

$$\binom{n}{m}\binom{m}{k} = \binom{n}{k}\binom{n-k}{m-k}, \quad n \geq m \geq k \geq 0.$$

Identity 8.

$$(n-k)\binom{n}{k} = n\binom{n-1}{k}, \quad n-1 \geq k \geq 0.$$

Identity 9.

$$\binom{n}{m}\binom{n-m}{k} = \binom{n}{m+k}\binom{m+k}{m}, \quad n \geq m+k, \, m \geq 0, \, k \geq 0.$$

Identity 10.

$$k(k-1)\binom{n}{k} = n(n-1)\binom{n-2}{k-2}, \quad n \geq k \geq 2.$$

Identity 11.

$$n\binom{n}{k} = k\binom{n}{k} + (k+1)\binom{n}{k+1}, \quad n-1 \geq k \geq 0.$$

2. Use the binomial theorem (Identity 3) to prove the following:

Identity 12.

$$\sum_{k=0}^{n}\binom{n}{k}(-1)^k = [n = 0].$$

3. Use Identities 4 and 12 to prove the following, the sums of the even and odd index binomial coefficients.

Identity 13.

$$\sum_{k \geq 0}\binom{n}{2k} = 2^{n-1} + \frac{1}{2}[n = 0].$$

Identity 14.

$$\sum_{k \geq 0}\binom{n}{2k+1} = 2^{n-1}[n \geq 1].$$

4. Give yet another proof of

Identity 4.

$$\sum_{k=0}^{n}\binom{n}{k} = 2^n,$$

this time using Pascal's recurrence $\binom{n}{k} = \binom{n-1}{k} + \binom{n-1}{k-1}$, with $\binom{0}{0} = 1$.

5. Use the recurrence $\binom{n}{k} + \binom{n}{k+1} = \binom{n+1}{k+1}$, together with the fact that $\binom{n}{0} = \binom{n+1}{0}$, to prove the following.

Identity 15 (Parallel Summation).

$$\sum_{k=0}^{m} \binom{n+k}{k} = \binom{n+m+1}{m}.$$

6. Prove the following, a binomial theorem-type generalization of Identity 13.

Identity 16.

$$\sum_{k\geq 0} \binom{n}{2k} x^{2k} y^{n-2k} = \frac{1}{2}\left((x+y)^n + (y-x)^n\right).$$

1.2 Notes

For those looking for a more comprehensive list of binomial identities and less emphasis on a variety of proof techniques, I suggest Gould's *Combinatorial Identities* [30] or Riordan's *Combinatorial Identities* [57].

2

Basic Techniques

Before we dive into the wider variety of approaches for proving binomial identities that will occupy most of this book we should spend a chapter establishing a few basic properties and techniques. First, we consider the generalization of the binomial coefficients to noninteger and negative values. This is an important generalization—one we will use frequently throughout the rest of the text. As is often the case, generalization will sometimes produce simpler proofs. In addition, some identities just make more sense in a general setting.

In the second section we discuss the special case of binomial coefficients with negative upper indices. The last two sections of this chapter concern the absorption identity and binomial inversion, two particularly important algebraic techniques for proving binomial identities.

2.1 The Generalized Binomial Coefficient

Thus far we have treated the binomial coefficient $\binom{n}{k}$ as if n and k must be nonnegative integers with $n \geq k$. In this section we generalize the binomial coefficient $\binom{n}{k}$ to the case where n is real and k is an integer. This not only gives us Newton's generalization of the binomial theorem, it also makes proofs of some binomial identities easier.

In Chapter 1 we gave four equivalent definitions of the binomial coefficients for integers n and k with $n \geq k \geq 0$. These are:

1. *The combinatorial definition:* $\binom{n}{k}$ is the number of subsets of size k that can be formed from a set of n elements.

2. *The recursive definition:* $\binom{n}{k} = \binom{n-1}{k} + \binom{n-1}{k-1}$ when $n - 1 \geq k \geq 1$, with boundary conditions $\binom{n}{0} = \binom{n}{n} = 1$.

3. *The binomial theorem definition:* $\binom{n}{k}$ is the coefficient of x^k in the expansion of $(x + 1)^n$ in powers of x.

4. *The factorial definition:* $\binom{n}{k} = \frac{n!}{k!(n-k)!}$.

If we remove the restriction $n \geq k$ while still requiring n and k to be nonnegative integers, the combinatorial and binomial theorem definitions immediately imply $\binom{n}{k} = 0$ for $n < k$. The recursive definition does as well, provided we also drop the $n - 1 \geq k$ restriction on the recurrence relation; see Exercise 1.

Removing the restriction $n \geq k$ on Definition 4 is trickier, however, as it appears to require that we make sense of $(-1)!, (-2)!, (-3)!$, and so forth. We will sidestep this difficult question[1] via a slight modification of the factorial definition of the binomial coefficients. We need to define some new notation and terminology first, though.

The *falling factorial* $n^{\underline{k}}$, valid for real n and nonnegative integer k, is given by $n^{\underline{k}} = n(n-1)(n-2)\cdots(n-k+1)$. Similarly, the *rising factorial* $n^{\overline{k}}$ is given by $n^{\overline{k}} = n(n+1)\cdots(n+k-1)$. (For the definition of $n^{\underline{k}}$ when k is negative, see Chapter 7.)

Now we're ready for a preliminary version of the falling factorial definition of the binomial coefficients:

$$\binom{n}{k} = \frac{n^{\underline{k}}}{k!} = \frac{n(n-1)(n-2)\cdots(n-k+1)}{k(k-1)\cdots(1)}.$$

It is easy to see that this definition does generalize the factorial definition of the binomial coefficients: $n^{\underline{k}} = \frac{n!}{(n-k)!}$ when $n \geq k \geq 0$ and n and k are both integers. Moreover, this definition gives $\binom{n}{k} = 0$ when $n < k$, in agreement with the other three definitions.

There is nothing about this definition that requires n to be nonnegative, though, or even an integer. Thus we could use it to generalize the binomial coefficient $\binom{n}{k}$ even further, to the case where n is any real number. If we make this extension, though, how well does it fit with the other definitions? Well, the combinatorial definition does not work for real n, as we do not have a working definition for a negative, fractional, or irrational number of subsets. (However, attempts have been made in this direction; see the notes.) The recursive definition can actually be extended to the case where n is negative, although real n in general would require different boundary conditions. Finally, the binomial theorem definition can be generalized to all real n through the use of Maclaurin series. We will explore both of these extensions (the recursive one and the binomial theorem one).

Before we do so, though, let's go ahead and make one more generalization by removing the restriction that k be nonnegative. None of our four definitions seem to give any reason that $\binom{n}{k}$ should be anything other than zero when $k < 0$, so let's force that.

Definition 4 (improved version). *The falling factorial definition*:

---

[1] While there is a way to generalize $n!$ to real values of n using the gamma function (see Chapter 4), the gamma function unfortunately turns out to be defined for all real numbers *except* negative integers n: precisely the ones for which we need to make sense of $n!$.

Identity 17. *For real n and integers k,*

$$\binom{n}{k} = \begin{cases} \dfrac{n^{\underline{k}}}{k!} = \dfrac{n(n-1)(n-2)\cdots(n-k+1)}{k(k-1)\cdots(1)}, & k \geq 0; \\ 0, & k < 0. \end{cases}$$

We will sometimes refer to the binomial coefficient defined in Identity 17 as the *generalized binomial coefficient*.

With Identity 17 in mind, let's now look at how the recursive and binomial theorem definitions generalize.

First, the recursive definition. As a reminder, this says that $\binom{n}{k} = \binom{n-1}{k} + \binom{n-1}{k-1}$ when $n - 1 \geq k \geq 1$, with boundary conditions $\binom{n}{0} = \binom{n}{n} = 1$. How well does this agree with Identity 17? As we have already mentioned, Exercise 1 shows that just removing the restriction $n - 1 \geq k$ on the recurrence relation produces $\binom{n}{k} = 0$ when $n < k$, in keeping with Identity 17.

Removing the nonnegativity restrictions on n and k while keeping the boundary conditions would cause a bit of a problem, though. For example, if $\binom{n}{n} = 1$ for all integers n then we would have $\binom{-1}{-1} = 1$, which directly contradicts Identity 17's assertion that $\binom{n}{k} = 0$ when $k < 0$. Instead, let's change the boundary condition $\binom{n}{n} = 1$ to $\binom{0}{0} = 1$ and $\binom{0}{k} = 0$ for $k \neq 0$. It turns out that this change both reproduces Pascal's triangle and forces $\binom{n}{k} = 0$ when $k < 0$. (See Exercise 2.) The boundary conditions are more compactly represented using the Iverson bracket notation, though, so we will use that in reformulating Pascal's recurrence.

Sir Isaac Newton (1642–1727) is considered to be one of the greatest mathematicians of all time. He is most famous for discovering (or, depending on your philosophical point of view, inventing) calculus, as well as formulating the theory of gravity. He also built the first reflecting telescope and contributed to the theory of light. In later years Newton served two terms as a member of the English Parliament, as well as master of the Royal Mint from 1700 until his death. Speaking of his own achievements, Newton once said, "If I have seen further, it is by standing on the shoulders of giants." Newton helped kickstart the Scientific Revolution and was admired throughout Europe by the time of his death. Alexander Pope summed up how many felt when he wrote the following: "Nature and nature's laws lay hid in night; God said 'Let Newton be' and all was light."

Identity 1 (Pascal's Recurrence). *For integers n and k,*

$$\binom{n}{k} = \binom{n-1}{k} + \binom{n-1}{k-1}, \quad \binom{n}{0} = 1, \quad \binom{0}{k} = [k = 0].$$

This is the form of Identity 1 that we will use throughout the rest of the text.

What does this say about values like $\binom{-1}{1}$, though? Well, by the boundary conditions, $\binom{-1}{0} = 1$ and $\binom{0}{1} = 0$. The recurrence $\binom{n}{k} = \binom{n-1}{k} + \binom{n-1}{k-1}$ with $n = 0$ and $k = 1$ then yields $\binom{-1}{1} = -1$. From there we can continue to generate binomial coefficients for negative values of n and nonnegative values of k, yielding the numbers in Figure 2.1.

n	$\binom{n}{0}$	$\binom{n}{1}$	$\binom{n}{2}$	$\binom{n}{3}$	$\binom{n}{4}$	$\binom{n}{5}$
-4	1	-4	10	-20	35	-56
-3	1	-3	6	-10	15	-21
-2	1	-2	3	-4	5	-6
-1	1	-1	1	-1	1	-1
0	1	0	0	0	0	0

FIGURE 2.1
First few rows of Pascal's triangle, extended to negative values of n

Generating the numbers in Figure 2.1 from Pascal's recurrence seems like a reasonable thing to do, but do these numbers actually agree with the generalized binomial coefficient defined in Identity 17? In other words, if we allow n and k to be any integers, does the definition of $\binom{n}{k}$ in Identity 17 actually satisfy $\binom{n}{k} = \binom{n-1}{k} + \binom{n-1}{k-1}$, with $\binom{n}{0} = 1$ and $\binom{0}{k} = [k = 0]$? This question is answered in the affirmative in Exercise 3. Thus Identity 1 gives us a definition of $\binom{n}{k}$ that is equivalent to that in Identity 17 when n and k are integers.

In addition to Pascal's recurrence, the binomial theorem definition of the binomial coefficients is also consistent with Identity 17. Recall that in Chapter 1 we used the Maclaurin series for $f(x) = (x+1)^n$ to prove that the factorial definition of $\binom{n}{k}$ follows from the definition of $\binom{n}{k}$ as the coefficient of x^k in the expansion of $(x+1)^n$ in powers of x. There is nothing about the definition of f nor about the Maclaurin series formula that requires n to be a nonnegative integer. If we drop that assumption and apply the Maclaurin series formula we obtain the generalization of the binomial theorem (Identity 3) known as the binomial series, first discovered by Newton.

Identity 18 (Binomial Series).

$$(x + 1)^n = \sum_{k=0}^{\infty} \binom{n}{k} x^k,$$

where n is a real number, $\binom{n}{k} = \dfrac{n^{\underline{k}}}{k!}$, *and* $|x| < 1$.

Proof. The Maclaurin series formula for $f(x)$ is given by

$$M(x) = \sum_{k=0}^{\infty} \frac{f^{(k)}(0)x^k}{k!}.$$

We see that $f(0) = (0+1)^n = 1$, $f'(0) = n(0+1)^{n-1} = n$, $f''(0) = n(n-1)(0+1)^{n-2} = n(n-1)$, and, in general, $f^{(k)}(0) = n(n-1)(n-2)\cdots(n-k+1)(0+1)^{n-k} = n^{\underline{k}}$. Therefore,

$$M(x) = \sum_{k=0}^{\infty} \frac{n^{\underline{k}}x^k}{k!} = \sum_{k=0}^{\infty} \binom{n}{k} x^k.$$

To complete the proof we need to show that $M(x)$ converges to $(x+1)^n$ for $|x| < 1$. This is done in Exercise 32. □

Summarizing, we have shown that the definitions of $\binom{n}{k}$ given in Figure 2.2 are equivalent, subject to the respective stated restrictions on n and k. These are the four definitions we will use throughout the remainder of the text. (We do mention a further generalization involving the gamma function in Chapter 4, but it is used only in a few places in that chapter.)

2.2 Negative Upper Indices

Let's take a closer look now at the negative upper index binomial coefficients in Figure 2.1. These numbers appear to be the same as the positive upper index binomial coefficients appearing in Pascal's triangle in Figure 1.1. The differences are that the triangle has been rotated and some of the values are negative. There is clearly a relationship, though. The next identity establishes precisely what that relationship is.

Identity 19. *For real n,*

$$\binom{-n}{k} = (-1)^k \binom{n+k-1}{k}.$$

Proof.

$$\binom{-n}{k} = \frac{-n(-n-1)(-n-2)\cdots(-n-k+1)}{k!}$$

$$= (-1)^k \frac{n(n+1)(n+2)\cdots(n+k-1)}{k!} = (-1)^k \frac{(n+k-1)^{\underline{k}}}{k!}$$

$$= (-1)^k \binom{n+k-1}{k}.$$

□

Four equivalent definitions of the binomial coefficients

1. *The combinatorial definition:* $\binom{n}{k}$ is the number of subsets of size k that can be formed from a set of n elements. It is valid for nonnegative integers n and k.

2. *The recursive definition* (Pascal's Recurrence): $\binom{n}{k} = \binom{n-1}{k} + \binom{n-1}{k-1}$, with boundary conditions $\binom{n}{0} = 1$ and $\binom{0}{k} = [k = 0]$. This is Identity 1, and it is valid for integers n and k.

3. *The binomial series definition:* $\binom{n}{k}$ is the coefficient of x^k in the Maclaurin series expansion of $(x+1)^n$ in powers of x. This is Identity 18, and it is valid for real n, integers k, and $|x| < 1$. (If we view the expansion as a formal power series, we do not even need the restriction $|x| < 1$ in Identity 18; see Chapter 6.)

4. *The falling factorial definition:*

$$\binom{n}{k} = \begin{cases} \dfrac{n^{\underline{k}}}{k!} = \dfrac{n(n-1)(n-2)\cdots(n-k+1)}{k(k-1)\cdots(1)}, & k \geq 0; \\ 0, & k < 0. \end{cases}$$

This is Identity 17, and it is valid for real n and integers k.

FIGURE 2.2
Improved versions of our four definitions of the binomial coefficients

Identity 19 has a nice application involving alternating row sums of Pascal's triangle. In Chapter 1 we proved one of the most basic of binomial coefficient identities, the row sum formula, Identity 4: $\sum_{k=0}^{n} \binom{n}{k} = 2^n$. Identity 19, together with the parallel summation formula $\sum_{k=0}^{m} \binom{n+k}{k} = \binom{n+m+1}{m}$ (Identity 15), gives us a proof for the alternating row sum.

Identity 12.

$$\sum_{k=0}^{n} \binom{n}{k}(-1)^k = [n = 0].$$

Proof.

$$\sum_{k=0}^{n} \binom{n}{k}(-1)^k = \sum_{k=0}^{n} \binom{-n+k-1}{k} = \binom{-n+n}{n} = \binom{0}{n} = [n = 0].$$

In the first step we use Identity 19, and in the second step we use Identity 15. (Identity 15 holds for negative values of n because its proof in Exercise 5 in Chapter 1 is based on the recurrence $\binom{n}{k} = \binom{n-1}{k} + \binom{n-1}{k-1}$, and we prove in Exercise 3 that this recurrence holds for negative values of n.) $\square$

What happens if we add only part of the way across row n of Pascal's triangle , say, from 0 to m, where $m < n$? It turns out that there is no closed-form expression for the *partial* row sum $\sum_{k=0}^{m} \binom{n}{k}$ [32, p. 165]. However, the proof of Identity 12 we just gave is easily modified to obtain a formula for the partial alternating row sum.

Identity 20.

$$\sum_{k=0}^{m} \binom{n}{k}(-1)^k = (-1)^m \binom{n-1}{m}.$$

Proof.

$$\sum_{k=0}^{m} \binom{n}{k}(-1)^k = \sum_{k=0}^{m} \binom{-n+k-1}{k} = \binom{-n+m}{m} = (-1)^m \binom{n-1}{m}.$$

$\square$

The binomial coefficient can be further extended to all but negative integer values of k; see Chapter 4, where we discuss the gamma function.

2.3 The Absorption Identity

The absorption identity is arguably the most useful of all the basic binomial identities—enough so that it gets its own section here. We have actually already seen this identity; it is Identity 6, part of Exercise 1 in Chapter 1.

Identity 6 (Absorption Identity). *For real n,*

$$k\binom{n}{k} = n\binom{n-1}{k-1}.$$

Proof. Using the falling factorial definition of the binomial coefficients, we have that both sides of the identity are 0 when $k \le 0$. If $k > 0$, then

$$\binom{n}{k}k = k\frac{n^{\underline{k}}}{k!} = \frac{n(n-1)(n-2)\cdots(n-k+1)}{(k-1)!} = n\frac{(n-1)^{\underline{k-1}}}{(k-1)!} = n\binom{n-1}{k-1}.$$

$\square$

(For a combinatorial proof, see Chapter 3.)

The absorption identity is particularly useful when proving binomial identities involving sums. Our next identity gives the flavor of the approach.

Identity 21.

$$\sum_{k=0}^{n} \binom{n}{k}k = n2^{n-1}.$$

Proof.

$$\sum_{k=0}^{n} \binom{n}{k} k = n \sum_{k=0}^{n} \binom{n-1}{k-1}, \qquad \text{by the absorption identity,}$$

$$= n2^{n-1}, \qquad \text{by Identity 4,}$$

remembering that $\binom{n-1}{-1} = 0$.   □

Here, we have a sum over k that includes both a binomial coefficient and a factor of k. The absorption identity allows us to swap the factor of k for a factor of n, leaving us with a binomial sum we already know. In general, this is the kind of situation in which the absorption identity is most useful: Replacing a factor on which a sum depends with a factor on which it does not will make the sum easier to evaluate.

For a second look at this approach, here is the alternating version of Identity 21.

Identity 22.

$$\sum_{k=0}^{n} \binom{n}{k} k(-1)^k = -[n = 1].$$

Proof. When $n = 0$ both sides of the identity are 0, and so the identity is clearly true. For $n \geq 1$, we have

$$\sum_{k=0}^{n} \binom{n}{k} k(-1)^k = n \sum_{k=0}^{n} \binom{n-1}{k-1}(-1)^k$$

$$= n \sum_{k=0}^{n-1} \binom{n-1}{k}(-1)^{k+1}, \qquad \text{switching indices,}$$

$$= -n[n-1 = 0] \qquad \text{by Identity 12,}$$

$$= -n[n = 1]$$

$$= -[n = 1].$$

□

(Here we see the usefulness of the Iverson bracket notation; the proof of Identity 22 would have been more complicated without it.)

The absorption identity can be used to prove more complicated identities as well. For example, we can use Identity 6 to prove the following, although that may not appear to be the case at first glance.

Identity 23.

$$\sum_{k=0}^{n} \binom{n}{k} \frac{1}{k+1} = \frac{2^{n+1} - 1}{n+1}.$$

Proof. If we replace k and n with $k+1$ and $n+1$, respectively, in Identity 6 and rearrange, we can obtain

$$\binom{n+1}{k+1}\frac{1}{n+1} = \binom{n}{k}\frac{1}{k+1}.$$

Then we have

$$\sum_{k=0}^{n}\binom{n}{k}\frac{1}{k+1} = \sum_{k=0}^{n}\binom{n+1}{k+1}\frac{1}{n+1} = \frac{1}{n+1}\sum_{k=1}^{n+1}\binom{n+1}{k}$$

$$= \frac{1}{n+1}\left(\sum_{k=0}^{n+1}\binom{n+1}{k} - \binom{n+1}{0}\right) = \frac{2^{n+1}-1}{n+1},$$

via Identity 4. □

A similar approach proves the alternating version of Identity 23 as well:

Identity 24.

$$\sum_{k=0}^{n}\binom{n}{k}\frac{(-1)^k}{k+1} = \frac{1}{n+1}.$$

Proof.

$$\sum_{k=0}^{n}\binom{n}{k}\frac{(-1)^k}{k+1} = \sum_{k=0}^{n}\binom{n+1}{k+1}\frac{(-1)^k}{n+1} = \frac{1}{n+1}\sum_{k=1}^{n+1}\binom{n+1}{k}(-1)^{k-1}$$

$$= \frac{1}{n+1}\left(\sum_{k=0}^{n+1}\binom{n+1}{k}(-1)^{k-1} + \binom{n+1}{0}\right) = \frac{1}{n+1},$$

via Identity 12. □

As our final example using the absorption identity, here is a variant on Identity 23 whose proof requires an additional idea: finite differences.

Identity 25.

$$\sum_{k=1}^{n}\binom{n}{k}\frac{1}{k} = \sum_{k=1}^{n}\frac{2^k-1}{k}.$$

Proof. We cannot use Identity 6 here directly, as the factor of $\frac{1}{k}$ cannot be absorbed into $\binom{n}{k}$. However, by combining finite differences with the recursion formula $\binom{n+1}{k} = \binom{n}{k} + \binom{n}{k-1}$ (Identity 1) we can find the sum.

Let $f(n) = \sum_{k=1}^{n}\binom{n}{k}\frac{1}{k}$. Then the finite difference $f(n+1) - f(n)$ is given by

$$f(n+1) - f(n) = \sum_{k=1}^{n+1}\binom{n+1}{k}\frac{1}{k} - \sum_{k=1}^{n}\binom{n}{k}\frac{1}{k}$$

$$= \sum_{k=1}^{n+1} \binom{n}{k-1} \frac{1}{k}, \qquad \text{using Pascal's recurrence,}$$

$$= \sum_{k=0}^{n} \binom{n}{k} \frac{1}{k+1}$$

$$= \frac{2^{n+1} - 1}{n+1}, \qquad \text{by Identity 23.}$$

Since finite differences telescope nicely, and using the fact that $f(0)$ is an empty sum and is thus 0, we have

$$\sum_{k=1}^{n} \binom{n}{k} \frac{1}{k} = f(n) = \sum_{k=1}^{n} (f(k) - f(k-1)) = \sum_{k=1}^{n} \frac{2^k - 1}{k}.$$

$\square$

Generalizations of most of these identities are given in the exercises. In addition, they may be proved using calculus techniques; see Chapter 4. The alternating identities have nice probabilistic proofs, too; see Chapter 5. We revisit finite differences as a tool for proving binomial coefficient identities in Chapter 7.

2.4   Binomial Inversion

The last of the basic techniques we consider in this chapter is binomial inversion. Given a binomial coefficient identity, binomial inversion produces a second, new, binomial coefficient identity paired with it (other than a few exceptions in which an identity is its own binomial inverse). This greatly increases the number of binomial coefficient identities we can prove.

Theorem 1 (Binomial Inversion).

$$f(n) = \sum_{k=0}^{n} \binom{n}{k} g(k)(-1)^k \iff g(n) = \sum_{k=0}^{n} \binom{n}{k} f(k)(-1)^k.$$

Proof. It suffices to prove only one direction of the implication, as the roles of f and g in the theorem statement are symmetric.

$$\sum_{k=0}^{n} \binom{n}{k} f(k)(-1)^k$$

$$= \sum_{k=0}^{n} \binom{n}{k} (-1)^k \sum_{j=0}^{k} \binom{k}{j} g(j)(-1)^j$$

$$= \sum_{k=0}^{n} \sum_{j=0}^{k} \binom{n}{j} \binom{n-j}{k-j} g(j)(-1)^{k+j}, \qquad \text{by Identity 7,}$$

$$= \sum_{j=0}^{n} \binom{n}{j} g(j)(-1)^{j} \sum_{k=j}^{n} \binom{n-j}{k-j}(-1)^{k}, \qquad \text{swapping the summation order,}$$

$$= \sum_{j=0}^{n} \binom{n}{j} g(j)(-1)^{j} \sum_{k=0}^{n-j} \binom{n-j}{k}(-1)^{k+j}, \qquad \text{switching indices,}$$

$$= \sum_{j=0}^{n} \binom{n}{j} g(j) \sum_{k=0}^{n-j} \binom{n-j}{k}(-1)^{k}.$$

Now, the inner sum is, by Identity 12, 1 if $n - j = 0$ and 0 otherwise. This means that the outer sum is nonzero only when $n = j$, and we end up with

$$\sum_{k=0}^{n} \binom{n}{k} f(k)(-1)^{k} = \binom{n}{n} g(n) = g(n).$$

$\square$

Let's take a look at some identities we can generate via binomial inversion. Alternating identities are an obvious place to start, although (as we shall see) binomial inversion can be applied to non-alternating binomial identities, too. Identity 12,

$$\sum_{k=0}^{n} \binom{n}{k}(-1)^{k} = [n = 0],$$

is the simplest alternating binomial identity we have seen thus far. However, it was used in the proof of binomial inversion, and it turns out that applying binomial inversion to this identity does not yield anything interesting. With $g(k) = 1$, $f(0) = 1$, and $f(k) = 0$ if $k \geq 1$, we obtain

$$1 = \sum_{k=0}^{n} \binom{n}{k} f(k)(-1)^{k} = \sum_{k=0}^{n} \binom{n}{k} [k = 0](-1)^{k} = \binom{n}{0}(-1)^{0} = 1,$$

which is not exactly a candidate for the mathematical breakthrough of the year.

However, this example does give us a hint for how we can use binomial inversion to prove something more interesting. Since binomial inversion is reversible, letting $g(0) = 1$ and $g(k) = 0$ if $k \geq 1$ yields Identity 12. This can be thought of as letting g be the indicator function of the number 0. If we make g the indicator function of different numbers we can generate different binomial coefficient identities. For example, if g is the indicator function of the number 1 then binomial inversion gives us the negative of Identity 22.

Identity 22.

$$\sum_{k=0}^{n} \binom{n}{k} k(-1)^k = -[n = 1].$$

Proof. Let

$$g(k) = [k = 1].$$

Then

$$\sum_{k=0}^{n} \binom{n}{k} g(k)(-1)^k = \binom{n}{1}(-1) = -n.$$

Applying binomial inversion, we have

$$\sum_{k=0}^{n} \binom{n}{k}(-k)(-1)^k = [n = 1].$$

Multiplying by -1 gives the result.   □

In fact, we can generalize Identity 22 by letting g be the indicator function on the number m, $0 \le m \le n$. This yields the following.

Identity 26.

$$\sum_{k=0}^{n} \binom{n}{k}\binom{k}{m}(-1)^k = (-1)^m[n = m].$$

Proof. Let

$$g(k) = [k = m].$$

Then

$$\sum_{k=0}^{n} \binom{n}{k} g(k)(-1)^k = \binom{n}{m}(-1)^m.$$

Applying binomial inversion produces

$$\sum_{k=0}^{n} \binom{n}{k}\binom{k}{m}(-1)^m(-1)^k = [n = m].$$

Multiplying by $(-1)^m$ produces the identity.   □

Let's take a look at a couple more examples using binomial inversion before we end the chapter. Since binomial inversion itself involves an alternating binomial sum, those types of identities are a good place to start. The only other alternating identity we have yet seen is Identity 24,

$$\sum_{k=0}^{n} \binom{n}{k}\frac{(-1)^k}{k+1} = \frac{1}{n+1}.$$

Unfortunately, this is one of the few cases in which an identity is its own binomial inverse! (This is because $f(n) = g(n) = \frac{1}{n+1}$.)) However, applying binomial inversion to the similar Identity 43,

$$\sum_{k=1}^{n} \binom{n}{k} \frac{(-1)^k}{k} = -H_n,$$

(see Exercise 16) does give us a new identity. (Here, H_n is the nth harmonic number, given by $H_n = \sum_{k=1}^{n} \frac{1}{k}$.)

Identity 27.

$$\sum_{k=0}^{n} \binom{n}{k} H_k(-1)^k = -\frac{1}{n}[n \geq 1].$$

Proof. In Identity 43 the summation begins at $k = 1$. This implicitly defines

$$g(k) = \frac{1}{k}[k \geq 1].$$

Binomial inversion, followed by multiplication by -1, does the rest. □

Binomial inversion also works on sums that do not alternate; in this case we just have to define $g(k)$ appropriately. For example, with Identity 21,

$$\sum_{k=0}^{n} \binom{n}{k} k = n2^{n-1},$$

binomial inversion produces the following:

Identity 28.

$$\sum_{k=0}^{n} \binom{n}{k} k 2^{k-1}(-1)^k = (-1)^n n.$$

Proof. We can rewrite Identity 21 as

$$\sum_{k=0}^{n} \binom{n}{k} k(-1)^k(-1)^k = n2^{n-1}.$$

This can be thought of as an alternating sum with $g(k) = k(-1)^k$. Applying binomial inversion gives Identity 28. □

Theorem 2 generalizes this idea of applying binomial inversion to sums that do not alternate. (See Exercise 27.)

2.5   Exercises

1. Prove that the recurrence $\binom{n}{k} = \binom{n-1}{k} + \binom{n-1}{k-1}$, valid for integers $n \geq 1$, $k \geq 1$, and with boundary conditions $\binom{n}{0} = \binom{n}{n} = 1$, yields $\binom{n}{k} = 0$ when $n < k$, $n \geq 0$, and $k \geq 0$.

2. Show that, when n and k are nonnegative integers, the following recurrences have the same solution:

   (a) $\binom{n}{k} = \binom{n-1}{k} + \binom{n-1}{k-1}$ for positive integers n and k, with $\binom{0}{0} = \binom{n}{n} = 1$ for nonnegative integers n;

   (b) $\binom{n}{k} = \binom{n-1}{k} + \binom{n-1}{k-1}$ for positive integers n and k, with $\binom{n}{0} = 1$ and $\binom{0}{k} = [k = 0]$ for nonnegative integers n and k.

   Then show that allowing n and k to be any integers in the latter recurrence implies $\binom{n}{k} = 0$ for $k < 0$.

3. Prove that the definition of $\binom{n}{k}$ given in Identity 17,

$$\binom{n}{k} = \begin{cases} \dfrac{n^{\underline{k}}}{k!} = \dfrac{n(n-1)(n-2)\cdots(n-k+1)}{k(k-1)\cdots(1)}, & k \geq 0; \\ 0, & k < 0; \end{cases}$$

   satisfies $\binom{n}{k} = \binom{n-1}{k} + \binom{n-1}{k-1}$ for all integers n and k, as well as boundary conditions $\binom{n}{0} = 1$ and $\binom{0}{k} = [k = 0]$.

4. Prove

   Identity 29.

$$\binom{-1}{k} = (-1)^k.$$

5. Prove

   Identity 30.

$$\binom{-1/2}{n} = \left(-\frac{1}{4}\right)^n \binom{2n}{n}.$$

6. Prove the following relationships between rising and falling factorial powers, for $k \geq 0$.

   (a)

   Identity 31.

$$n^{\overline{k}} = (n + k - 1)^{\underline{k}}.$$

(b)

Identity 32.

$$n^{\overline{k}} = (-1)^k (-n)^{\underline{k}}.$$

7. The falling and rising factorial powers satisfy their own versions of the binomial theorem. Prove them. (Hint: You will need Identity 57, which we have not seen yet.)

(a)

Identity 33.

$$\sum_{k=0}^{n} \binom{n}{k} x^{\underline{k}} y^{\underline{n-k}} = (x+y)^{\underline{n}}.$$

(b)

Identity 34.

$$\sum_{k=0}^{n} \binom{n}{k} x^{\overline{k}} y^{\overline{n-k}} = (x+y)^{\overline{n}}.$$

8. Use the absorption identity twice to prove

Identity 35.

$$\sum_{k=0}^{n} \binom{n}{k} k(k-1) = n(n-1)2^{n-2}.$$

9. (A generalization of Identity 21.) Prove

Identity 36.

$$\sum_{k=0}^{n} \binom{n}{k} k^{\underline{m}} = n^{\underline{m}} 2^{n-m}.$$

10. Use the absorption identity twice to prove

Identity 37.

$$\sum_{k=0}^{n} \binom{n}{k} k(k-1)(-1)^k = 2[n=2].$$

11. Prove

Identity 38.

$$\sum_{k=0}^{n} \binom{n}{k} k^{\underline{m}} (-1)^k = (-1)^m m! [n=m].$$

12. Prove

   Identity 39.

   $$\sum_{k=0}^{n} \binom{n}{k} \frac{1}{(k+1)(k+2)} = \frac{2^{n+2}-1}{(n+1)(n+2)} - \frac{1}{n+1}.$$

13. Prove

   Identity 40.

   $$\sum_{k=0}^{n} \binom{n}{k} \frac{1}{(k+1)^{\overline{m}}} = \frac{1}{(n+1)^{\overline{m}}} \left(2^{n+m} - \sum_{k=0}^{m-1} \binom{n}{k} \right).$$

14. Prove

   Identity 41.

   $$\sum_{k=0}^{n} \binom{n}{k} \frac{(-1)^k}{(k+1)(k+2)} = \frac{1}{n+2}.$$

15. Prove

   Identity 42.

   $$\sum_{k=0}^{n} \binom{n}{k} \frac{(-1)^k}{(k+1)^{\overline{m}}} = \frac{1}{(n+m)(m-1)!}.$$

16. Use the absorption identity and an argument like that used to prove Identity 25 to prove the following:

   Identity 43.

   $$\sum_{k=1}^{n} \binom{n}{k} \frac{(-1)^k}{k} = -H_n,$$

   where H_n is the nth harmonic number: $H_n = \sum_{k=1}^{n} \frac{1}{k}$.

17. Prove the following, a generalization of Identity 24.

   Identity 44.

   $$\sum_{k=0}^{m} \binom{n}{k} \frac{(-1)^k}{k+1} = \frac{1}{n+1} \binom{n}{m+1} (-1)^m + \frac{1}{n+1}.$$

18. Prove

Identity 45.

$$\sum_{k=0}^{n} \binom{n}{k} \frac{1}{(k+1)^2} = \frac{1}{n+1} \sum_{k=0}^{n} \frac{2^{k+1}-1}{k+1}.$$

19. Prove

Identity 46.

$$\sum_{k=0}^{n} \binom{n}{k} \frac{(-1)^k}{(k+1)^2} = \frac{H_{n+1}}{n+1}.$$

20. Prove

Identity 47.

$$\sum_{k \geq 0} \binom{n}{2k} \frac{1}{2k+1} = \frac{2^n}{n+1}.$$

21. Prove

Identity 48.

$$\sum_{k \geq 0} \binom{n}{2k+1} \frac{1}{k+1} = \frac{2^{n+1}-2}{n+1}.$$

22. While in this chapter we have focused on examples using the absorption identity, the other basic binomial coefficient identities in Exercise 1 of Chapter 1 can also be used to prove more sophisticated binomial coefficient identities. For example, use trinomial revision, Identity 7,

$$\binom{n}{m}\binom{m}{k} = \binom{n}{k}\binom{n-k}{m-k},$$

to prove

Identity 49.

$$\sum_{k=0}^{n} \binom{n}{k}\binom{k}{m} = 2^{n-m}\binom{n}{m}.$$

23. Use trinomial revision (Identity 7) to generalize Identity 49 to the following:

Identity 50.

$$\sum_{k=0}^{n} \binom{n}{k}\binom{k}{m} x^{n-k} y^{k-m} = (x+y)^{n-m}\binom{n}{m}.$$

24. Use trinomial revision (Identity 7) and parallel summation (Identity 15),

$$\sum_{k=0}^{m} \binom{n+k}{k} = \binom{n+m+1}{m},$$

to prove

Identity 51.

$$\sum_{k=0}^{m} \frac{\binom{m}{k}}{\binom{n}{k}} = \frac{\binom{n+1}{m}}{\binom{n}{m}} = \frac{n+1}{n+1-m}.$$

25. Prove

Identity 52.

$$\sum_{k=0}^{m} \frac{\binom{m}{k}\binom{k}{r}}{\binom{n}{k}} = \frac{\binom{n+1}{m-r}}{\binom{n}{m}}.$$

(Hint: While similar to Exercise 24, you will also need to use Identity 77, which we have not seen yet.)

26. What identity do we obtain if we apply binomial inversion to Identity 20? (Don't do too much work after applying binomial inversion.)

27. Prove the following variation of the binomial inversion formula in Theorem 1.

Theorem 2.

$$f(n) = \sum_{k=0}^{n} \binom{n}{k} g(k) \iff g(n) = \sum_{k=0}^{n} \binom{n}{k} f(k)(-1)^{n-k}.$$

28. Prove

Identity 53.

$$\sum_{k=0}^{n} \binom{n}{k} \frac{(-1)^k}{k+2} = \frac{1}{(n+1)(n+2)}.$$

29. Prove

Identity 54.

$$\sum_{k=0}^{n} \binom{n}{k} \frac{H_{k+1}}{k+1}(-1)^k = \frac{1}{(n+1)^2}.$$

30. Prove

Identity 55.

$$\sum_{k=0}^{n} \binom{n}{k} k^{\underline{m}} 2^{k-m}(-1)^k = (-1)^n n^{\underline{m}}.$$

31. Prove

Identity 56.

$$\sum_{k=0}^{n} \binom{n}{k} \frac{2^k}{k+1}(-1)^k = \frac{1}{n+1}[n \text{ is even}].$$

32. (This exercise completes the proof of Identity 18.) Let

$$M_n(x) = \sum_{k=0}^{\infty} \binom{n}{k} x^k.$$

(a) Show that $M_n(x)$ has a radius of convergence of 1.

(b) Show that $y = M_n(x)$ satisfies the differential equation

$$(1+x)\frac{dy}{dx} = ny.$$

(c) Finally, show that $\dfrac{d}{dx}\dfrac{M_n(x)}{(1+x)^n} = 0$, and use this to prove that
$M_n(x) = (1+x)^n$ for $|x| < 1$.

2.6   Notes

Propp [54] uses the Euler characteristic to generalize the combinatorial interpretation of $\binom{n}{k}$ to values of n other than nonnegative integers. In particular,
his work shows how a negative integer value for n can be interpreted combinatorially in a way that makes sense of Identity 19,

$$\binom{-n}{k} = (-1)^k \binom{n+k-1}{k}.$$

As we shall see in Chapter 3 (specifically, Identity 62), $\binom{n+k-1}{k}$ can be thought
of as the number of ways to choose k elements from a set containing n elements
when the same element is allowed to be chosen more than once.

There are many more inverse relations involving the binomial coefficients
than those given in Theorems 1 and 2. In fact, Riordan devotes two chapters
of his *Combinatorial Identities* [57] to this topic.

3

Combinatorics

A combinatorial proof of an identity tends to involve some kind of counting argument. Because of this, combinatorial proofs are some of the most beautiful proofs in mathematics: They frequently give more insight into why an identity is true than an argument that relies more on symbolic manipulations. Finding a good combinatorial proof is often not as straightforward as proving a result by other methods, however; while you get better at constructing combinatorial proofs with experience, it does tend to be something of an art. This chapter, in addition to proving binomial coefficient identities, is an introduction to some basic combinatorial proof techniques. We will primarily use the interpretation of $\binom{n}{k}$ as the number of subsets of size k that can be formed from n elements, but we will also see how this way of thinking about binomial coefficients leads to other combinatorial interpretations that may not be obvious at first. For example, we will see that $\binom{n}{k}$ counts the number of ways to choose a committee of size k from a group of n people, as well as the number of sequences of n coin flips in which exactly k flips are heads. We also discuss how binomial coefficients can be used to count certain kinds of paths in a lattice, as well as the number of selections possible when objects can be chosen more than once. In the final two sections we look at alternating sum binomial identities and two techniques for proving them: involutions and the principle of inclusion-exclusion. Finally, Chapter 8 contains several additional combinatorial proofs involving Fibonacci, Stirling, and other kinds of numbers.

While we will generally not take the time to point this out for each proof, it may be helpful to keep in mind two primary archetypes for combinatorial proofs of binomial identities. First, there are *double-counting proofs*, which show that the two sides of an identity count the number of elements in a particular set but in different ways. Since the underlying set is the same the two sides of the identity must be equal. There are also *bijective proofs*. In these the two sides of an identity are shown to count different sets, but then these sets are demonstrated to have the same size via a bijection from one set to the other. Most of our combinatorial proofs will be double-counting arguments, but we will see some bijective arguments as well.

3.1 Basic Arguments

Let's review the binomial coefficient identities we've already proved combinatorially.

- Identity 1 (Pascal's Recurrence):

$$\binom{n}{k} = \binom{n-1}{k} + \binom{n-1}{k-1},$$

with $\binom{n}{0} = \binom{n}{n} = 1$.

- Identity 2:

$$\binom{n}{k} = \frac{n!}{k!(n-k)!}.$$

(Well, technically, we proved the combinatorial definition of the binomial coefficients from this formula, but the argument is reversible.)

- Identity 3 (Binomial Theorem):

$$(x+y)^n = \sum_{k=0}^{n} \binom{n}{k} x^k y^{n-k}.$$

- Identity 4:

$$\sum_{k=0}^{n} \binom{n}{k} = 2^n.$$

Probably the most important binomial coefficient identity not among this list is the symmetry one.

Identity 5 (Symmetry).

$$\binom{n}{k} = \binom{n}{n-k}.$$

Proof. The left side of this identity counts the number of subsets of size k chosen from n elements, while the right side counts the number of subsets of size $n-k$ chosen from n elements. Given a subset S of size k chosen from n elements, there is exactly one subset S' of size $n-k$ consisting of the $n-k$ elements *not* in S. Thus there as many subsets of size k from n elements as there are subsets of size $n-k$. □

This proof of Identity 5 is an example of a bijective proof.

The next six identities (not counting Identity 59, which is a special case of the identity just before it) can be grouped in pairs. Each pair illustrates a somewhat different combinatorial approach.

Identity 6 (Absorption Identity).

$$k\binom{n}{k} = n\binom{n-1}{k-1}.$$

Proof. Let's think about what the left side might be counting. We know that $\binom{n}{k}$ is the number of subsets of size k chosen from $\{1, 2, \ldots, n\}$. If we single out one of the elements from that subset, we have k choices, and so that gives us a combinatorial interpretation of $k\binom{n}{k}$. Rephrasing this more concretely, $k\binom{n}{k}$ is the number of ways to choose a committee of size k from n people and give the committee a chair.

Alternatively, we could choose the chair of the committee first. There are n ways to do this. Then there are $\binom{n-1}{k-1}$ ways to choose the remaining members of the committee. So there are also $n\binom{n-1}{k-1}$ ways to choose a committee of size k from n people and give the committee a chair. □

This proof of Identity 6 is an example of a double-counting proof. It also illustrates a useful technique for constructing such proofs: Swap the order in which something is being counted. The next identity gives another example.

Identity 7 (Trinomial Revision).

$$\binom{n}{m}\binom{m}{k} = \binom{n}{k}\binom{n-k}{m-k}.$$

Proof. Again, let's think about what the left side might be counting. If we use the committee interpretation again, $\binom{n}{m}$ is the number of committees of size m from n people, and $\binom{m}{k}$ could be interpreted as the number of ways to choose a subcommittee of size k from that committee. So $\binom{n}{m}\binom{m}{k}$ would be the number of ways to choose a committee of size m containing a subcommittee of size k from n total people.

We could also choose the subcommittee first. There are $\binom{n}{k}$ ways to do this. Then there are $\binom{n-k}{m-k}$ ways to choose the members of the committee who will not be on the subcommittee. Thus $\binom{n}{k}\binom{n-k}{m-k}$ is also the number of ways to choose a committee of size m containing a subcommittee of size k from n total people. □

Many binomial coefficient identities involve sums. These can often be interpreted combinatorially by finding a combinatorial interpretation of the summand for a fixed k, and then summing up over all possible values of k. For simpler sums over the lower index in the binomial coefficient, the committee interpretation often works. The next two identities illustrate this idea. (As a reminder, all identities are listed in the index of identities and theorems starting on page 343.)

Identity 21.

$$\sum_{k=0}^{n} \binom{n}{k} k = n2^{n-1}.$$

Proof. We already have a combinatorial interpretation of $k\binom{n}{k}$—namely, the number of chaired committees of size k that can be formed from n people. Summing over all values of k must give the total number of chaired committees that could be chosen from n people.

For the right side, the n could come from choosing the chair first. Then, as in the combinatorial proof of Identity 4, any of the remaining $n-1$ people could be chosen or not chosen for the committee. With two choices for each person, there are 2^{n-1} committees that could be formed from the remaining $n-1$ people. Thus $n2^{n-1}$ also gives the number of chaired committees that could be chosen from n people. $\square$

The next identity is called *Vandermonde's identity.*

Identity 57 (Vandermonde's Identity).

$$\sum_{k=0}^{r}\binom{n}{k}\binom{m}{r-k} = \binom{n+m}{r}.$$

Proof. The summand on the left side pairs a committee of size k from a group of n people with a committee of size $r-k$ from a second group of m people. If we combine the two committees we get one of size r in which k are from the first group and $r-k$ are from the second group. Summing over all possible values of k must yield the total number of ways to choose a committee of size r from the two groups of $n+m$ people; i.e., $\binom{n+m}{r}$. $\square$

For simpler binomial sums where the summation is over the upper index it sometimes helps to interpret the summand as the number of ways a set of objects could be placed in certain positions. We will see this in the next few identities.

Born in Paris, **Alexandre-Théophile Vandermonde** (1735–1796) spent his early adulthood as a violinist, turning to mathematics only at age 35. He published just four mathematics papers, but they include contributions to combinatorics and the theory of determinants, as well as problems in areas that today we would call abstract algebra and topology. Interestingly enough, the Vandermonde determinant, despite being named for him, does not actually appear anywhere in his published work. Vandermonde also worked on music theory, arguing (perhaps somewhat surprisingly for a mathematician) that music should be considered more of an art than a science.

Identity 58.

$$\sum_{k=0}^{n}\binom{k}{m} = \binom{n+1}{m+1}.$$

Proof. Imagine we flip a coin $n+1$ times. The right side $\binom{n+1}{m+1}$ can be thought of as the number of those sequences of coin flips in which we obtain $m+1$

heads, as choosing $m+1$ elements from $\{1, 2, \ldots, n+1\}$ can be thought of as choosing on which flips the heads occur.

For the left side, we're choosing m positions rather than $m+1$, and the total number of positions from which we're allowed to choose is the variable of summation. If the *last* head occurs on flip $k+1$, then the number of ways that could happen would be the number of ways to choose m positions from the k possible ones before the last head. This is $\binom{k}{m}$. Thus summing $\binom{k}{m}$ from 0 to n also gives the number of sequences of $n+1$ coin flips in which $m+1$ heads are obtained. $\qquad\square$

A special case of Identity 58 is the famous formula for the sum of the first n positive integers.

Identity 59.
$$\sum_{k=0}^{n} k = \binom{n+1}{2} = \frac{n(n+1)}{2}.$$

We actually give two proofs of Identity 59. The first is a straightforward consequence of Identity 58. For the second, though, we give a different combinatorial interpretation of the identity—one that generalizes nicely to other power sums.

Proof 1. Take $m = 1$ in Identity 58. $\qquad\square$

Proof 2. One way to interpret the right side of Identity 59 is that you are choosing two numbers out of $n+1$ possibilities. One can think of that in terms of functions f from $\{0, 1\}$ (a set of size two) to $\{0, 1, \ldots, n\}$ (a set of size $n+1$). However, the number of such functions is actually $(n+1)^2$, as there are $n+1$ choices for $f(0)$ and for $f(1)$. Thus $\binom{n+1}{2}$ must count the number of such functions with some restrictions. One of these must be that $f(0)$ and $f(1)$ are different, as the binomial coefficient necessarily chooses different things. The other is that once the two numbers in $\{0, 1, \ldots, n\}$ are selected to form the range, there is only one way to assign these numbers as $f(0)$ and $f(1)$. This can be accomplished by requiring $f(0) > f(1)$. To sum up, the right side of Identity 59 counts the number of functions from $\{0, 1\}$ to $\{0, 1, \ldots, n\}$ in which $f(0) > f(1)$.

The left side must condition on something; a natural choice is the value of $f(0)$. If $f(0) = k$, then the restriction $f(0) > f(1)$ means that $f(1)$ can be any of $\{0, 1, \ldots, k-1\}$, for a total of k choices. Summing over the possible values of $f(0)$, then, means that the left side also counts the number of functions from $\{0, 1\}$ to $\{0, 1, \ldots, n\}$ in which $f(0) > f(1)$. $\qquad\square$

(The sum of the first n squares is Exercise 20. The generalization to the sum of the first n mth powers is more difficult; we will see it as Identity 227 in Chapter 8.)

Going back to the coin-flipping interpretations, here's another identity that's a little harder.

Identity 15 (Parallel Summation).

$$\sum_{k=0}^{m} \binom{n+k}{k} = \binom{n+m+1}{m}.$$

Proof. The right side can be interpreted as the number of sequences of $n+m+1$ coin flips that produce m tails and $n+1$ heads. With the proof of Identity 58 in mind, the summand $\binom{n+k}{k}$ on the left side can be thought of as the number of sequences in which exactly k of the tails occur before the last head. Summing k from 0 to m thus yields the number of sequences of $n+m+1$ coin flips in which m tails and $n+1$ heads are obtained. □

Here's a slightly more difficult variation on the previous identity.

Identity 60.

$$\sum_{k=0}^{n} \binom{n+k}{k} 2^{n-k} = 4^n.$$

Proof. As in the proof we just did, we can think of $\binom{n+k}{k}$ as counting the number of ways to have k tails occur before head $n+1$. But then the additional factor of 2^{n-k} means that after head $n+1$ we could have either heads or tails occur with each of $n-k$ additional flips. Thus the left side is counting sequences of $2n+1$ coin flips in which *at least* $n+1$ heads occur. This is equivalent to counting sequences of $2n+1$ coin flips in which there are more heads than tails.

Now the right side becomes clear as well. Of the 2^{2n+1} possible sequences, half would have more heads than tails. There are $\frac{1}{2}(2^{2n+1}) = 4^n$ of these. □

Instead of conditioning on the last event of some particular kind occurring, as we did in three prior proofs, sometimes it is helpful to think of that special "last" event as a separate kind of event. Thus, for example, we could also prove Identity 15 by supposing we have n blue balls, m red balls, and 1 black ball and counting the number of sequences of balls that have the black ball occurring after all of the blue balls. Here's a more explicit example.

Identity 61.

$$\sum_{k=0}^{n} \binom{k}{m} \binom{n-k}{r} = \binom{n+1}{m+r+1}.$$

Proof. Using the colored balls interpretation, the right side looks like the number of ways to place m blue balls, r red balls, 1 black ball, and $n - m - r$ green balls in a certain configuration. Since the right side consists solely of $\binom{n+1}{m+r+1}$, though, the ordering of the balls must be determined once the positions of the blue and red balls and the black ball are chosen.

Since the black ball is the "special ball," the left side is probably conditioning on its position. Then $\binom{k}{m}$ would be the number of ways to place the blue balls before the black ball if the black ball falls in position $k + 1$. The factor $\binom{n-k}{r}$ would be the number of ways to place the red balls in the remaining $n - k$ positions. So the left side counts sequences of these colored balls in which all the blue balls occur before the black ball and all the red balls occur after the black ball, with the green balls interspersed in the remaining $n - m - r$ positions. This fits with the right side interpretation as well, as once $m + r + 1$ positions are chosen, the blue balls must go in the first m of these positions, the black ball in the next position, and the red balls in the last r chosen positions. The green balls fill out the remaining (unchosen) positions. □

3.2 Lattice Path Counting

Another place binomial coefficients appear is in a classic counting problem involving paths in a grid. An $n \times m$ grid of squares is called a *lattice*. See Figure 3.1, for example. A path in the lattice that only uses steps of one unit to the right or one unit up is called a *lattice path*. How many lattice paths are there from $(0, 0)$ to (n, m)? The answer turns out to be the binomial coefficient $\binom{n+m}{n}$. Let's prove that.

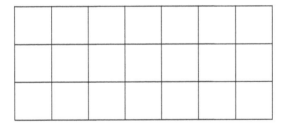

FIGURE 3.1
A 7×3 lattice

Theorem 3. *There are $\binom{n+m}{n}$ lattice paths from $(0, 0)$ to (n, m).*

Proof. Regardless of the path you take you must take a total of n steps to the right and m steps up, for a total of $n + m$ steps. Out of these $n + m$ steps, the steps to the right could be the first n, or the last n, or any other combination of n steps. In fact, the path is determined by which of the $n + m$ steps are the n steps to the right. This is because once the right steps are chosen, the remaining m steps must be up steps. (Alternatively, one could choose which m steps are to be up steps. Then the remaining steps must be steps to the right.) For example, the path in Figure 3.2 is the result of choosing steps 1, 3, 4, 7, 8, 9, and 10 as the right steps. This forces steps 2, 5, and 6 to be the up steps. Choosing a path determines the steps, too, of course, so that each selection of right steps corresponds to exactly one path, and vice versa. There are $\binom{n+m}{n}$ ways to choose which steps will be the right steps in a $n \times m$ lattice, and thus there are $\binom{n+m}{n}$ lattice paths from $(0,0)$ to (n,m). $\qquad\square$

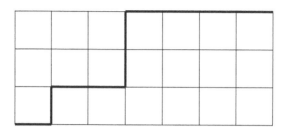

FIGURE 3.2
A 7×3 lattice with a path

The lattice path interpretation gives a different take on some binomial coefficient identities we have already seen. In fact, we have already made a lattice path argument for the following.

Identity 5.
$$\binom{n}{k} = \binom{n}{n-k}.$$

Proof. Both sides count the number of lattice paths from $(0,0)$ to $(k, n-k)$. There are $\binom{n}{k}$ choices for the right steps, and then the up steps are determined by that choice. Alternatively, there are $\binom{n}{n-k}$ choices for the up states, and then the right steps are determined by that choice. $\qquad\square$

One way to obtain an identity from lattice path counting is to condition on where the path crosses a certain line in the lattice. For example, every lattice path from $(0,0)$ to (n,m) must touch the top of the lattice at some point. If we condition on the *first* time a path touches the top, we get a new proof of Identity 58.

Identity 58.
$$\sum_{k=0}^{n} \binom{k}{m} = \binom{n+1}{m+1}.$$

Proof. By Theorem 3, by counting up steps we know there are $\binom{n+1}{m+1}$ lattice paths from $(0,0)$ to $(n-m, m+1)$. As we argued above, each lattice path must touch the line $y = m+1$ somewhere for the first time. Touching the line $y = m+1$ for the first time at coordinate $(j, m+1)$ means precisely that the path uses the line segment from (j, m) to $(j, m+1)$ as part of its path, as any path that does not use this line segment does not hit $y = m+1$ for the first time at $(j, m+1)$. Another way to think of it is that the path hitting $y = m+1$ for the first time at $(j, m+1)$ is equivalent to the path crossing $y = m + 1/2$ at $x = j$. See, for example, Figure 3.3, where the path crossing the dashed line $y = 2.5$ at $x = 3$ is equivalent to the path touching $y = 3$ for the first time at $(3, 3)$.

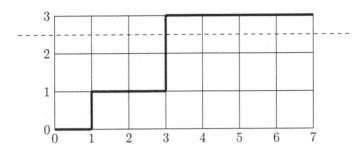

FIGURE 3.3
This lattice path touches the top for the first time at (3,3).

If k is the number of steps taken before the path crosses $y = m+1/2$, then there are $\binom{k}{m}$ paths from $(0,0)$ to (j, m), where m is the number of up steps and $k = j + m$. Every path that touches $y = m + 1$ first at $(j, m + 1)$ then traverses the segment (j, m) to $(j, m + 1)$. From $(j, m + 1)$ to $(n - m, m + 1)$ the path can only move to the right. Thus there are $\binom{k}{m}$ paths from $(0,0)$ to $(n - m, m + 1)$ that touch the line $y = m + 1$ for the first time at $(j, m + 1)$, where $k = j + m$. Since k can run from 0 to $n - m + m = n$, summing over all values of k produces

$$\sum_{k=0}^{n} \binom{k}{m} = \binom{n+1}{m+1}.$$

$\square$

While we did say that this is a new proof of Identity 58, in fact it is structurally quite similar to the combinatorial proof of Identity 58 we presented earlier in the chapter: In that proof, heads correspond to steps up, and tails correspond to steps right.

We can modify the lattice path argument for Identity 58 to prove its generalization in Identity 61.

Identity 61.

$$\sum_{k=0}^{n} \binom{k}{m}\binom{n-k}{r} = \binom{n+1}{m+r+1}.$$

Proof. By Theorem 3, the right side is the number of lattice paths from $(0,0)$ to $(n - m - r, m + r + 1)$. For the left side, condition on the number k of steps the path takes before the line $y = m + 1/2$. The path must take exactly m steps up before crossing the line, and there are $\binom{k}{m}$ possible ways to choose them. Then the path takes the step that crosses the line $y = m + 1/2$. There are $n - k$ steps remaining, of which r steps must be up, for a total of $\binom{n-k}{r}$ possible paths after crossing $y = m + 1/2$. See, for example, Figure 3.4, where $n = 9$, $m = 1$, $r = 1$, and $k = 4$. Multiplying together and summing over all possible values of k, we have

$$\sum_{k=0}^{n} \binom{k}{m}\binom{n-k}{r} = \binom{n+1}{m+r+1}.$$

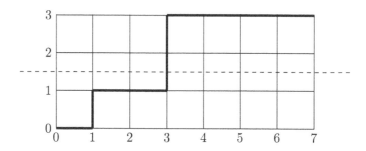

FIGURE 3.4
This lattice path crosses the line $y = 1.5$ after completing four steps.

□

Again, notice the structural similarities between this proof and the colored balls proof we presented earlier in the chapter. The black ball is the step that crosses the line $y = m + 1/2$, the blue balls are the up steps before crossing the line, the red balls are the up steps after crossing the line, and the green balls are the right steps.

As a final example, conditioning on where the path crosses one of the downward-sloping diagonals gives us a lattice path proof of Vandermonde's identity.

Identity 57 (Vandermonde's Identity).

$$\sum_{k=0}^{r} \binom{n}{k}\binom{m}{r-k} = \binom{n+m}{r}.$$

Proof. Unlike the previous two proofs, the path crosses the dividing line in question at a lattice point (see Figure 3.5) as opposed to in the middle of a line segment.

Let's let n and m be the number of steps taken by the path before and after reaching a given downward-sloping diagonal, respectively. As we can

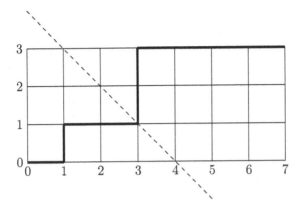

FIGURE 3.5
This lattice path crosses the diagonal after three right and four total steps.

see in Figure 3.5, n and m are functions of the diagonal chosen and do not change from path to path. Let r be the number of right steps in the lattice. In Figure 3.5, we have $n = 4$, $m = 6$, and $r = 7$. The total number of lattice paths from the lower left diagonal to the upper right corner is clearly $\binom{n+m}{r}$.

Let k be the number of right steps before reaching the diagonal. For the path in Figure 3.5, $k = 3$. There are $\binom{n}{k}$ ways to choose these and thus $\binom{n}{k}$ paths before touching the diagonal. Then there are $r - k$ remaining rightward steps out of m more to be taken. Thus there are $\binom{m}{r-k}$ possible paths after touching the diagonal. Multiplying and summing over all values of k, we have Vandermonde's identity,

$$\sum_{k=0}^{n} \binom{n}{k}\binom{m}{r-k} = \binom{n+m}{r}.$$

$\square$

Counting lattice paths with respect to touching an upward-sloping diagonal is more complicated, as the paths don't necessarily cross the diagonal exactly once—or even at all! We'll consider this problem in Chapter 8 when we look at Catalan numbers.

3.3 Choosing with Replacement

The binomial coefficient $\binom{n}{k}$ counts the number of ways to choose k items, without replacement, from a group of n elements. In other words, once an item is chosen, it cannot be chosen again. What if we allow replacement, though, so that an item can be chosen more than once? Instead of constructing a set, we would be constructing what is called a *multiset*. For example, $\{1,1,2,3\}$ and $\{2,2,2,3\}$ are multisets of size 4 chosen from the elements $\{1,2,3\}$. Let $\left(\binom{n}{k}\right)$ denote the number of multisets of size k chosen from n elements.

Another perspective on $\left(\binom{n}{k}\right)$ is that it counts the number of ways that k identical candies can be distributed among n people. For example, if we had four candies to give to three people, $\{1,1,2,3\}$ would correspond to person 1 receiving two candies, person 2 receiving one candy, and person 3 receiving one candy. From the multiset standpoint, the number of candies person i receives is the number of times i appears in the corresponding multiset. For still another interpretation of $\left(\binom{n}{k}\right)$, see Exercise 5.

The next identity shows that $\left(\binom{n}{k}\right)$ can also be expressed as a binomial coefficient.

Identity 62.

$$\left(\binom{n}{k}\right) = \binom{n+k-1}{k}.$$

Proof. The classic way to show this is with a "stars and bars" argument. Multichoosing k items from a set of n is equivalent to creating a sequence of $n-1$ bars and k stars. The idea is that the bars separate the stars into groups. For example, here is the representation of the multiset $\{1,1,2,3\}$ as a sequence of stars and bars:

$$\star\star \mid \star \mid \star.$$

The number of stars before the first bar indicates the number of times the element 1 is chosen. The number of stars between the first and second bars indicates the number of times the element 2 is chosen, and so forth. We only need $n-1$ bars rather than n because the number of stars after bar $n-1$ indicates the number of times element n is chosen.

Placing a sequence of $n-1$ bars and k stars, though, is equivalent to having $n+k-1$ positions and then choosing which k of these positions will contain stars. Once the stars are placed, the bars must go in the remaining $n-1$ positions. Thus $\left(\binom{n}{k}\right) = \binom{n+k-1}{k}$. □

The interpretation of $\left(\binom{n}{k}\right)$ as a binomial coefficient can lead to combinatorial proofs of binomial coefficient identities. For example, Identity 15 has a combinatorial proof in terms of multichoosing.

Identity 15 (Parallel Summation).

$$\sum_{k=0}^{m} \binom{n+k}{k} = \binom{n+m+1}{m}.$$

Proof. Identity 15 rewritten using multiset notation is the following.

Identity 63.
$$\sum_{k=0}^{m} \left(\binom{n+1}{k}\right) = \left(\binom{n+2}{m}\right).$$

Both sides count the number of ways to choose m objects with replacement from $n+2$ total objects. For the right side, this is just the combinatorial interpretation of $\left(\binom{n+2}{m}\right)$. The summand on the left side counts the number of ways to choose k objects with replacement from the first $n+1$ objects, as the remaining $m-k$ must then be chosen from item $n+2$. Summing over all values of k yields the total number of ways to choose m objects with replacement from $n+2$ total objects. $\square$

Some binomial identities look more elegant in multichoose form. Identity 64, a multichoose version of Vandermonde's identity (Identity 57), is such an example.

Identity 64.
$$\sum_{k=0}^{r} \left(\binom{n}{k}\right) \left(\binom{m}{r-k}\right) = \left(\binom{n+m}{r}\right).$$

Proof. Suppose we have r identical pieces of candy to give to n boys and m girls. The right side is the number of ways to distribute the candies to the $n+m$ children.

For the left side, condition on the number k of candies distributed among the n boys. The number of ways to do this is $\left(\binom{n}{k}\right)$. The remaining $r-k$ candies must be distributed among the m girls, in $\left(\binom{m}{r-k}\right)$ ways. Multiplying together and summing up yields the identity. $\square$

The interested reader is invited to rewrite Identity 64 in terms of binomial coefficients for comparison. (It turns out to be a special case of Identity 77.)

3.4 Alternating Binomial Sums and Involutions

In the previous chapter we used binomial inversion to prove a few identities featuring alternating binomial sums. At first glance it may seem like such identities cannot possibly have combinatorial proofs, since the sums in the identities contain both addition and subtraction. However, in this chapter we're going to look at two methods for proving identities featuring alternating sums. The first method uses what are called *sign-reversing involutions*. Let's take a look at how this technique works on a proof of the simplest alternating binomial identity we've seen, Identity 12.

Identity 12.

$$\sum_{k=0}^{n} \binom{n}{k}(-1)^k = [n = 0].$$

Proof. First, if $n = 0$ then the identity simplifies to $1 = 1$, which is obviously true. For $n \geq 1$ let's take a closer look at what this identity is claiming. When $n = 4$, for example, it says that

$$\binom{4}{0} - \binom{4}{1} + \binom{4}{2} - \binom{4}{3} + \binom{4}{4} = 0.$$

We could rewrite this as

$$\binom{4}{0} + \binom{4}{2} + \binom{4}{4} = \binom{4}{1} + \binom{4}{3}.$$

The left side of this equation counts the number of subsets of $\{1, 2, \ldots, 4\}$ that have an even number of elements. Similarly, the right side counts the number of subsets that have an odd number of elements. From a combinatorial point of view, then, Identity 12 is saying that there are as many subsets of $\{1, 2, \ldots, n\}$ that have an even number of elements as that have an odd number of elements. To find a combinatorial proof of the identity, then, we could simply show that the set E_n of even-sized subsets of $\{1, 2, \ldots, n\}$ has the same number of elements as O_n, the set of odd-sized subsets. The following function will help us with that.

Given a subset S of $\{1, 2, \ldots, n\}$, define ϕ by

$$\phi(S) = \begin{cases} S \cup \{n\}, & n \notin S; \\ S - \{n\}, & n \in S. \end{cases}$$

Basically, ϕ looks to see whether the set S contains the largest element, element n. If S doesn't contain n, ϕ adds it to S. If S does contain n, ϕ removes it from S. For example, in the $n = 4$ case ϕ maps the following subsets of $\{1, 2, 3, 4\}$ to each other:

$\emptyset \Leftrightarrow \{4\}$	$\{1, 2\} \Leftrightarrow \{1, 2, 4\}$
$\{1\} \Leftrightarrow \{1, 4\}$	$\{1, 3\} \Leftrightarrow \{1, 3, 4\}$
$\{2\} \Leftrightarrow \{2, 4\}$	$\{2, 3\} \Leftrightarrow \{2, 3, 4\}$
$\{3\} \Leftrightarrow \{3, 4\}$	$\{1, 2, 3\} \Leftrightarrow \{1, 2, 3, 4\}$

The important thing to note is that ϕ pairs up even-sized subsets with odd-sized subsets. The fact that such a pairing exists proves that E_4 and O_4 have the same number of elements, which in turn proves Identity 12 in the $n = 4$ case.

The function ϕ is what's called an *involution*: ϕ is its own inverse. We can see this, for example, with $S = \{1, 2\}$. Applying ϕ to S yields $\{1, 2, 4\}$, and applying ϕ again gives us $\{1,2\}$ back. Mathematically, we say $\phi(\phi(S)) = S$. The fact that ϕ exists proves $|E_4| = |O_4|$ because involutions form a bijection between their domains and codomains, and the only way finite sets can have a bijection between them is if they have the same number of elements.

We're now ready to move to the general n case. If $S \subseteq \{1, 2, \ldots, n\}$ then $\phi(S)$ either has one more or one fewer elements than S. This means that if S has an even number of elements then $\phi(S)$ has an odd number of elements. In other words, if $S \in E_n$ then $\phi(S) \in O_n$. Similarly, if $S \in O_n$ then $\phi(S) \in E_n$. In addition (as we argued before), $\phi(\phi(S)) = S$. This means that ϕ is an involution between E_n and O_n, and so $|E_n| = |O_n|$.

The preceding discussion has been under the assumption that $n \geq 1$, but the interpretation also holds in the case where $n = 0$. Here, ϕ is not defined, as there is no largest element n. Thus we must go back to what Identity 12 is saying: There is one more even-sized subset of the empty set than there are odd-sized subsets. Since the only subset of the empty set is itself, and it has even size, the sum is 1 when $n = 0$. $\qquad \square$

The adjective *sign-reversing* is often added to functions such as ϕ because they form an involution between a set whose elements contribute positively to the sum in question (the even-sized subsets of $\{1, 2, \ldots, n\}$ in the statement of Identity 12) and a set whose elements contribute negatively. Sign-reversing involutions can often be used to prove alternating binomial sum identities.

Identity 12 also says that half of the subsets of $\{1, 2, \ldots, n\}$ have an even number of elements and half have an odd number of elements. With 2^n total subsets, that translates into a combinatorial explanation of the following two identities, which we have already seen. (The presence of the Iverson bracket in Identity 13 is just to ensure that the sum evaluates to 1 when $n = 0$.)

Identity 13.

$$\sum_k \binom{n}{2k} = 2^{n-1} + \frac{1}{2}[n = 0].$$

Identity 14.

$$\sum_{k \geq 0} \binom{n}{2k+1} = 2^{n-1}[n \geq 1].$$

An involution similar to the one in the proof of Identity 12 works on the next identity.

Identity 22.

$$\sum_{k=0}^{n} \binom{n}{k} k(-1)^k = -[n = 1].$$

Proof. We know that $\binom{n}{k}k$ counts the number of chaired committees of size k that can be formed from a group of n people. From this point of view, the identity is saying that there are as many chaired committees with an even number of members as there are with an odd number of members, except in the case $n = 1$, where there is one more committee with an odd number of members. When $n = 0$, there can be no chaired committees, and so the identity holds in this case.

For $n \geq 1$, let's construct a sign-reversing involution on the set of chaired committees that can be formed from n people. Number the people from 1 to n. Let (S, x) denote a set S of committee members, together with its chair x. If we try to use the same involution as with Identity 12, though, we're going to run into problems, as the involution must map chaired committees to chaired committees. The difficulty is that if the chair x is also person n, removal of person n would result in an unchaired committee.

We can get around this by modifying the involution in the proof of Identity 12. Given S and x, let y denote the person with the largest number who is not x. If we remove y from S then we still have a chaired committee, so we're fine there. Define

$$\phi(S, x) = \begin{cases} (S \cup \{y\}, x), & y \notin S; \\ (S - \{y\}, x), & y \in S. \end{cases}$$

Since (S, x) and $\phi(S, x)$ have different parities (the number of people on each committee is increased or decreased by one upon application of ϕ) and $\phi(\phi(S, x)) = (S, x)$, ϕ is a sign-reversing involution on the set of all chaired committees. The chaired committees paired up by the involution cancel each other out in the sum.

This doesn't quite prove that $\sum_{k=0}^{n} \binom{n}{k} k (-1)^k = 0$ in all cases, though. It turns out that there are some chaired committees for which ϕ is not defined. These committees are thus not paired up with any others by ϕ, and so we still have to count these leftover committees in order to determine the value of $\sum_{k=0}^{n} \binom{n}{k} (-1)^k$. Which committees are these? Well, the only way in which ϕ might not be defined is for y not to exist. As long as $n \geq 2$, though, y is defined, as there is guaranteed to be a person with largest number who is not x. Thus $\sum_{k=0}^{n} \binom{n}{k} k (-1)^k = 0$ when $n \geq 2$. When $n = 1$, however, the chaired committee consisting of person 1 has (obviously) person $x = 1$ as the chair. There is no person with largest number who is not x, and so ϕ is not defined in this case. This single chaired committee when $n = 1$ has odd parity because 1 is odd. Thus $\sum_{k=0}^{n} \binom{n}{k} k (-1)^k = -1$ when $n = 1$. $\qquad\square$

As we can see with this example, sometimes the involution is not defined on part of the set under consideration. When this happens we must count separately the leftover elements. In Identity 22 this was easy to do. Here is a slightly more difficult example.

Identity 65.

$$\sum_{k=0}^{n} \binom{n}{k}^2 (-1)^k = (-1)^{n/2} \binom{n}{n/2} [n \text{ is even}].$$

Proof.

Since $\binom{n}{k} = \binom{n}{n-k}$, we can rewrite the sum as $\sum_{k=0}^{n} \binom{n}{k}\binom{n}{n-k}(-1)^k$. Then $\binom{n}{k}\binom{n}{n-k}$ can be thought of as counting ordered pairs (A, B), each of which is a subset of $\{1, 2, \ldots, n\}$, such that $|A| = k$ and $|B| = n - k$. The sum, then, is taken over all pairs such that $|A| + |B| = n$.

Given (A, B), let x denote the largest element in the *symmetric difference* $A \oplus B = (A - B) \cup (B - A)$. Effectively, x is the largest element that is in either A or B but not both. Let's define ϕ so that it moves x from A to B if x is in A and from B to A if x is in B. In other words,

$$\phi(A, B) = \begin{cases} (A - \{x\}, B \cup \{x\}), x \in A; \\ (A \cup \{x\}, B - \{x\}), x \in B. \end{cases}$$

Since the parity of (A, B) is determined by the number of elements in A, and $\phi(A, B)$ has one more or one fewer elements in its first set than does A, ϕ is sign-reversing. Since $\phi(\phi(A, B)) = (A, B)$, ϕ is also an involution. To complete the proof we need to count the pairs (A, B) for which ϕ is not defined. Let C denote the set of such pairs. Then C consists of those (A, B) for which there is no largest element in $A \oplus B$; i.e., pairs (A, B) for which $A \oplus B = \emptyset$. However, the only way to have $A \oplus B = \emptyset$ is to have $A = B$. If n is odd then the requirement $|A| + |B| = n$ means that we cannot have $A = B$, and so $|C| = 0$. If n is even then $|C|$ is just the number of subsets of $\{1, 2, \ldots, n\}$ of size $n/2$; i.e., $\binom{n}{n/2}$, and the parity is determined by whether $|A| = n/2$ is odd or even. □

In addition to the exercises in this chapter, see Chapter 8 for another example of using a sign-reversing involution to prove an identity.

3.5   The Principle of Inclusion-Exclusion

Another useful tool for proving alternating binomial sum identities is the principle of inclusion-exclusion. While the formal statement of the principle can appear frightening (especially the first time you see it), it encodes a rather basic idea about counting by first overcounting and then subtracting off the part that was overcounted. Let's look at an example to see how this works.

Let S be the set of positive integers up to 60 that are divisible by 2 or 3. How big is S? There are 30 numbers up to 60 that are divisible by 2 (the even numbers) and 20 that are divisible by 3 (the multiples of 3). However, it's not true that $|S| = 30 + 20 = 50$, because that double-counts the numbers that are multiples of 2 and of 3. These double-counted numbers are the multiples of 6, and there are 10 of them. Subtracting them off removes the effect of double-counting, and we get $|S| = 30 + 20 - 10 = 40$. (The reader can verify this by listing these numbers if he or she wishes.)

Let's add a level of complexity by letting S now be the set of positive integers up to 60 that are divisible by 2, 3, or 5. We know that $|S|$ is not $30 + 20 + 12$ because of the overcounting. We can compensate as before for the overcounting by subtracting off the multiples of 6, 10, and 15, as doing so removes those numbers that are multiples of both 2 and 3, 2 and 5, and 3 and 5, respectively. However, if we do this we're subtracting off too much: Those numbers that are multiples of 2, 3, and 5 don't get counted at all! They are counted once with the multiples of 2, once with the multiples of 3, and once with the multiples of 5, yet they also get "uncounted" once with the multiples of 6, once with the multiples of 10, and once with the multiples of 15. To fix this we must add back the multiples of 30. Putting all of this together, we get $|S| = 30 + 20 + 12 - 10 - 6 - 4 + 2 = 44$.

We can express this formally as follows: Let A_1, A_2, and A_3 be the positive integers up to 60 that are multiples of 2, 3, and 5, respectively. Our argument in the previous paragraph says that

$$|A_1 \cup A_2 \cup A_3| = |A_1| + |A_2| + |A_3| - |A_1 \cap A_2| - |A_1 \cap A_3| - |A_2 \cap A_3| + |A_1 \cap A_2 \cap A_3|.$$

This should hold true regardless of the interpretation we place on A_1, A_2, and A_3.

What about the size of the union of four sets, or five, or even n sets? The principle of inclusion-exclusion tells how to deal with that. We will give it in two forms: The first is the generalization of the idea we've been discussing to n sets, and the second (in somewhat simplified form) is more useful to us for proving binomial identities.

Theorem 4 (Principle of Inclusion-Exclusion). *Given finite sets $A_1, A_2, \ldots, A_n$,*

$$\left| \bigcup_{i=1}^{n} A_i \right| = \sum_{i=1}^{n} |A_i| - \sum_{1 \leq i < j \leq n} |A_i \cap A_j| + \sum_{1 \leq i < j < k \leq n} |A_i \cap A_j \cap A_k| - \cdots + (-1)^{n-1} |A_1 \cap \cdots \cap A_n|,$$

and, if $A_i \subseteq A$ for each i and A is also finite,

$$\left| \bigcap_{i=1}^{n} A_i \right| = |A| - \sum_{i=1}^{n} |\bar{A}_i| + \sum_{1 \leq i < j \leq n} |\bar{A}_i \cap \bar{A}_j| - \sum_{1 \leq i < j < k \leq n} |\bar{A}_i \cap \bar{A}_j \cap \bar{A}_k| + \cdots + (-1)^{n} |\bar{A}_1 \cap \cdots \cap \bar{A}_n|.$$

Again, the first formula says that to find the number of elements in the union of the A_i's, we can start by summing the number of elements in each of the A_i's. However, any element in two particular sets is double-counted, and so to correct that we need to subtract the number of elements in the intersections of each pair of the A_i's. Unfortunately, any element in the intersection of any three sets is subtracted off too many times, and so we have to add those back, and so forth. The second formula is just the first rewritten using DeMorgan's law for distributing complements over unions.

Proof. Let $x \in \bigcup_{i=1}^{n} A_i$. The proof will show that the left and right sides of the first formula in Theorem 4 each count x exactly once. For the left side, this is clear, so the work is in showing that the right side counts x a total of one time as well. First, if A_i does not contain x then any intersection with A_i will count x zero times. So we only need to consider the A_i's that do contain x. Suppose there are m of these A_i's that contain x. (Since x must be in at least one of the A_i's, $m \geq 1$.) How many times will x be counted in the intersection of j of these m subsets? The intersection of any particular j of these will contain x once, and there are $\binom{m}{j}$ ways to choose j of the m sets to intersect, so for a fixed value of j the contribution to the number of times the right side counts x is $\binom{m}{j}(-1)^{j-1}$. Summing up over all possible values of j, we have $\sum_{j=1}^{m} \binom{m}{j}(-1)^{j-1} = 1 - \sum_{j=0}^{m} \binom{m}{j}(-1)^{j} = 1$, by Identity 12. Thus the left and right sides both count x a total of one time. As this must be true for any element in the union of the A_i's, the first formula is proved.

For the second formula, we have

$$\left| \bigcap_{i=1}^{n} A_i \right| = \left| \overline{\bigcup_{i=1}^{n} \bar{A}_i} \right|$$

$$= |A| - \left| \bigcup_{i=1}^{n} \bar{A}_i \right|$$

$$= |A| - \sum_{i=1}^{n} |\bar{A}_i| + \sum_{1 \leq i < j \leq n} |\bar{A}_i \cap \bar{A}_j|$$

$$- \sum_{1 \leq i < j < k \leq n} |\bar{A}_i \cap \bar{A}_j \cap \bar{A}_k| + \cdots + (-1)^n |\bar{A}_1 \cap \cdots \cap \bar{A}_n|,$$

where the first step is DeMorgan's law for distributing complements over unions. (See Exercise 1.) □

There are no binomial coefficients in the statement of Theorem 4. However, for many applications the intersection of any k of the sets $A_1, A_2, \ldots, A_n$ does not depend on the particular choice of the subsets. In that simplified case Theorem 4 reduces to the following two formulas, which do contain binomial coefficients.

Corollary 1. *Given finite sets $A_1, A_2, \ldots, A_n$, if the intersection of any k of these has cardinality a_k independent of the choice of the k sets, then*

$$\left| \bigcup_{k=1}^{n} A_k \right| = \sum_{k=1}^{n} \binom{n}{k} a_k (-1)^{k-1}.$$

If $A_i \subseteq A$ for each i, A is finite, the intersection of any k of the $\bar{A}_i$'s has cardinality $\bar{a}_k$ independent of the choice of the k sets, and we define $\bar{a}_0 = |A|$ (so that an empty intersection is the common superset), then

$$\left| \bigcap_{k=1}^{n} A_k \right| = \sum_{k=0}^{n} \binom{n}{k} \bar{a}_k (-1)^k.$$

Proof. If the intersection of any k of the sets $A_1, A_2, \ldots, A_n$ has cardinality a_k, then the sum over all such intersections reduces to a_k times the number $\binom{n}{k}$ of ways of choosing k of the sets to intersect. Then the first formula in Theorem 4 simplifies to

$$\left| \bigcup_{k=1}^{n} A_k \right| = \sum_{k=1}^{n} \binom{n}{k} a_k (-1)^{k-1}.$$

The argument for the second formula is virtually identical.   □

The second of the two formulas in Corollary 1 is generally the more useful for proving binomial coefficient identities, as the sum in the second formula is over all possible values of k.

Our proof of Theorem 4 used Identity 12. However, Identity 12 turns out to be a consequence of Theorem 4 as well. Thus Identity 12 is equivalent to Theorem 4. Let's take a look at that proof.

Identity 12.

$$\sum_{k=0}^{n} \binom{n}{k} (-1)^k = [n = 0].$$

Proof. As with our involution proof, the identity is clearly true when $n = 0$. Otherwise, suppose we have n copies of the empty set; i.e., $A_1 = A_2 = \cdots = A_n = \emptyset$. Each of these can be considered a subset of $A = \{a\}$. Since the intersection of any k of the sets $\bar{A}_1, \bar{A}_2, \ldots, \bar{A}_n$ is just $\{a\}$, as is A itself, by Corollary 1 we have

$$0 = \left| \bigcap_{k=1}^{n} A_k \right| = \sum_{k=0}^{n} \binom{n}{k} (-1)^k.$$

□

Here are two more examples of the use of Corollary 1.

Identity 66.

$$\sum_{k=0}^{n} \binom{n}{k} 2^{n-k} (-1)^k = 1.$$

Proof. Let A_k, $1 \le k \le n$, be the collection of all subsets of $\{1, 2, \ldots, n\}$ that do not contain the element k. We have $|\bigcap_{k=1}^{n} A_k| = 1$, as the intersection of all of the A_k's consists only of the empty set. The intersection of $\bar{A}_{i_1}, \bar{A}_{i_2}, \ldots, \bar{A}_{i_k}$ is the collection of all subsets that do contain all of $i_1, i_2, \ldots, i_k$. Since any such subset must contain these k elements, and any of the remaining $n - k$ elements can independently be either in or out of these subsets, there are $1^k 2^{n-k}$ such subsets. Also, there are 2^n total subsets of $\{1, 2, \ldots, n\}$. By Corollary 1, then, we have

$$1 = \sum_{k=0}^{n} \binom{n}{k} 2^{n-k} (-1)^k.$$

$\square$

Identity 67.

$$\sum_{k=0}^{n} \binom{n}{k} (n - k)^n (-1)^k = n!.$$

Proof. Let A_k be the set of all functions f from the set $\{1, 2, \ldots, n\}$ to itself for which $f(i) = k$ for at least one i. Thus the intersection of the A_k's is the set of onto functions from $\{1, 2, \ldots, n\}$ to itself. Since an onto function from a finite set to itself must be one-to-one as well, the intersection of the A_k's is the set of bijections on $\{1, 2, \ldots, n\}$. Since there are n choices for $f(1)$, followed by $n - 1$ choices for $f(2)$ given $f(1)$, and so forth, and there are $n!$ bijections on $\{1, 2, \ldots, n\}$.

The functions in the intersection of $\bar{A}_{i_1}, \bar{A}_{i_2}, \ldots, \bar{A}_{i_k}$ are those that map from the set $\{1, 2, \ldots, n\}$ to the set $\{1, 2, \ldots, n\} - \{i_1, i_2, \ldots i_k\}$. There are $n - k$ elements in this latter set, and so there are $(n - k)^n$ functions in $\bigcap_{j=1}^{k} \bar{A}_{i_j}$. In addition, there are n^n total functions from $\{1, 2, \ldots, n\}$ to itself. By Corollary 1, then,

$$n! = \sum_{k=0}^{n} \binom{n}{k} (n - k)^n (-1)^k.$$

$\square$

We end this section by discussing what is probably the classic use of inclusion-exclusion: deriving the formula for the number of derangements on n elements.

The derangement problem is often described using a scenario like the following. Suppose each of n students writes his name on a piece of paper and drops it into a hat. After all the names are in the hat each student pulls out a slip of paper from the hat. In how many ways can the students pull out the slips of paper so that no student gets his own name?

More formally, a *fixed point* of a permutation is one in which the same element is mapped to itself. For example, the permutation $(3, 2, 1, 4)$ contains two fixed points: 2 and 4, as 2 occurs in the second position and 4 in the fourth. A *derangement* is a permutation with no fixed points. We denote the number of derangements on n elements by D_n. For example, $D_4 = 9$, as there are nine permutations on four elements that have no fixed points:

$$(2, 1, 4, 3), \quad (2, 3, 4, 1), \quad (2, 4, 1, 3), \quad (3, 1, 4, 2), \quad (3, 4, 1, 2),$$
$$(3, 4, 2, 1), \quad (4, 1, 2, 3), \quad (4, 3, 1, 2), \quad (4, 3, 2, 1).$$

Identity 68. *Let D_n denote the number of derangements on n elements. Then*

$$D_n = \sum_{k=0}^{n} \binom{n}{k} (n-k)!(-1)^k = n! \sum_{k=0}^{n} \frac{(-1)^k}{k!}.$$

Proof. Let A_i denote the set of permutations on n elements in which element i is deranged (i.e., is not mapped to itself). The intersection of all n of the A_i's is the set of derangements on n elements and thus has cardinality D_n. The set $\bar{A}_i$ is the set of permutations on n elements in which i is mapped to itself. Thus the intersection of any k of these is the set of permutations in which k specific elements are mapped to themselves. If these k elements must be mapped to themselves, that leaves us with $n - k$ elements to be permuted; this can be done in $(n-k)!$ ways. Thus $\bar{a}_k = (n-k)!$. By Corollary 1, we have

$$D_n = \sum_{k=0}^{n} \binom{n}{k} (n-k)!(-1)^k = n! \sum_{k=0}^{n} \frac{(-1)^k}{k!}.$$

$\square$

We will see the probabilistic version of the principle of inclusion-exclusion in Chapter 5. We will also see inclusion-exclusion in Chapter 8, where we use it to prove an identity involving the Stirling numbers of the second kind.

3.6 Exercises

For the combinatorial proofs, feel free to think in terms of any of the scenarios presented in this chapter, such as committees, coin flips, colored balls, and lattice paths. Proofs of alternating sum identities will require either sign-reversing involutions or the principle of inclusion-exclusion.

1. Prove DeMorgan's laws for finite sets $A_1, A_2, \ldots, A_n$, each of which is a subset of another finite set A.

   (a)

   $$\overline{\bigcap_{i=1}^{n} A_i} = \bigcup_{i=1}^{n} \bar{A}_i.$$

(b)

$$\overline{\bigcup_{i=1}^{n} A_i} = \bigcap_{i=1}^{n} \bar{A}_i.$$

2. Show that $\binom{n+k-1}{k-1}$ counts the number of solutions in nonnegative integers to the equation $x_1 + x_2 + \cdots + x_k = n$.

3. Show that $\binom{n-1}{k-1}$ counts the number of solutions in positive integers to the equation $x_1 + x_2 + \cdots + x_k = n$.

4. Show that $\binom{n}{k}$ counts the number of increasing sequences of length k that consist only of numbers from $\{1, 2, \ldots, n\}$.

5. Show that $\left(\binom{n}{k}\right)$ counts the number of nondecreasing sequences of length k that consist only of numbers from $\{1, 2, \ldots, n\}$.

6. Show that, for $0 \le m \le k - 1$, $\binom{n+m}{k}$ counts the number of nondecreasing sequences $(a_i)_{i=1}^{k}$ of length k that consist only of numbers from $\{1, 2, \ldots, n\}$ and for which there are m fixed values of i such that $a_i = a_{i-1}$ is allowed. For example, if $n = 10$ and $a_i = a_{i-1}$ is allowed only for $i \in \{2, 4, 6\}$ then $(1, 1, 3, 4, 5, 9)$ is an acceptable sequence (as the only duplicate occurs in position 2) but $(1, 2, 2, 2, 8, 8)$ is not (the duplicates in positions 4 and 6 are fine, but the duplicate in position 3 is not allowed). (This exercise is a generalization of Exercises 4 and 5; Exercise 4 is the special case $m = 0$, and Exercise 5 is the special case $m = k - 1$.)

7. Use a lattice path argument to prove Identity 15.

Identity 15 (Parallel Summation).

$$\sum_{k=0}^{m} \binom{n+k}{k} = \binom{n+m+1}{m}.$$

8. Identity 17 expresses the binomial coefficient in terms of the falling factorial:

$$\binom{n}{k} = \begin{cases} \dfrac{n^{\underline{k}}}{k!} = \dfrac{n(n-1)(n-2)\cdots(n-k+1)}{k(k-1)\cdots(1)}, & k \ge 0; \\ 0, & k < 0. \end{cases}$$

Show that $\left(\binom{n}{k}\right)$ can be expressed in terms of the rising factorial:

Identity 69.

$$\left(\binom{n}{k}\right) = \begin{cases} \dfrac{n^{\overline{k}}}{k!} = \dfrac{n(n+1)(n+2)\cdots(n+k-1)}{k(k-1)\cdots(1)}, & k \ge 0; \\ 0, & k < 0. \end{cases}$$

9. Prove

   Identity 63.

   $$\sum_{k=0}^{m} \left(\!\!\binom{n+1}{k}\!\!\right) = \left(\!\!\binom{n+2}{m}\!\!\right),$$

   the multiset version of Identity 15, using the interpretation of $\left(\!\!\binom{n}{k}\!\!\right)$ as the number of ways to distribute k identical candies to n people.

10. Give a combinatorial proof for the recurrence relation for the multichoose coefficients $\left(\!\!\binom{n}{k}\!\!\right)$.

    Identity 70.

    $$\left(\!\!\binom{n}{k}\!\!\right) = \left(\!\!\binom{n}{k-1}\!\!\right) + \left(\!\!\binom{n-1}{k}\!\!\right),$$

    valid for integers $n > 0, k > 0$, with boundary conditions $\left(\!\!\binom{n}{0}\!\!\right) = 1$ for $n \geq 0$ and $\left(\!\!\binom{0}{k}\!\!\right) = 0$ for $k > 0$.

    Find combinatorial proofs for the identities in Exercises 11 through 38.

11.

   Identity 8.

   $$(n-k)\binom{n}{k} = n\binom{n-1}{k}.$$

12.

   Identity 9.

   $$\binom{n}{m}\binom{n-m}{p} = \binom{n}{m+p}\binom{m+p}{m}.$$

13.

   Identity 10.

   $$k(k-1)\binom{n}{k} = n(n-1)\binom{n-2}{k-2}.$$

14.

   Identity 11.

   $$n\binom{n}{k} = k\binom{n}{k} + (k+1)\binom{n}{k+1}.$$

15.

   Identity 35.

   $$\sum_{k=0}^{n}\binom{n}{k}k(k-1) = n(n-1)2^{n-2}.$$

16.

Identity 71.

$$\sum_{k=0}^{n} \binom{n}{k} k^2 = n(n+1)2^{n-2}.$$

17. (A generalization of Identity 21.)

Identity 36.

$$\sum_{k=0}^{n} \binom{n}{k} k^{\underline{m}} = n^{\underline{m}} 2^{n-m}.$$

18.

Identity 72.

$$\sum_{k=0}^{n} \binom{n}{k} k(n-k) = n(n-1)2^{n-2}.$$

19.

Identity 73.

$$\sum_{k=0}^{n} \binom{k}{m} k = n \binom{n+1}{m+1} - \binom{n+1}{m+2}.$$

20.

Identity 74.

$$\sum_{k=0}^{n} k^2 = \binom{n+1}{2} + 2\binom{n+1}{3}.$$

(Compare with Identity 203.)

21.

Identity 49.

$$\sum_{k=0}^{n} \binom{n}{k}\binom{k}{m} = 2^{n-m}\binom{n}{m}.$$

22. For positive integers x and y,

Identity 50.

$$\sum_{k=0}^{n} \binom{n}{k}\binom{k}{m} x^{n-k} y^{k-m} = (x+y)^{n-m}\binom{n}{m}.$$

23.

Identity 75.

$$\sum_{k=0}^{n} \binom{n}{k}\binom{m}{r-k}k = n\binom{n+m-1}{r-1}.$$

24.

Identity 76.

$$\sum_{k=0}^{n} \binom{n}{k}\binom{m}{r-k}k^2 = n\binom{n+m-1}{r-1} + n(n-1)\binom{n+m-2}{r-2}.$$

25. (A generalization of Identity 61. Try for both colored-balls and lattice path proofs.)

Identity 77.

$$\sum_{k=-m}^{n} \binom{m+k}{r}\binom{n-k}{s} = \binom{m+n+1}{r+s+1}.$$

26. (A generalization of Identity 12.)

Identity 20.

$$\sum_{k=0}^{m} \binom{n}{k}(-1)^k = (-1)^m\binom{n-1}{m}.$$

27.

Identity 37.

$$\sum_{k=0}^{n} \binom{n}{k}k(k-1)(-1)^k = 2[n=2].$$

28.

Identity 78.

$$\sum_{k=0}^{n} \binom{n}{k}k^2(-1)^k = (-1)^n n![n=1 \ or \ n=2].$$

29.

Identity 38.

$$\sum_{k=0}^{n} \binom{n}{k}k^{\underline{m}}(-1)^k = (-1)^m m![n=m].$$

30.

Identity 26.

$$\sum_{k=0}^{n} \binom{n}{k} \binom{k}{m} (-1)^k = (-1)^m [n = m].$$

31. (A generalization of Identity 26.)

Identity 79.

$$\sum_{k=0}^{n} \binom{n}{k} \binom{r+k}{m} (-1)^k = (-1)^n \binom{r}{m-n}.$$

32.

Identity 80.

$$\sum_{k=0}^{n} \binom{n}{k} \binom{k-1}{m} (-1)^k = (-1)^m [n \leq m].$$

33.

Identity 81.

$$\sum_{k=0}^{m} \binom{n}{k} \binom{n}{m-k} (-1)^k = (-1)^{m/2} \binom{n}{m/2} [m \text{ is even}].$$

34.

Identity 82.

$$\sum_{k=0}^{m} \left(\binom{n}{k}\right) \left(\binom{n}{m-k}\right) (-1)^k = \left(\binom{n}{m/2}\right) [m \text{ is even}].$$

35.

Identity 28.

$$\sum_{k=0}^{n} \binom{n}{k} k 2^{k-1} (-1)^k = (-1)^n n.$$

36.

Identity 83.

$$\sum_{k=0}^{n} \binom{n}{k}^2 k(-1)^k = \begin{cases} (-1)^{(n+1)/2} n \binom{n-1}{(n-1)/2}, & n \text{ is odd;} \\ (-1)^{n/2} n \binom{n-1}{(n-2)/2}, & n \text{ is even.} \end{cases}$$

37.

Identity 84. *If $m < n$,*

$$\sum_{k=0}^{n} \binom{n}{k}(n-k)^m(-1)^k = 0.$$

38.

Identity 85.

$$\sum_{k=0}^{n} \binom{n}{k}(n-k)^{n+1}(-1)^k = \frac{n(n+1)!}{2}.$$

39. Use an inclusion-exclusion argument to prove the following identity.

Identity 86.

$$\sum_{k=0}^{n} \binom{n}{k}\binom{2n-k}{n}(-1)^k = 1.$$

40. Prove the following identity using an inclusion-exclusion argument.

Identity 87.

$$\sum_{k=0}^{n} \binom{n}{k}\binom{n+m-k-1}{m}(-1)^k = \binom{m-1}{n-1}.$$

41. Identity 68 contains a formula for the number of derangements D_n on n elements. Here is a different proof. First, give a combinatorial proof of the following:

Identity 88.

$$\sum_{k=0}^{n} \binom{n}{k}D_{n-k} = n!.$$

Then apply binomial inversion to Identity 88 to obtain Identity 68:

$$D_n = \sum_{k=0}^{n} \binom{n}{k}(n-k)!(-1)^k = n!\sum_{k=0}^{n}\frac{(-1)^k}{k!}.$$

3.7   Notes

A standard reference for proving binomial coefficient identities via combinatorial arguments is Benjamin and Quinn's *Proofs that Really Count* [7]. Several of the proofs in this chapter, or variations of them, appear in this text.

Lattice path proofs have a long history in combinatorics. Feller's text [26] is a classic; see also Mohanty's [49] monograph and Humphreys's survey article [37].

The principle of inclusion-exclusion can be described in more generality than we give here. Chapter 2 of Stanley's *Enumerative Combinatorics* [75] has a nice discussion. See also Chapter 5 of Aigner's *A Course in Enumeration* [1] for a more sophisticated discussion of inclusion-exclusion and involution arguments.

Lutgen [45] gives four proofs of Identity 82, including a combinatorial one, as well as a couple of extensions.

4

Calculus

Most people who have a passing familiarity with binomial identities probably do not think of calculus when they think of techniques for proving them. In this chapter we'll show that calculus is actually a very powerful tool for proving certain kinds of binomial identities. We'll see how simply differentiating and integrating the binomial theorem can yield quick proofs of many identities. We'll prove Leibniz's generalized product rule for derivatives and note that it involves the binomial coefficients, and we'll derive some identities from the rule as well. In the last section we'll go a little deeper by considering two special functions that can be defined in terms of integrals, the gamma and beta functions. We'll see how the binomial coefficient can be represented in terms of the beta integral, and we'll use this representation to prove several binomial identities we have not yet seen.

4.1   Differentiation

As a reminder, the binomial theorem (Identity 3) is

$$\sum_{k=0}^{n} \binom{n}{k} x^k y^{n-k} = (x+y)^n.$$

Differentiating the binomial theorem in different ways is the primary focus of this section. As we shall see, it can be used to prove a surprisingly large number of binomial identities.

For example, if we differentiate the binomial theorem with respect to x, we get

Identity 89.

$$\sum_{k=0}^{n} \binom{n}{k} k x^{k-1} y^{n-k} = n(x+y)^{n-1}.$$

Choosing different values of x and y in Identity 89 produces different identities, including some we have already seen. However, we do have to be a little careful in the case $n = 0$, as choosing values for x and y such that $x = -y$ would yield division by zero. In situations like this, the formula assumes that the n is evaluated first (which is, of course, consistent with the power rule we used to produce it). The same care needs to be taken with subsequent identities involving differentiation (such as Identities 90, 91, 92, and 93).

Two simple examples with Identity 89 produce identities we have already seen. If we let $x = y = 1$ then we get Identity 21:

Identity 21.

$$\sum_{k=0}^{n} \binom{n}{k} k = n2^{n-1}.$$

Substituting $x = -1, y = 1$ yields Identity 22:

Identity 22.

$$\sum_{k=0}^{n} \binom{n}{k} k(-1)^k = -[n = 1].$$

We can take further derivatives as well. For instance, a second derivative of the binomial theorem with respect to x yields

Identity 90.

$$\sum_{k=0}^{n} \binom{n}{k} k(k-1)x^{k-2}y^{n-k} = n(n-1)(x+y)^{n-2}.$$

Identity 35, which we have already seen, is the special case $x = y = 1$:

Identity 35.

$$\sum_{k=0}^{n} \binom{n}{k} k(k-1) = n(n-1)2^{n-2}.$$

We can continue taking derivatives with respect to x, and it is not too hard to see how the pattern goes, so let's just jump to the general case. Let $x^{\underline{m}}$ denote the falling factorial $x^{\underline{m}} = x(x-1)\cdots(x-m+1)$, which we first saw in Chapter 2. If we differentiate the binomial theorem m times with respect to x then we have the following.

Identity 91.

$$\sum_{k=0}^{n} \binom{n}{k} k^{\underline{m}} x^{k-m} y^{n-k} = n^{\underline{m}}(x+y)^{n-m}.$$

For example, Identity 36 is the special case $x = y = 1$.

Identity 36.

$$\sum_{k=0}^{n} \binom{n}{k} k^{\underline{m}} = n^{\underline{m}} 2^{n-m}.$$

Identity 36 gives what we might call a binomial falling factorial sum, and we can obtain variations on this by substituting different values of x and y into Identity 91. But what if we want a binomial power sum, say $\sum_{k=0}^{n} \binom{n}{k} k^m$?

Let's start with the k^2 case. Looking at Identity 89,

$$\sum_{k=0}^{n} \binom{n}{k} k x^{k-1} y^{n-k} = n(x+y)^{n-1},$$

what we need is to introduce another factor of k by differentiating x again (or possibly y somehow). However, the exponent on x is not right; it's $k-1$ rather than k. For the exponent to be k, we would need another factor of x. So let's put it in there. This requires multiplying both sides by x, and we get

$$\sum_{k=0}^{n} \binom{n}{k} k x^{k} y^{n-k} = nx(x+y)^{n-1}.$$

Differentiating the right side is slightly more complicated now because we need the product rule, but we can do it. We get $n(x+y)^{n-1} + n(n-1)x(x+y)^{n-2} = n(x+y)^{n-2}(x+y+(n-1)x) = n(x+y)^{n-2}(nx+y)$. This yields the following identity.

Identity 92.

$$\sum_{k=0}^{n} \binom{n}{k} k^2 x^{k-1} y^{n-k} = n(x+y)^{n-2}(nx+y).$$

Now we can obtain our power sum by letting $x = y = 1$. We get Identity 71.

Identity 71.

$$\sum_{k=0}^{n} \binom{n}{k} k^2 = n(n+1)2^{n-2}.$$

If we repeat the process that we used to prove Identity 92—multiplying by x and then differentiating with respect to x—we can get power sums for k^3, k^4, etc. This gets more complicated as the power on k increases, though. There's a better way to do this with the Stirling numbers of the second kind, which we shall see in Chapter 8.

We can also differentiate with respect to y, although that is equivalent to differentiating with respect to x and reindexing the sum. We can mix x and y derivatives as well. For example, taking one x and one y derivative of the binomial theorem gives us

Identity 93.

$$\sum_{k=0}^{n} \binom{n}{k} k(n-k) x^{k-1} y^{n-k-1} = n(n-1)(x+y)^{n-2}.$$

Then, for example, we can produce Identity 72 by setting $x = y = 1$:

Identity 72.

$$\sum_{k=0}^{n} \binom{n}{k} k(n-k) = n(n-1)2^{n-2}.$$

A less obvious calculus argument allows us to prove the next identity.

Identity 67.

$$\sum_{k=0}^{n} \binom{n}{k} (n-k)^n (-1)^k = n!.$$

Proof. By the binomial theorem, we have

$$\sum_{k=0}^{n} \binom{n}{k} e^{(n-k)x} (-1)^k = (e^x - 1)^n.$$

Now, differentiate both sides n times. The left side is

$$\sum_{k=0}^{n} \binom{n}{k} (n-k)^n e^{(n-k)x} (-1)^k.$$

The right side is (after a derivative or two)

German **Gottfried Wilhelm Leibniz** (1646–1716) was one of the most brilliant polymaths of the 17th century, with contributions to philosophy, psychology, law, politics, ethics, philology, geology, and even library science. Mathematically, he is most known for having discovered (or invented) calculus independently of Isaac Newton. Unfortunately, the argument between their supporters over who first developed calculus is one of the uglier incidents in the history of mathematics. While it is now agreed that Newton's work on calculus preceded Leibniz's, the latter's influence lives on through notation we use for derivatives and integrals today.

$$\frac{d^{n-1}}{dx^{n-1}} \left(n(e^x-1)^{n-1} e^x \right) = \frac{d^{n-2}}{dx^{n-2}} \left(n(n-1)(e^x-1)^{n-2} e^{2x} + n(e^x-1)^{n-1} e^x \right).$$

After n derivatives we will have n terms.

One of these will be the term $n! e^{nx}$. All of the others will contain at least one factor of $e^x - 1$. Now, let $x = 0$. The $e^{(n-k)x}$ factor vanishes on the left side, and everything on the right goes to zero except for $n! e^0$. The identity follows. $\qquad\square$

Since $\frac{d}{dx} e^x = e^x$, judicious use of the exponential function can prove some binomial coefficient identities, as we see with this calculus proof of Identity 67.

The binomial theorem is not the only tool we have for proving binomial identities via differentiation. Leibniz's generalized product rule for derivatives is also useful.

Identity 94 (Generalized Product Rule). *For sufficiently differentiable functions f and g,*

$$(fg)^{(n)} = \sum_{k=0}^{n} \binom{n}{k} f^{(k)} g^{(n-k)}.$$

Proof. We can prove this by induction. When $n = 0$, the formula claims $(fg)(x) = f(x)g(x)$, which of course is true. Now assume that the formula is true for a given n. We have

$$(fg)^{(n+1)}(x) = \frac{d}{dx}(fg)^{(n)}(x) = \frac{d}{dx}\sum_{k=0}^{n}\binom{n}{k}f^{(k)}(x)g^{(n-k)}(x)$$

$$= \sum_{k=0}^{n}\binom{n}{k}f^{(k+1)}(x)g^{(n-k)}(x) + \sum_{k=0}^{n}\binom{n}{k}f^{(k)}(x)g^{(n+1-k)}(x)$$

$$= \sum_{k=1}^{n+1}\binom{n}{k-1}f^{(k)}(x)g^{(n+1-k)}(x) + \sum_{k=0}^{n}\binom{n}{k}f^{(k)}(x)g^{(n+1-k)}(x)$$

$$= \sum_{k=0}^{n+1}\left(\binom{n}{k-1} + \binom{n}{k}\right)f^{(k)}(x)g^{(n+1-k)}(x)$$

$$= \sum_{k=0}^{n+1}\binom{n+1}{k}f^{(k)}(x)g^{(n+1-k)}(x),$$

where the last step follows from Identity 1. $\qquad\square$

The binomial theorem itself is actually a corollary to Leibniz's generalized product rule.

Identity 3 (Binomial Theorem).

$$(x+y)^n = \sum_{k=0}^{n}\binom{n}{k}x^k y^{n-k}.$$

Proof. Take $f(z) = e^{xz}$ and $g(z) = e^{yz}$. Then $(fg)(z) = e^{(x+y)z}$. Subbing into Identity 94, we obtain

$$(x+y)^n e^{(x+y)z} = \sum_{k=0}^{n}\binom{n}{k}x^k e^{xz}y^{n-k}e^{yz}$$

$$e^{(x+y)z}\sum_{k=0}^{n}\binom{n}{k}x^k y^{n-k}.$$

Dividing both sides by $e^{(x+y)z}$ yields the binomial theorem. $\qquad\square$

We can also prove Vandermonde's identity from the generalized product rule.

Identity 57 (Vandermonde's Identity).

$$\sum_{k=0}^{r}\binom{n}{k}\binom{m}{r-k} = \binom{n+m}{r}.$$

Proof. Let $f(x) = x^n$ and $g(x) = x^m$. We have that $f^{(k)}(x) = n(n-1)\cdots(n-k+1) = n^{\underline{k}}x^{n-k}$. Similarly, $g^{(r-k)}(x) = m^{\underline{r-k}}x^{m-r+k}$, and $(fg)^{(r)}(x) = (n+m)^{\underline{r}}x^{n+m-r}$. Applying Identity 94 and differentiating r times, we have

$$(n+m)^{\underline{r}}x^{n+m-r} = \sum_{k=0}^{r}\binom{r}{k}n^{\underline{k}}x^{n-k}m^{\underline{r-k}}x^{m-r+k}$$

$$\implies (n+m)^{\underline{r}}x^{n+m-r} = \sum_{k=0}^{r}\frac{r!}{(r-k)!\,k!}n^{\underline{k}}m^{\underline{r-k}}x^{n+m-r}$$

$$\implies \frac{(n+m)^{\underline{r}}}{r!} = \sum_{k=0}^{r}\frac{n^{\underline{k}}}{k!}\frac{m^{\underline{r-k}}}{(r-k)!}$$

$$\implies \binom{n+m}{r} = \sum_{k=0}^{r}\binom{n}{k}\binom{m}{r-k}.$$

$\square$

See Exercises 8 and 9 for some variations or special cases of Vandermonde's identity.

4.2 Integration

In the previous section we proved binomial identities by differentiating the binomial theorem. Now let's see what sorts of identities we can derive by integrating it. Indefinite integration would leave an additional constant to deal with, so let's just work with definite integration. Another advantage of definite integration is that we can choose bounds; that increases the number of variables we can use. (See Exercise 11 for applying indefinite integration to the binomial theorem.)

Our first identity is the following.

Identity 95.

$$\sum_{k=0}^{n}\binom{n}{k}\frac{(x_2^{k+1} - x_1^{k+1})y^{n-k}}{k+1} = \frac{(x_2+y)^{n+1} - (x_1+y)^{n+1}}{n+1}.$$

Proof. The two sides come from integrating the binomial theorem with respect to x, from x_1 to x_2:

$$\int_{x_1}^{x_2}\sum_{k=0}^{n}\binom{n}{k}x^k y^{n-k}\,dx = \int_{x_1}^{x_2}(x+y)^n\,dx$$

$$\implies \sum_{k=0}^{n}\binom{n}{k}\frac{x^{k+1}y^{n-k}}{k+1}\Big|_{x_1}^{x_2} = \frac{(x+y)^{n+1}}{n+1}\Big|_{x_1}^{x_2}$$

$$\implies \sum_{k=0}^{n} \binom{n}{k} \frac{(x_2^{k+1} - x_1^{k+1})y^{n-k}}{k+1} = \frac{(x_2 + y)^{n+1} - (x_1 + y)^{n+1}}{n+1}.$$

$\square$

With its five variables, Identity 95 may not look particularly elegant. For the most part, though, Identity 95 is less important in practice than some of its special cases are. That is the pattern that we will continue to see in this section: We will derive general identities that may not be all that aesthetically appealing, but—via substitution—the general identity immediately gives us multiple special cases that are more useful.

One such special case of Identity 95 occurs when $y = x_2 = 1$ and $x_1 = 0$:

Identity 23.

$$\sum_{k=0}^{n} \binom{n}{k} \frac{1}{k+1} = \frac{2^{n+1} - 1}{n+1}.$$

Another is when $y = 1$, $x_1 = -1$, $x_2 = 0$.

Identity 24.

$$\sum_{k=0}^{n} \binom{n}{k} \frac{(-1)^k}{k+1} = \frac{1}{n+1}.$$

We can also get an identity involving the even-numbered coefficients from the left side of Identity 23 by letting $y = x_2 = 1$ and $x_1 = -1$ in Identity 95.

Identity 47.

$$\sum_{k \geq 0} \binom{n}{2k} \frac{1}{2k+1} = \frac{2^n}{n+1}.$$

Proof. As mentioned, with $y = x_2 = 1$ and $x_1 = -1$ in Identity 95, we have

$$\sum_{k=0}^{n} \binom{n}{k} \frac{1 + (-1)^k}{k+1} = \frac{2^{n+1}}{n+1}.$$

The numerator on the left side, though, is 0 when k is odd and 2 when k is even. Summing over only even values, then, and dividing by 2, we get Identity 47. $\square$

If we integrate a second time with respect to x we need to pick one of x_1 or x_2 in Identity 95 to be the variable of integration. The variable x_2 seems the more natural choice. With x_1, x_2, and x_3 as our three x-variable parameters, we get the following.

Identity 96.

$$\sum_{k=0}^{n} \binom{n}{k} \left(\frac{(x_3^{k+2} - x_2^{k+2})y^{n-k}}{(k+1)(k+2)} - \frac{(x_3 - x_2)x_1^{k+1}y^{n-k}}{k+1} \right)$$
$$= \frac{(x_3 + y)^{n+2} - (x_2 + y)^{n+2}}{(n+1)(n+2)} - \frac{(x_3 - x_2)(x_1 + y)^{n+1}}{n+1}.$$

Proof. The two sides come from

$$\int_{x_2}^{x_3} \sum_{k=0}^{n} \binom{n}{k} \frac{(x^{k+1} - x_1^{k+1})y^{n-k}}{k+1} \, dx = \int_{x_2}^{x_3} \frac{(x+y)^{n+1} - (x_1+y)^{n+1}}{n+1} \, dx.$$

□

One of the most important special cases of Identity 96 is when $x_1 = x_2 = 0$, $x_3 = y = 1$:

Identity 39.

$$\sum_{k=0}^{n} \binom{n}{k} \frac{1}{(k+1)(k+2)} = \frac{2^{n+2} - 1}{(n+1)(n+2)} - \frac{1}{n+1},$$

Another occurs with $x_1 = x_2 = 0$, $x_3 = -1$, $y = 1$:

Identity 41.

$$\sum_{k=0}^{n} \binom{n}{k} \frac{(-1)^k}{(k+1)(k+2)} = \frac{1}{n+2}.$$

Proof. Identity 96 with $x_1 = x_2 = 0$, $x_3 = -1$, $y = 1$ yields

$$\sum_{k=0}^{n} \binom{n}{k} \frac{(-1)^k}{(k+1)(k+2)} = \frac{1}{n+1} - \frac{1}{(n+1)(n+2)}.$$

However, $\frac{1}{(n+1)(n+2)} = \frac{1}{n+1} - \frac{1}{n+2}$ (via partial fractions decomposition), and so the right side simplifies to $\frac{1}{n+2}$. □

As with differentiation, integrating solely with respect to y is equivalent to integrating with respect to x and reindexing the sum. Also as with differentiation, we can integrate first with respect to x and then with respect to y. For example, integrating Identity 95 with respect to y from y_1 to y_2 produces

Identity 97.

$$\sum_{k=0}^{n} \binom{n}{k} \frac{(x_2^{k+1} - x_1^{k+1})(y_2^{n+1-k} - y_1^{n+1-k})}{(k+1)(n+1-k)}$$
$$= \frac{(x_2+y_2)^{n+2} - (x_2+y_1)^{n+2} - (x_1+y_2)^{n+2} + (x_1+y_1)^{n+2}}{(n+1)(n+2)}.$$

The most important special cases of Identity 97 are $x_2 = y_2 = 1$, $x_1 = y_1 = 0$,

Identity 98.

$$\sum_{k=0}^{n} \binom{n}{k} \frac{1}{(k+1)(n+1-k)} = \frac{2^{n+2} - 2}{(n+1)(n+2)},$$

and $x_1 = -1$, $y_2 = 1$, $x_2 = y_1 = 0$,

Identity 99.

$$\sum_{k=0}^{n} \binom{n}{k} \frac{(-1)^k}{(k+1)(n+1-k)} = \frac{1+(-1)^n}{(n+1)(n+2)} = \frac{2}{(n+1)(n+2)} [n \text{ is even}].$$

What about an identity for $\sum_{k=1}^{n} \binom{n}{k} \frac{1}{k}$? (We proved an expression for this using the absorption identity in Chapter 2, but what about a calculus proof?) This sum is a little harder than the ones we have seen thus far in this section. To introduce a factor of k into the denominator we need to *divide* the binomial theorem by x before performing the integration. Unfortunately, this makes the resulting integral more difficult to evaluate, and we get another sum on the right, too. Let's take a look at the identity, and then we'll go through the derivation.

Identity 100.

$$\sum_{k=1}^{n} \binom{n}{k} \frac{(x_2^k - x_1^k)y^{n-k}}{k} = \sum_{k=1}^{n} \frac{((x_2+y)^k - (x_1+y)^k)y^{n-k}}{k}.$$

Proof. If we simply divide the binomial theorem by x and integrate, the term y^n/x means that an integral from x_1 to x_2 will not be defined if $0 \in [x_1, x_2]$. To avoid this problem, let's subtract y^n from $(x+y)^n$ before dividing by x. We get

$$\int_{x_1}^{x_2} \frac{(x+y)^n - y^n}{x} dx = \int_{x_1}^{x_2} \sum_{k=1}^{n} \binom{n}{k} x^{k-1} y^{n-k} dx = \sum_{k=1}^{n} \binom{n}{k} \frac{(x_2^k - x_1^k)y^{n-k}}{k}.$$

Evaluating the integral a second way would be simpler with something like z^n rather than $(x+y)^n$, so let's try the substitution $z = x + y$. We get

$$\int_{x_1}^{x_2} \frac{(x+y)^n - y^n}{x} dx = \int_{x_1+y}^{x_2+y} \frac{z^n - y^n}{z - y} dz.$$

Fortunately, $z - y$ divides $z^n - y^n$ (see Exercise 12) leaving $\sum_{k=1}^{n} z^{k-1} y^{n-k}$. Thus we have

$$\int_{x_1}^{x_2} \frac{(x+y)^n - y^n}{x} dx = \int_{x_1+y}^{x_2+y} \sum_{k=1}^{n} z^{k-1} y^{n-k} dz$$

$$= \sum_{k=1}^{n} \frac{((x_2+y)^k - (x_1+y)^k)y^{n-k}}{k}.$$

$\square$

Now we have our identity for $\sum_{k=1}^{n} \binom{n}{k} \frac{1}{k}$:

Identity 25.

$$\sum_{k=1}^{n} \binom{n}{k} \frac{1}{k} = \sum_{k=1}^{n} \frac{2^k - 1}{k}.$$

Proof. Take $x_1 = 0, x_2 = y = 1$ in Identity 100. $\square$

Unfortunately, the right side of Identity 25 does not appear to simplify.

Perhaps the simplest special case of Identity 100 is an alternating one, with $x_1 = -1, x_2 = 0, y = 1$. Here, $H_n = \sum_{k=1}^{n} \frac{1}{k}$, the nth harmonic number.

Identity 43.

$$\sum_{k=1}^{n} \binom{n}{k} \frac{(-1)^{k+1}}{k} = H_n.$$

We end this section on integration with an identity involving sums of the form $\sum_{k=0}^{n} \binom{n}{k} \frac{1}{(k+1)^2}$.

Identity 101.

$$\sum_{k=0}^{n} \binom{n}{k} \frac{(x_2^{k+1} - x_1^{k+1}) y^{n-k}}{(k+1)^2} = \frac{1}{n+1} \sum_{k=0}^{n} \frac{((x_2 + y)^{k+1} - (x_1 + y)^{k+1}) y^{n-k}}{k+1}.$$

For the proof, see Exercise 15.

We get a couple of nice corollaries from Identity 101. Not surprisingly, they parallel Identities 25 and 43.

Identity 45.

$$\sum_{k=0}^{n} \binom{n}{k} \frac{1}{(k+1)^2} = \frac{1}{n+1} \sum_{k=0}^{n} \frac{2^{k+1} - 1}{k+1}.$$

Proof. Take $x_1 = 0, x_2 = y = 1$ in Identity 101. $\square$

Identity 46.

$$\sum_{k=0}^{n} \binom{n}{k} \frac{(-1)^k}{(k+1)^2} = \frac{H_{n+1}}{n+1}.$$

Proof. Take $x_1 = -1, x_2 = 0, y = 1$ in Identity 101. $\square$

4.3   The Beta Integral and the Gamma Function

The binomial coefficient $\binom{n}{k}$ also has a representation in terms of Euler's *beta integral*, or *beta function*. This representation can be exploited to prove some new classes of binomial coefficient identities, such as infinite sums involving reciprocals of the binomial coefficients. These identities look quite different from any we have encountered thus far; they are also fairly difficult to prove using techniques we have seen up to this point in the text.

There are multiple ways to prove the representation of the binomial coefficient as a beta integral. The proof in this section uses another important special function, the *gamma function*, whose properties were also first studied by Euler. (For some other methods, see the exercises, as well as Section 5.4.)

Let's take a look at the gamma function. It is denoted $\Gamma(n)$ and defined, for $n > 0$, by

$$\Gamma(n) = \int_0^\infty e^{-t} t^{n-1} dt.$$

The most important property of the gamma function is that when n is a positive integer, it is almost the factorial function. In fact, the integral defining the gamma function is the standard way to generalize the factorial function to positive noninteger values. (To get negative values you have to use analytic continuation in the complex plane; this works for all but the negative integers and zero.) To prove the relationship between the gamma function and the factorial function let's first look at a simple recurrence that the gamma function obeys.

Identity 102. *For $n > 1$, $\Gamma(n) = (n-1)\Gamma(n-1)$.*

Proof. Integration by parts works beautifully here. With the formula $\int u\, dv = uv - \int v\, du$, let $u = t^{n-1}$, $dv = e^{-t} dt$. Then $du = (n-1)t^{n-2} dt$, $v = -e^{-t}$, and we have, for $n > 1$,

$$\Gamma(n) = -t^{n-1} e^{-t} \Big|_0^\infty + (n-1) \int_0^\infty e^{-t} t^{n-2} dt$$

$$= (n-1)\Gamma(n-1).$$

(The expression $\lim_{t\to\infty}(-t^{n-1}e^{-t})$ can be shown to equal 0 by repeated application of l'Hôpital's rule.) $\qquad\square$

With Identity 102 in hand we are now ready to prove that the gamma function generalizes the factorial function.

Identity 103. *For positive integer $n > 0$, $\Gamma(n) = (n-1)!$.*

Proof. All we really need to do to prove Identity 103 is unroll the recurrence in Identity 102 and evaluate $\Gamma(1)$ at the base of the recursion. This is

$$\Gamma(1) = \int_0^\infty e^{-t} dt = -e^{-t} \Big|_0^\infty = 0 - (-1) = 1.$$

Thus we have

$$\Gamma(n) = (n-1)(n-2)\cdots(1)\Gamma(1) = (n-1)!.$$

$\qquad\square$

Since the binomial coefficient can be represented as a ratio of factorials, Identity 103 implies that it can also be expressed as a ratio of gamma functions. This representation is our next step in expressing the binomial coefficient in terms of the beta integral.

Identity 104. *For nonnegative integer k and real n with $n - k > -1$,*

$$\binom{n}{k} = \frac{\Gamma(n+1)}{\Gamma(k+1)\Gamma(n-k+1)}.$$

Proof. If n and k are nonnegative integers with $n \geq k$, this follows directly from Identity 103 and the factorial representation of the binomial coefficients (Identity 2). For other values of n, the recurrence $\Gamma(n) = (n-1)\Gamma(n-1)$ implies that

$$\frac{\Gamma(n+1)}{\Gamma(k+1)\Gamma(n-k+1)} = \frac{n(n-1)\cdots(n-k+1)}{k!},$$

which is consistent with our definition of the generalized binomial coefficient $\binom{n}{k}$ for real n and nonnegative integer k (Identity 17). $\qquad\square$

Now that we've seen how to express the binomial coefficient as a ratio of gamma functions, let's take a look at the relationship between the beta integral and the gamma function. Thanks to Identity 104, this will give us the binomial coefficient $\binom{n}{k}$ in terms of the beta integral.

The *beta integral* is denoted $B(a,b)$ and is defined, for $a > 0$, $b > 0$, to be

$$B(a,b) = \int_0^1 x^{a-1}(1-x)^{b-1}\,dx.$$

(Note, as with the gamma function, the $a - 1$ and $b - 1$ exponents.)

Identity 105. *For nonnegative real a and b,*

$$B(a,b) = \frac{\Gamma(a)\Gamma(b)}{\Gamma(a+b)}.$$

Proof. The product $\Gamma(a)\Gamma(b)$ can be rewritten as a double integral:

$$\Gamma(a)\Gamma(b) = \left(\int_0^\infty e^{-x}x^{a-1}\,dx\right)\left(\int_0^\infty e^{-y}y^{b-1}\,dy\right)$$

$$= \int_0^\infty \int_0^\infty e^{-x-y}x^{a-1}y^{b-1}\,dx\,dy.$$

Now, use the change of variables $x = uv$, $y = u(1-v)$. The Jacobian of the transformation is $\begin{vmatrix} v & u \\ 1-v & -u \end{vmatrix} = -uv - (u - uv) = -u$, and so we get

$$\int_0^\infty \int_0^\infty e^{-x-y}x^{a-1}y^{b-1}\,dx\,dy = \int_0^\infty \int_0^1 e^{-u}(uv)^{a-1}u^{b-1}(1-v)^{b-1}u\,dv\,du$$

$$= \left(\int_0^\infty e^{-u} u^{a+b-1} du \right) \left(\int_0^1 v^{a-1} (1-v)^{b-1} dv \right)$$

$$= \Gamma(a+b) B(a,b).$$

$\square$

Finally, we're ready to state and prove the relationship between the binomial coefficients and the beta integral.

Identity 106. *For nonnegative integer k and real $n - k > -1$,*

$$\binom{n}{k}^{-1} = (n+1) \int_0^1 x^k (1-x)^{n-k} dx.$$

Proof. Thanks to Identities 104 and 105 and the recurrence $\Gamma(n) = (n-1)\Gamma(n-1)$, we have

$$\binom{n}{k}^{-1} = \frac{\Gamma(k+1)\Gamma(n-k+1)}{\Gamma(n+1)} = \frac{(n+1)\Gamma(k+1)\Gamma(n-k+1)}{\Gamma(n+2)}$$

$$= (n+1)B(k+1, n-k+1) = (n+1) \int_0^1 x^k (1-x)^{n-k} dx.$$

$\square$

With the usual notation for the beta integral, Identity 106 says that

$$\binom{n}{k} = \frac{1}{(n+1)B(k+1, n-k+1)}.$$

Now that we have Identity 106, let's look at some binomial identities that can be proved using the beta integral. Some of the more interesting ones are those involving sums of reciprocals of binomial coefficients. For example, the next identity gives the sum of the reciprocals of the binomial coefficients in a particular column of Pascal's triangle.

Identity 107. *For $k \geq 2$,*

$$\sum_{n=k}^{\infty} \frac{1}{\binom{n}{k}} = \frac{k}{k-1}.$$

Proof. Identity 106 gives us

$$\sum_{n=k}^{\infty} \frac{1}{\binom{n}{k}} = \sum_{n=k}^{\infty} (n+1) \int_0^1 x^k (1-x)^{n-k} dx$$

$$= \int_0^1 \frac{x^k}{(1-x)^k} \sum_{n=k}^{\infty} (n+1)(1-x)^n dx.$$

At this point it would be really nice if our summand was something like y^n. Then we would have a geometric series, and this would be easy to evaluate. The infinite sum here almost fits this pattern; the problem is the extra factor of $n + 1$. However, $n + 1$ is only one larger than the exponent on the variable factor $(1 - x)^n$. This means that the summand is (up to sign) the *derivative* of $(1 - x)^{n+1}$, and $(1 - x)^{n+1}$ as a summand would fit the form of a geometric series. This observation leads to a clever technique: Represent the summand as a derivative, and then swap the differentiation and the summation. Applying this idea and working through the resulting manipulations yields us the rest of the proof of the identity.

$$\int_0^1 \frac{x^k}{(1-x)^k} \sum_{n=k}^{\infty} (n+1)(1-x)^n \, dx$$

$$= \int_0^1 \frac{x^k}{(1-x)^k} \sum_{n=k}^{\infty} \frac{d}{dx}(-(1-x)^{n+1}) \, dx$$

$$= -\int_0^1 \frac{x^k}{(1-x)^k} \frac{d}{dx} \left(\sum_{n=k}^{\infty} (1-x)^{n+1} \right) dx$$

$$= -\int_0^1 \frac{x^k}{(1-x)^k} \frac{d}{dx} \left(\frac{(1-x)^{k+1}}{1-(1-x)} \right) dx$$

$$= -\int_0^1 \frac{x^k}{(1-x)^k} \frac{d}{dx} \left(\frac{(1-x)^{k+1}}{x} \right) dx$$

$$= -\int_0^1 \frac{x^k}{(1-x)^k} \frac{x(k+1)(1-x)^k(-1) - (1-x)^{k+1}}{x^2} \, dx$$

$$= \int_0^1 \left((k+1)x^{k-1} + (1-x)x^{k-2} \right) dx$$

$$= \int_0^1 \left(kx^{k-1} + x^{k-2} \right) dx$$

$$= 1 + \frac{1}{k-1}$$

$$= \frac{k}{k-1}.$$

$\square$

A more complicated example involves the sum of the reciprocals of the central binomial coefficients $\binom{2n}{n}$. While in the previous example the sum is a nice, simple fraction, the sum here involves the irrational numbers π and $\sqrt{3}$.

Identity 108.

$$\sum_{n=0}^{\infty} \frac{1}{\binom{2n}{n}} = \frac{4}{3} + \frac{2\pi\sqrt{3}}{27}.$$

Proof. By Identity 106, we have

$$\sum_{n=0}^{\infty} \frac{1}{\binom{2n}{n}} = \sum_{n=0}^{\infty} (2n+1) \int_0^1 x^n (1-x)^n \, dx$$

$$= \int_0^1 \sum_{n=0}^{\infty} (2n+1)(x(1-x))^n \, dx.$$

Now we have a situation similar to that in the proof of Identity 107. By making the variable switch $y = \sqrt{x(1-x)}$ we can proceed exactly as in that proof.

$$\int_0^1 \sum_{n=0}^{\infty} (2n+1)(x(1-x))^n \, dx = \int_{x=0}^1 \sum_{n=0}^{\infty} (2n+1)y^{2n} \, dx$$

$$= \int_{x=0}^1 \sum_{n=0}^{\infty} \frac{d}{dy} \left(y^{2n+1} \right) \, dx$$

$$= \int_{x=0}^1 \frac{d}{dy} \left(\sum_{n=0}^{\infty} y^{2n+1} \right) \, dx$$

$$= \int_{x=0}^1 \frac{d}{dy} \left(\frac{y}{1-y^2} \right) \, dx$$

$$= \int_{x=0}^1 \frac{1+y^2}{(1-y^2)^2} \, dx$$

$$= \int_0^1 \frac{1+x-x^2}{(1-x+x^2)^2} \, dx.$$

At this point we have the integral of a rational function, which can be evaluated via standard integration techniques. (See Exercise 17.) We have

$$\int_0^1 \frac{1+x-x^2}{(1-x+x^2)^2} \, dx = \frac{2\sqrt{3}}{9} \arctan \left(\frac{2x-1}{\sqrt{3}} \right) + \frac{2(2x-1)}{3(1-x+x^2)} \Big|_0^1$$

$$= \frac{2\sqrt{3}}{9} \arctan \left(\frac{1}{\sqrt{3}} \right) + \frac{2}{3} - \frac{2\sqrt{3}}{9} \arctan \left(\frac{-1}{\sqrt{3}} \right) - \frac{-2}{3}$$

$$= \frac{4}{3} + \frac{2\sqrt{3}}{9} \left(\frac{\pi}{6} - \frac{-\pi}{6} \right)$$

$$= \frac{4}{3} + \frac{2\pi\sqrt{3}}{27}.$$

$\square$

See the exercises for more examples finding sums of reciprocals of binomial coefficients.

One of the basic integration techniques taught to calculus students is partial fractions decomposition. For example, $\frac{1}{x(x+1)}$ decomposes fairly quickly, using the Heaviside cover-up method, as $\frac{1}{x} - \frac{1}{x+1}$. Decomposing $\frac{1}{x(x+1)(x+2)}$ this way would be more challenging, though, and attempting something like $(x(x+1)\cdots(x+10))^{-1}$ using Heaviside would only be for those with a special love of algebra. However, $x(x+1)\cdots(x+10) = (x+10)^{\underline{11}}$, and this representation as a falling factorial means the fraction can be thought of as the reciprocal of a generalized binomial coefficient. We use this idea to prove the general case, the decomposition of $(x(x+1)\cdots(x+n))^{-1}$. Identity 109 generalizes Identity 24 as well.

Identity 109. *For nonnegative integer n and real x such that $x \notin \{0, -1, \ldots, -n\}$,*

$$\sum_{k=0}^{n} \binom{n}{k} \frac{(-1)^k}{k+x} = \frac{n!}{x(x+1)\cdots(x+n)}.$$

Proof. We have

$$\frac{n!}{x(x+1)\cdots(x+n)} = \frac{n!}{(x+n-1)^{\underline{n}}(x+n)} = \binom{x+n-1}{n}^{-1} \frac{1}{x+n}$$

$$= \int_0^1 t^n (1-t)^{x-1} dt = \int_0^1 u^{x-1}(1-u)^n du$$

$$= \int_0^1 \sum_{k=0}^{n} \binom{n}{k} u^{k+x-1}(-1)^k du = \sum_{k=0}^{n} \binom{n}{k} \frac{(-1)^k}{k+x},$$

where we use the substitution $u = 1 - t$ in the fourth step and then expand $(1-u)^n$ via the binomial theorem in the fifth step. □

Another property of the gamma function and the beta integral is that they can be used to extend the definition of $\binom{n}{k}$ to all real—or even complex—values n and k for which the gamma function or beta integral exist. For example, see Exercise 23 for the value of $\binom{0}{1/2}$ under this definition.

Finally, the ratio of $\Gamma(n + 1/2)$ to $\Gamma(n)$ is related to the central binomial coefficient $\binom{2n}{n}$; see Exercise 26.

4.4 Exercises

1. Use a calculus argument to prove

Identity 28.

$$\sum_{k=0}^{n} \binom{n}{k} k 2^{k-1} (-1)^k = n(-1)^n.$$

2. Use a calculus argument to prove

Identity 78.

$$\sum_{k=0}^{n} \binom{n}{k} k^2 (-1)^k = (-1)^n n! [n = 1 \text{ or } n = 2].$$

3. Use a calculus argument to prove

Identity 110.

$$\sum_{k=0}^{n} \binom{n}{k} k^3 = n^2 (n + 3) 2^{n-3}.$$

4. Use a calculus argument to prove

Identity 111.

$$\sum_{k=0}^{n} \binom{n}{k} k(k + 1) = n(n + 3) 2^{n-2}.$$

5. Use a calculus argument to prove this generalization of Identity 111:

Identity 112.

$$\sum_{k=0}^{n} \binom{n}{k} k(k + m) = n(n + 2m + 1) 2^{n-2}.$$

6. Suppose we differentiate the binomial theorem j times with respect to x and $n - j$ times with respect to y. What identity do we obtain?

7. Expand $(e^x - e^{-x})^n$ via the binomial theorem and use an argument like that in our calculus proof of Identity 67 to prove

Identity 113.

$$\sum_{k=0}^{n} \binom{n}{k} (n - 2k)^n (-1)^k = 2^n n!.$$

8. Use the generalized product rule with $f(x) = x^{-(n+1)}, g(x) = x^{-(m+1)}$, $n, m \geq 0$, to prove

Identity 114.

$$\sum_{k=0}^{r} \binom{n+k}{n} \binom{m+r-k}{m} = \binom{n+m+r+1}{r}.$$

9. Use the generalized product rule with $f(x) = x^n, g(x) = x^{-(m+1)}$, $n > m \geq 0$, to prove

Identity 115.

$$\sum_{k=0}^{r} \binom{n}{r-k}\binom{m+k}{m}(-1)^k = \binom{n-m-1}{r}.$$

10. Prove these more general versions of Identities 114 and 115 by relabeling variables and indices.

   (a)

   Identity 61.

   $$\sum_{k=-m}^{n} \binom{m+k}{r}\binom{n-k}{s} = \binom{m+n+1}{r+s+1}.$$

   (b)

   Identity 116.

   $$\sum_{k=-s}^{r} \binom{n}{r-k}\binom{s+k}{m}(-1)^k = (-1)^{m+s}\binom{n-m-1}{r+s-m}.$$

11. What identity do we obtain if we use indefinite integration on the binomial theorem, rather than the definite integration we used to prove Identity 95? Show that this identity is a special case of Identity 95.

12. Show that

$$\frac{z^n - y^n}{z-y} = \sum_{k=1}^{n} z^{k-1} y^{n-k}.$$

13. Use a calculus argument to prove the following two identities (which you should find to be special cases of a single identity).

   (a)

   Identity 117.

   $$\sum_{k=0}^{n} \binom{n}{k} \frac{1}{(k+1)(k+2)(k+3)}$$
   $$= \frac{2^{n+3}-1}{(n+1)(n+2)(n+3)} - \frac{1}{(n+1)(n+2)} - \frac{1}{2(n+1)}.$$

   (b)

   Identity 118.

   $$\sum_{k=0}^{n} \binom{n}{k} \frac{(-1)^k}{(k+1)(k+2)(k+3)} = \frac{1}{2(n+3)}.$$

14. Use a calculus argument to prove the following.

Identity 119.
$$\sum_{k=0}^{n} \binom{n}{k} \frac{1}{k+2} = \frac{n2^{n+1}+1}{(n+1)(n+2)}.$$

15. Prove Identity 101:

$$\sum_{k=0}^{n} \binom{n}{k} \frac{(x_2^{k+1} - x_1^{k+1})y^{n-k}}{(k+1)^2} = \frac{1}{n+1} \sum_{k=0}^{n} \frac{((x_2+y)^{k+1} - (x_1+y)^{k+1})y^{n-k}}{k+1}.$$

16. Show that
$$\int_0^1 x^k(1-x)^{n-k}dx = \frac{k}{n+1} \int_0^1 x^{k-1}(1-x)^{n-1-(k-1)}dx$$

when $k > 0$ and $n - k > -1$. Unroll this recurrence and prove the base case $\int_0^1 (1-x)^{n-k}dx = \frac{1}{n-k+1}$, thereby giving a proof of Identity 106 that does not involve the gamma function.

17. Finish the derivation of Identity 108 by completing the following two steps.

(a) Use partial fractions decomposition to show that
$$\frac{1+x-x^2}{(1-x+x^2)^2} = \frac{-1}{1-x+x^2} + \frac{2}{(1-x+x^2)^2}.$$

(b) Then, use trigonometric substitution to show that
$$\int \left(\frac{-1}{1-x+x^2} + \frac{2}{(1-x+x^2)^2} \right) dx$$
$$= \frac{2\sqrt{3}}{9} \arctan\left(\frac{2x-1}{\sqrt{3}} \right) + \frac{2(2x-1)}{3(1-x+x^2)} + C.$$

18. Prove

Identity 120.
$$\sum_{k=0}^{n} \frac{(-1)^k}{\binom{n}{k}} = \frac{2n+2}{n+2} [n \text{ is even}].$$

19. Prove

Identity 121.
$$\sum_{n=0}^{\infty} \frac{2^n}{(2n+1)\binom{2n}{n}} = \frac{\pi}{2}.$$

20. Prove

Identity 122. *For $k \geq 1$,*

$$\sum_{n=k}^{\infty} \frac{1}{\binom{n}{k}(n+1)} = \frac{1}{k}.$$

21. Prove an alternative integral representation of the beta function,

Identity 123.

$$B(a,b) = 2 \int_0^{\pi/2} \sin^{2a-1}\theta \cos^{2b-1}\theta \, d\theta,$$

thereby proving this counterpart to Identity 106:

Identity 124.

$$\binom{n}{k}^{-1} = 2(n+1) \int_0^{\pi/2} \sin^{2k+1}\theta \cos^{2n-2k+1}\theta \, d\theta.$$

22. Give an alternate proof of Identity 107 using Identity 124.

23. Show that the definition of $\binom{n}{k}$ as a beta integral (Identity 106) yields $\binom{0}{1/2} = \frac{2}{\pi}$.

24. Prove that $\Gamma(1/2) = \sqrt{\pi}$. (There are at least two methods that will work. One involves relating $\Gamma(1/2)$ to $B(1/2, 1/2)$ and then evaluating the latter. Another uses the substitution $u = t^{1/2}$, which results in a famous integral.)

25. Prove that $\Gamma(0)$ does not exist by showing that $\int_0^{\infty} \frac{e^{-t}}{t} \, dt$ diverges.

26. Prove the following formula relating the ratio of $\Gamma(n+1/2)$ and $\Gamma(n)$ to the central binomial coefficients:

Identity 125. *For $n > 0$,*

$$\frac{\Gamma(n+1/2)}{\Gamma(n+1)} = \binom{2n}{n} \frac{\sqrt{\pi}}{4^n}.$$

27. In this exercise we'll work through the steps to prove the famous *Wallis product*. (In the next exercise we'll use one of the intermediate results in this derivation to help give a bound on the value of the central binomial coefficient $\binom{2n}{n}$.)

Identity 126 (Wallis Product).

$$\frac{2}{1} \cdot \frac{2}{3} \cdot \frac{4}{3} \cdot \frac{4}{5} \cdot \frac{6}{5} \cdot \frac{6}{7} \cdots = \prod_{n=1}^{\infty} \left(\frac{2n}{2n-1} \cdot \frac{2n}{2n+1} \right) = \frac{\pi}{2}.$$

(a) Use integration by parts to prove the sine reduction formula, valid for $n \geq 2$:

$$\int \sin^n x \, dx = -\frac{1}{n} \sin^{n-1} x \cos x + \frac{n-1}{n} \int \sin^{n-2} x \, dx.$$

(b) Let $I_n = \int_0^{\pi/2} \sin^n x \, dx$, and use the sine reduction formula repeatedly to find explicit formulas for I_{2n} and I_{2n+1}.

(c) Show that

$$\frac{I_{2n+1}}{I_{2n}} = \frac{2}{\pi} \prod_{i=1}^{n} \left(\frac{2i}{2i-1} \cdot \frac{2i}{2i+1} \right)$$

$$= \frac{2}{\pi} \left(\frac{2}{1} \cdot \frac{2}{3} \cdot \frac{4}{3} \cdot \frac{4}{5} \cdot \frac{6}{5} \cdot \frac{6}{7} \cdots \frac{2n \cdot 2n}{(2n-1) \cdot (2n+1)} \right).$$

(d) Use the fact that $0 < \sin x < 1$ for $x \in (0, \pi/2)$ to show that

$$\frac{2n}{2n+1} < \frac{I_{2n+1}}{I_{2n}} < 1.$$

(e) Finally, take limits and apply (c) and (d) to complete the proof.

28. Use the inequality in part (d) of Exercise 27 to help prove

Identity 127.

$$\frac{4^n}{\sqrt{\pi(n+1/2)}} < \binom{2n}{n} < \frac{4^n}{\sqrt{\pi n}}.$$

As a side note, Identity 127 implies a common approximation for the central binomial coefficients:

$$\binom{2n}{n} \approx \frac{4^n}{\sqrt{\pi n}}.$$

4.5   Notes

The proof of Vandermonde's identity (Identity 57) in this chapter appears in Spivey [69].

5

Probability

In this chapter we will see how some binomial identities may be viewed through a probabilistic lens. Generally speaking, probabilistic proofs are a close relative of combinatorial proofs. In fact, many combinatorial proofs can be easily converted into probabilistic ones, and vice versa. However, there are some distinct ways in which probabilistic concepts can be used to prove binomial identities.

The first section of this chapter examines three important distributions whose probability mass functions feature binomial coefficients: the binomial distribution, the negative binomial distribution, and the hypergeometric distribution. We will also look at the use of expected values and indicator variables to prove some binomial identities. Near the end of the chapter we'll see the probabilistic version of inclusion-exclusion, as well as a probabilistic proof of the beta integral representation of $\binom{n}{k}$ that we saw in Chapter 4.

Some basic familiarity with terminology and notation used in discrete probability is required for this chapter, although otherwise the material is largely self-contained. The one exception is the last section involving the beta integral, which does require the reader to know how to find probabilities using the continuous uniform distribution.

5.1 Binomial, Negative Binomial, and Hypergeometric Distributions

Let's start our discussion of probabilistic methods for proving binomial identities with the three major probability distributions that involve the binomial coefficients: the binomial distribution, the negative binomial distribution, and the hypergeometric distribution. What we'll see is that we can use these distributions to set up expressions involving the binomial coefficients. By thinking probabilistically, we will be able to evaluate these expressions fairly easy, yielding binomial identities.

The first two probability distributions—the binomial and negative binomial distributions—can be expressed in terms of coin flipping. Suppose we have a coin that has a probability p of turning up heads on a single flip.

Two of the many probabilistic questions we could ask about this coin are the following.

1. If we flip the coin n times, how many of the n flips are heads?

2. If we flip the coin until we obtain r heads, how many flips does it take?

If, in the first question, X is the number of flips that are heads, then X has what is called the *binomial distribution* with parameters n and p. Before we find the probabilities associated with particular values of X let's get a little more formal about what is going on here.

Flipping a coin is an example of a *Bernoulli trial*, which is just an event in which the outcome is either success or failure. (We can define "success" as "the coin lands head-side up," although we could just as easily have defined "success" as "the coin lands tails-side up.") A *Bernoulli process* is a sequence of independent Bernoulli trials that all have the same success probability p. So flipping the same coin repeatedly is an example of a Bernoulli process. More generally, then, a random variable X has the binomial distribution with parameters n and p if X counts the number of successes in a Bernoulli process consisting of exactly n trials in which each trial has success probability p.

Let's find a formula for $P(X = k)$, the probability that there are exactly k successes in n trials for a binomial (n, p) distribution. This formula is called the *probability mass function* for the binomial distribution.

The Bernoulli family of Basel, Switzerland, produced several famous mathematicians and scientists during the 17th and 18th centuries. **Jacob Bernoulli** (1655–1705) was the oldest of these. He made contributions to calculus, differential equations, and infinite series; discovered the constant e (in the context of continuously compounded interest); and derived an early version of the law of large numbers. The Bernoulli numbers appeared in Jacob's posthumously published *Ars Conjectandi*, where he used them to construct the sum-of-powers formula in Identity 248. Jacob collaborated with his brother Johann for many years, but rivalry between the two led to a rather public falling-out near the end of Jacob's life.

Theorem 5. *If X has a binomial (n, p) distribution, then, for $k = 0, 1, \ldots, n$,*

$$P(X = k) = \binom{n}{k} p^k (1 - p)^{n-k}.$$

Proof. One way to obtain exactly k successes in n events is for the first k events to be successes and the remaining $n - k$ events to be failures. The probability that this happens is $p^k(1-p)^{n-k}$ (a factor of p for each of the k successes and a factor of $1-p$ for each of the $n-k$ failures). However, there are other ways to obtain exactly k successes besides the first k events being successes. For example, the last k events could be the successes. In fact, any subset of size k of the n trials could be used to define which k events are to be successes. There are $\binom{n}{k}$ such subsets, and so we have

$$P(X = k) = \binom{n}{k}p^k(1-p)^{n-k}.$$

$\square$

Since the sum over all possible values of X must be 1, we immediately obtain

$$\sum_{k=0}^{n} \binom{n}{k}p^k(1-p)^{n-k} = 1, \tag{5.1}$$

which is just a special case of the binomial theorem when $x = p$ and $y = 1-p$. In fact, this property of Theorem 5 can actually be used to give an alternate proof of the binomial theorem for the case in which x and y are both positive. Let's take a look at that now.

Identity 3. *If* $x, y > 0$,

$$(x + y)^n = \sum_{k=0}^{n} \binom{n}{k}x^k y^{n-k}.$$

Proof. Let $p = \dfrac{x}{x+y}$ in Equation (5.1). Then $1 - p = \dfrac{y}{x+y}$. Therefore,

$$\sum_{k=0}^{n} \binom{n}{k}\left(\frac{x}{x+y}\right)^k \left(\frac{y}{x+y}\right)^{n-k} = 1$$

$$\implies \sum_{k=0}^{n} \binom{n}{k}x^k y^{n-k}(x+y)^{-n} = 1$$

$$\implies \sum_{k=0}^{n} \binom{n}{k}x^k y^{n-k} = (x+y)^n.$$

$\square$

(As it turns out, the probabilistic proof we saw in Chapter 1—that of Identity 4—uses the same argument we used to prove Theorem 5 and Identity 3 for the special case $p = 1/2$.)

Let's turn to the second of our two questions about coin flipping: Suppose we have a coin that has a probability p of turning up heads on a single flip. If we flip the coin until we obtain a total of r heads, how many total flips does it take? If X denotes the number of flips required, then X has what is called a *negative binomial distribution* with parameters r and p. In terms of our Bernoulli process formulation, X is the number of trials required until the rth success occurs in the Bernoulli process. If $r = 1$ then X is said to have a *geometric distribution* with parameter p.

What are the probabilities associated with the different values of X in a negative binomial distribution? We answer that now.

Theorem 6. *If X has a negative binomial (r, p) distribution, then, for $k = r, r + 1, \ldots$,*

$$P(X = k) = \binom{k-1}{r-1} p^r (1-p)^{k-r}.$$

Proof. If the rth success occurs on exactly the kth trial, then trial k must be a success, and the previous $k - 1$ trials must consist of exactly $r - 1$ successes and $k - r$ failures. The probability of the latter happening is, by Theorem 5, $\binom{k-1}{r-1} p^{r-1} (1-p)^{r-k}$. Since trial k being a success has probability p, we have

$$P(X = k) = \binom{k-1}{r-1} p^r (1-p)^{k-r}.$$

$\square$

At this point we cannot quite assert yet that, if X is negative binomial (r, p), then $\sum_{k=r}^{\infty} P(X = k) = 1$. The problem is that there are scenarios in which we could continue to flip a coin without ever achieving r heads. (For example, suppose every flip is a tail.) This means that there might be positive probability of never achieving r heads, in which case $\sum_{k=r}^{\infty} P(X = k)$ evaluates to some number strictly less than 1. However, this turns out not to be the case, as we shall see in the next identity.

Identity 128. *If $0 < p \le 1$ and r is a positive integer, then*

$$\sum_{k=r}^{\infty} \binom{k-1}{r-1} p^r (1-p)^{k-r} = 1.$$

Proof. We have

$$\sum_{k=r}^{\infty} \binom{k-1}{r-1} p^r (1-p)^{k-r} = p^r \sum_{k=0}^{\infty} \binom{k+r-1}{r-1} (1-p)^k, \quad \text{switching indices}$$

$$= p^r \sum_{k=0}^{\infty} \binom{k+r-1}{k} (1-p)^k, \quad \text{by Identity 5}$$

$$= p^r \sum_{k=0}^{\infty} \binom{-r}{k} (-1)^k (1-p)^k, \quad \text{by Identity 19}$$

$$= p^r \sum_{k=0}^{\infty} \binom{-r}{k} (p-1)^k$$

$$= p^r(p-1+1)^{-r}, \qquad \text{by Identity 18}$$
$$= 1.$$

$\square$

This implies that the probability of flipping a coin without end and achieving tails on every flip is zero. Also, the proof of Identity 128 explains the reason for the name *negative binomial distribution*: As we can see here, $P(X = k+r) = \binom{-r}{k}(p-1)^k p^r$, which is analogous to $P(X = k)$ for a binomial distribution. (Note that the requirement $p > 0$ is necessary in the second-to-last step for the infinite series in Identity 18 to converge. This makes sense with the coin-flipping problem modeled by the negative binomial distribution, too: If there is zero probability of obtaining heads on a single flip then one cannot ever obtain r heads.)

Rewriting Identity 128 slightly yields a variant of the binomial theorem that sums over the upper index in the binomial coefficient rather than the lower.

Identity 129. *If $0 \le x < 1$, then*

$$\sum_{n=k}^{\infty} \binom{n}{k} x^n = \frac{x^k}{(1-x)^{k+1}}.$$

Proof. Exercise 7. $\square$

The third of the three major distributions is often expressed in terms of selecting colored balls from a jar, rather than flipping coins. Suppose we have a jar containing m numbered red balls and n numbered blue balls. We draw r balls from the jar without replacing any of them. If X denotes the number of red balls chosen, then X has a *hypergeometric distribution* with parameters m, n, and r. The following theorem gives the probabilities associated with a hypergeometric distribution.

Theorem 7. *If X has a hypergeometric (m, n, r) distribution, then, for $k = 0, 1, \ldots, m$,*

$$P(X = k) = \frac{\binom{m}{k}\binom{n}{r-k}}{\binom{m+n}{r}}.$$

Proof. Drawing r balls from a jar and having k of those be red means that $r - k$ must be blue. There are $\binom{m}{k}$ ways to choose exactly k red balls from the m total red balls and $\binom{n}{r-k}$ ways to choose exactly $r - k$ blue balls from the n total blue balls. Multiplying these together and dividing by $\binom{m+n}{r}$, the total number of ways to choose r balls from the jar, gives the probability. $\square$

The hypergeometric distribution provides us with another way to prove Vandermonde's identity, Identity 57. (In fact, the hypergeometric distribution really is Vandermonde's identity in disguise. The proof of Theorem 7 is the direct probabilistic analogue of the combinatorial proof of identity 57.)

Identity 57.

$$\sum_{k=0}^{m} \binom{m}{k}\binom{n}{r-k} = \binom{m+n}{r}.$$

Proof. If X has a hypergeometric (m, n, r) distribution, then $\sum_{k=0}^{m} P(X = k) = 1$. Therefore,

$$\sum_{k=0}^{m} \frac{\binom{m}{k}\binom{n}{r-k}}{\binom{m+n}{r}} = 1$$

$$\implies \sum_{k=0}^{m} \binom{m}{k}\binom{n}{r-k} = \binom{m+n}{r}.$$

$\square$

5.2   Expected Values and Moments

In the last section we saw how we could produce binomial identities by using a probability mass function involving binomial coefficients and the fact that probabilities must sum to 1. There are other properties of probability distributions we can use to produce binomial identities, though—not just the probability mass functions. In this section we'll see how the expected value, or, more generally, the moments, of a probability distribution may also be used in this way.

Informally, the expected value of a random variable is its average or mean value. Formally, the *expected value* $E(X)$ of the discrete random variable X is given by

$$E(X) = \sum_{k=1}^{\infty} x_k p_k,$$

where X can take on the values $x_1, x_2, \ldots$, and $p_k = P(X = x_k)$. In many cases (as with binomial and hypergeometric random variables) the infinite sum is actually finite. This definition extends to functions of X as well; we have

$$E(f(X)) = \sum_{k=1}^{\infty} f(x_k) p_k.$$

Some classes of functions f are important enough that they have their own names. For example, $E(X^m)$ is called the *mth moment* of X.

To see how expected values can be used to prove binomial identities, let's look at the case of the binomial distribution. If the probability of achieving heads on a single flip is $1/2$, and we flip a coin 20 times, then on average we should get 10 heads. Generalizing, if X is binomial $(n, 1/2)$, then $E(X)$ should be $n/2$. Using the definition of expected value, together with Theorem 5, yields

$$\frac{n}{2} = \sum_{k=0}^{n} k \binom{n}{k} \left(\frac{1}{2}\right)^k \left(1 - \frac{1}{2}\right)^{n-k}$$

$$\implies \frac{n}{2} = \frac{1}{2^n} \sum_{k=0}^{n} \binom{n}{k} k$$

$$\implies \sum_{k=0}^{n} \binom{n}{k} k = n2^{n-1},$$

which is Identity 21.

There's something a little unsatisfying about this, however. We said that "$E(X)$ *should be* $n/2$," and "should be" doesn't really belong in a proof. We need to make this more rigorous. To do that we need the property that expectation is a linear operator, as well as the concept of an indicator variable.

Theorem 8 (Linearity of Expectation). *If X and Y are random variables and a and b are constants, then*

$$E(aX + bY) = aE(X) + bE(Y).$$

Proof. Let $Z = aX + bY$. If $z_1, z_2, \ldots$ denote the values that Z can take, then we have

$$E(aX + bY) = E(Z)$$

$$= \sum_{k=1}^{\infty} z_k P(Z = z_k)$$

$$= \sum_{k=1}^{\infty} (ax_k + by_k) P(Z = z_k)$$

$$= a \sum_{k=1}^{\infty} x_k P(Z = z_k) + b \sum_{k=1}^{\infty} y_k P(Z = z_k)$$

$$= aE(X) + bE(Y).$$

$\square$

One aspect of the proof of Theorem 8 perhaps requires explanation. In the last line we make the statement $\sum_{k=1}^{\infty} x_k P(Z = z_k) = E(X)$ (as well as the corresponding statement for $E(Y)$). In doing so we are using the probability space on Z to calculate the expected value of X. This is fine as long as X is defined on the same probability space as Z. The cases in which this is not true are not ones that we will encounter in our discussion, and so we will not concern ourselves with them. (For more details on probability spaces, see any standard probability text.)

An *indicator variable* is one that only takes on the values 0 or 1. For example, we could let the indicator variable X be 1 if a coin flip turns up heads and 0 if it turns up tails. This may seem like a simple concept, but it actually simplifies some expected value calculations dramatically. Let's take a look at how this works by giving the more rigorous proof of Identity 21 that we promised. We'll first do the case of general p since the argument is the same.

Identity 130. *If $0 \leq p \leq 1$, then*

$$\sum_{k=0}^{n} \binom{n}{k} p^k (1-p)^{n-k} k = np.$$

Proof. Let X be a binomial (n, p) random variable, and define the set of indicator variables $X_1, X_2, \ldots, X_n$ via

$$X_k = \begin{cases} 1, & \text{event } k \text{ is a success;} \\ 0, & \text{event } k \text{ is a failure.} \end{cases}$$

It's easy to calculate $E(X_k)$; for each k we have $E(X_k) = 1p + 0(1-p) = p$.

Now we define X in terms of the X_k indicator variables. Since X is the number of successes out of the n trials, we have

$$X = X_1 + X_2 + \cdots + X_n = \sum_{k=1}^{n} X_k.$$

Applying linearity of expectation, we get

$$E(X) = E\left(\sum_{k=1}^{n} X_k\right) = \sum_{k=1}^{n} E(X_k) = \sum_{k=1}^{n} p = np.$$

Thus, by Theorem 5,

$$np = E(X) = \sum_{k=0}^{n} k \binom{n}{k} p^k (1-p)^{n-k}.$$

$\square$

Now Identity 21 is immediate.

Identity 21.

$$\sum_{k=0}^{n} \binom{n}{k} k = n2^{n-1}.$$

Proof. Substitute $p = 1/2$ into Identity 130, and then multiply both sides by 2^n. $\qquad\square$

Indicator variables can also be used to calculate $E(X^2)$, the second moment, where X is binomial (n, p). This gives us another binomial identity.

Identity 131. *If* $0 \leq p \leq 1$, *then*

$$\sum_{k=0}^{n} \binom{n}{k} p^k (1-p)^{n-k} k^2 = n^2 p^2 + np(1-p).$$

Proof. As in the proof of Identity 130, let X be a binomial (n, p) random variable, and define $X_1, X_2, \ldots, X_n$ via

$$X_k = \begin{cases} 1, & \text{event } k \text{ is a success;} \\ 0, & \text{event } k \text{ is a failure.} \end{cases}$$

Then we have

$$E(X^2) = E\left((X_1 + \cdots + X_n)(X_1 + \cdots + X_n)\right) = \sum_{j=1}^{n}\sum_{k=1}^{n} E(X_j X_k).$$

We now have a double sum involving the indicator variables. Fortunately, the required expected values aren't too difficult to calculate. We have to break the terms in the double sum into two cases:

1. If j and k are the same, then $X_j X_k = X_k^2$. The variable X_k^2 can only take on two values: 1, if event k is a success, and 0, if event k is a failure. These two scenarios have probabilities p and 0, respectively. Thus $E(X_k^2) = 1p + 0(1-p) = p$.

2. If j and k are different, then $X_j X_k$ can also only take on two values: 1, if events j and k are both successes, and 0, otherwise. These two scenarios have probabilities p^2 and $1 - p^2$, respectively. Thus $E(X_j X_k) = 1p^2 + 0(1 - p^2) = p^2$ when $j \neq k$.

Now, the double sum $\displaystyle\sum_{j=1}^{n}\sum_{k=1}^{n} E(X_j X_k)$ contains n^2 total terms, n of which have $j = k$, leaving $n^2 - n$ for the $j \neq k$ case. Thus

$$\sum_{j=1}^{n}\sum_{k=1}^{n} E(X_j X_k) = np + (n^2 - n)p^2 = n^2 p^2 + np(1-p).$$

By the definition of $E(X^2)$ and Theorem 5, then, we get

$$\sum_{k=0}^{n} \binom{n}{k} p^k (1-p)^{n-k} k^2 = n^2 p^2 + np(1-p).$$

$\qquad\square$

Letting $p = 1/2$ in Identity 131 and rearranging, we have Identity 71:

Identity 71.

$$\sum_{k=0}^{n} \binom{n}{k} k^2 = n(n+1)2^{n-2}.$$

We can use indicator variables plus linearity of expectation to obtain the expected value of a hypergeometrically-distributed random variable, too, yielding a binomial identity we have not yet seen.

Identity 132.

$$\sum_{k=0}^{m} \binom{m}{k}\binom{n}{r-k} k = \binom{m+n}{r}\frac{rm}{m+n}.$$

Proof. A binomially-distributed random variable counts a number of successes. A hypergeometrically-distributed random variable counts a number of successes, too, where "success" means drawing a red ball (or, in general, selecting an item of a certain kind). So if X is hypergeometric (m, n, r), let

$$X_k = \begin{cases} 1, & \text{a red ball is selected on draw } k; \\ 0, & \text{a blue ball is selected on draw } k. \end{cases}$$

The probability of selecting a red ball on any one particular draw is just $\dfrac{m}{m+n}$, as there are m red balls and $m+n$ total balls. Thus

$$E(X_k) = (1)\frac{m}{m+n} + (0)\frac{n}{m+n} = \frac{m}{m+n},$$

and hence

$$E(X) = E\left(\sum_{k=1}^{r} X_k\right) = \sum_{k=1}^{r} E(X_k) = \frac{rm}{m+n}.$$

By Theorem 7 and the definition of expected value, though,

$$E(X) = \sum_{k=0}^{m} \frac{\binom{m}{k}\binom{n}{r-k}}{\binom{m+n}{r}} k.$$

Therefore,

$$\sum_{k=0}^{m} \frac{\binom{m}{k}\binom{n}{r-k}}{\binom{m+n}{r}} k = \frac{rm}{m+n}$$

$$\implies \sum_{k=0}^{m} \binom{m}{k}\binom{n}{r-k} k = \binom{m+n}{r}\frac{rm}{m+n}.$$

$\square$

Computing $E(X^2)$, where X is hypergeometric, yields another binomial identity. (See Exercise 16.)

One difference between the indicator variables in the proofs of Identities 130 and 131 and those in the proof of Identity 132 is that the former are independent of each other and the latter are not. (For example, the chances of obtaining heads on flip 10 are not affected by the result of flip 1. However, the chances of obtaining a red ball on draw 10 change depending on whether a red or blue ball was selected on draw 1, as the draw 1 selection changes the proportions of colored balls in the jar.) Lack of independence often makes for nasty probability calculations. It does not affect linearity of expectation, though, and so using indicator variables lets us avoid that potential messiness completely.

We can also obtain binomial identities by calculating the expected value of a random variable X with a negative binomial (r, p) distribution. The indicator variable approach that we have been using does not work in quite the same way, though, since a negative binomial random variable does not count the number of times that something happens. (The problem is that if we use indicator variables to keep track of the number of trials until the rth success, we don't know in advance how many indicator variables we will need.) Still, there are some different ways to calculate the expected value. We will look at two of them, as they each illustrate useful techniques in probability.

Identity 133. *If $0 < p \le 1$ and r is a positive integer, then*

$$\sum_{k=r}^{\infty} \binom{k-1}{r-1} p^r (1-p)^{k-r} k = \frac{r}{p}.$$

Proof 1. The first proof is a modification of the indicator variable method. If X is negative binomial, then X *can* be expressed as the sum of a fixed number of simpler random variables; they're just not indicator variables. Remember that if X is negative binomial (r, p) then X counts the number of trials required until the rth success is obtained. Alternatively, since the trials are independent, this process can be thought of as running a sequence of r experiments, each of which ends when the first success is obtained. The value of X is then the total number of trials over all r experiments. From this point of view we get $X = X_1 + X_2 + \cdots + X_r$, where X_k is the number of trials required for experiment k only. The X_k's are independent geometric (p) random variables.

To calculate $E(X)$ we're going to need $E(X_k)$. The most straightforward way to find this is to do so directly, via the definition of expected value and the negative binomial probabilities when $r = 1$. (See Exercise 17.) However, we're going to use a simpler and more clever method. We know that $E(X_k)$ is the expected number of trials until the first success. Condition on the result of the first trial. This is either a success, in which case we're done, or a failure, in which case the process of finding the first success essentially restarts itself.

These two outcomes have probabilities p and $1 - p$, respectively, and this means that $E(X_k)$ must satisfy the following equation:

$$E(X_k) = (p)1 + (1 - p)E(1 + X_k).$$

Solving, we obtain $E(X_k) = 1/p$.

Therefore,

$$E(X) = E\left(\sum_{k=1}^{r} X_k\right) = \sum_{k=1}^{r} E(X_k) = \frac{r}{p}.$$

By Theorem 6 and the definition of $E(X)$, then, we have

$$\sum_{k=r}^{\infty} \binom{k-1}{r-1} p^r (1-p)^{k-r} k = \frac{r}{p}.$$

$\square$

Proof 2. The second proof is even simpler; it uses a variation of the method we used in Proof 1. Let Y_r denote a negative binomial (r, p) random variable. Now, condition on the result of the first trial. This is a success with probability p. In this case, we've found our first success, and we only need to run $r - 1$ more experiments. On the other hand, the first trial is a failure with probability $1 - p$. In this case the process essentially restarts itself. This means that $E(Y_r)$ must satisfy

$$E(Y_r) = pE(1 + Y_{r-1}) + (1 - p)E(1 + Y_r).$$

Rearranging, we have

$$E(Y_r) = E(Y_{r-1}) + \frac{1}{p}.$$

Unrolling the resulting recurrence, while recalling that $E(Y_1) = 1/p$, we have

$$E(Y_2) = \frac{1}{p} + \frac{1}{p} = \frac{2}{p},$$

$$E(Y_3) = E(Y_2) + \frac{1}{p} = \frac{3}{p},$$

$$\vdots$$

$$E(Y_r) = \frac{r}{p}.$$

Thus

$$\sum_{k=r}^{\infty} \binom{k-1}{r-1} p^r (1-p)^{k-r} k = \frac{r}{p}.$$

$\square$

(See Exercise 19 for a somewhat cleaner-looking—at least on the left side—version of Identity 133.)

As with the binomial and hypergeometric distributions, we can obtain a binomial coefficient identity by finding $E(X^2)$ for a negative binomial random variable X.

Identity 134. *If $0 < p \le 1$ and r is a positive integer, then*

$$\sum_{k=r}^{\infty} \binom{k-1}{r-1} p^r (1-p)^{k-r} k^2 = \frac{r}{p} \left(\frac{r+1}{p} - 1 \right).$$

Proof. We use the variables we used in Proof 1 of Identity 133. (See Exercise 21 for the method using Proof 2 of Identity 133.) Let X be a negative binomial (r, p) random variable, and let $X_1, X_2, \ldots, X_r$ be independent geometric (p) random variables so that $X = \sum_{k=1}^{r} X_k$.

We're eventually going to need $E(X_k^2)$, so let's calculate it now using the recurrence method we used before. (See Exercise 18 for the direct method using the definition of expected value.) The first trial is either a success, in which case we stop, or a failure, in which case the process restarts itself. Since the probabilities of these events happening are p and $1 - p$, respectively, we have that

$$E(X_k^2) = p(1) + (1-p)E((1+X_k)^2)$$
$$= p + (1-p)(E(X_k^2) + 2E(X_k) + 1)$$
$$= 1 + (1-p)E(X_k^2) + 2(1-p)\frac{1}{p},$$

since $E(X_k) = 1/p$. (See Exercise 17 or the first proof of Identity 133.) A little algebra yields

$$pE(X_k^2) = 1 + \frac{2}{p} - 2$$
$$\implies E(X_k^2) = \frac{2-p}{p^2}.$$

Using this expression for $E(X_k^2)$, we obtain, as in the proof of Identity 131,

$$E(X^2) = E\left(\sum_{k=1}^{r} X_k \right)^2 = \sum_{k=1}^{r} E(X_k^2) + \sum_{j=1}^{r} \sum_{\substack{k=1 \\ k \ne j}}^{r} E(X_j X_k).$$

Since X_j and X_k are independent we have $E(X_j X_k) = E(X_j)E(X_k) = 1/p^2$. Putting all of this together, we have

$$E(X^2) = r\frac{2-p}{p^2} + r(r-1)\frac{1}{p^2} = \frac{r}{p} \left(\frac{r+1}{p} - 1 \right).$$

Theorem 6 and the definition of expected value then give us

$$\sum_{k=r}^{\infty} \binom{k-1}{r-1} p^r (1-p)^{k-r} k^2 = \frac{r}{p} \left(\frac{r+1}{p} - 1 \right).$$

$\square$

(See Exercise 20 for a version of Identity 134 that is simpler on the left side.)

In this section we have mostly focused on using indicator variables to calculate moments. However, the moments of the binomial, negative binomial, and hypergeometric distributions also satisfy recurrence relations that can be used to calculate their moments. These relations are given in Exercises 25, 27, and 29. In addition, in Chapter 8 we use the Stirling numbers of the second kind to find explicit formulas for the moments of these three distributions. (See Exercises 9, 10, and 11 in Chapter 8.)

5.3   Inclusion-Exclusion Revisited

The principle of inclusion-exclusion that we saw in Chapter 3 has a probabilistic version, and this can also be used to prove binomial identities. In many instances the combinatorial proof of an identity that uses inclusion-exclusion is virtually identical to its probabilistic version, but in other cases the probabilistic viewpoint results in a noticeably simpler proof. Let's look at some probabilistic variants of the inclusion-exclusion formula and then at a few examples using them.

Theorem 9 (Principle of Inclusion-Exclusion – Probabilistic Version). *Given events $A_1, A_2, \ldots, A_n$,*

$$P\left(\bigcup_{i=1}^{n} A_i \right) = \sum_{i=1}^{n} P(A_i) - \sum_{1 \le i < j \le n} P(A_i \cap A_j) + \sum_{1 \le i < j < k \le n} P(A_i \cap A_j \cap A_k)$$
$$- \cdots + (-1)^{n-1} P(A_1 \cap \cdots \cap A_n),$$

and, if there exists an event A such that $A_i \subseteq A$ for each i,

$$P\left(\bigcap_{i=1}^{n} A_i \right) = P(A) - \sum_{i=1}^{n} P(\bar{A}_i) + \sum_{1 \le i < j \le n} P(\bar{A}_i \cap \bar{A}_j)$$
$$- \sum_{1 \le i < j < k \le n} P(\bar{A}_i \cap \bar{A}_j \cap \bar{A}_k) + \cdots + (-1)^n P(\bar{A}_1 \cap \cdots \cap \bar{A}_n).$$

Proof. If the sample space S on which the A_i events are drawn is finite, then one can simply divide the equations in the combinatorial version of inclusion-exclusion (Theorem 4) by the size of S to complete the proof. If S is infinite, however, we need induction. See Exercise 31. $\qquad\square$

In the second equation in Theorem 9 we usually take A to be S. Corollary 1 has a probabilistic version as well; see Corollary 2.

Our first example using Theorem 9 is a probabilistic proof of Identity 24. As we work through the proof, we'll discuss why Theorem 9 is an effective tool here, as well as how one might go about using Theorem 9 in general to prove binomial identities.

Identity 24.

$$\sum_{k=0}^{n} \binom{n}{k} \frac{(-1)^k}{k+1} = \frac{1}{n+1}.$$

Proof. This identity looks like a good candidate for using inclusion-exclusion because of the alternating sum on the left. The sum also contains the binomial coefficient $\binom{n}{k}$ and a simple fractional $\frac{1}{k+1}$ expression. While not that conducive to a combinatorial proof because the numbers involved aren't integers, the simple unit fractions likely have a nice probabilistic interpretation. In fact, the right side looks like the probability of selecting one special item out of $n+1$ choices. With that in mind, the left side looks like it involves something having to do with selecting items out of $k+1$ choices, as k varies from 0 to n. We've been working with balls and jars with the hypergeometric distribution, so let's construct a scenario in those terms.

Suppose there are balls numbered 1 through $n+1$ placed in a jar. We select the balls, one-by-one, and remove them from the jar. What is the probability that ball $n+1$ is the first ball chosen? It's pretty clear that this is $\frac{1}{n+1}$, the right side of our identity.

To use inclusion-exclusion for the left side we have to interpret "$n+1$ is the first ball chosen" as the union or intersection of a collection of other events. Moreover, we have to do this in such a way as to work in the binomial coefficient and unit fraction $\frac{1}{k+1}$ as part of calculating the probabilities when these events are intersected. From the intersection standpoint, ball $n+1$ does have to come before ball 1 and before ball 2 and before ball 3 and so forth. So let's let A_i, $1 \leq i \leq n$, be the event that ball $n+1$ is drawn before ball i and use Theorem 9 to find an expression for $P\left(\bigcap_{i=1}^{n} A_i\right)$. First, $\bar{A}_i$ is the event that ball $n+1$ is drawn after ball i. Since ball $n+1$ is just as likely to be drawn before ball i as after ball i, we have, for any i, $P(\bar{A}_i) = \frac{1}{2}$. Similarly, $\bar{A}_i \cap \bar{A}_j$ is the event that ball $n+1$ is drawn after ball i and drawn after ball j, i.e., the event that ball $n+1$ comes last out of the three balls i, j, and $n+1$. Thus $P(\bar{A}_i \cap \bar{A}_j) = \frac{1}{3}$. Continuing this logic gives $\bar{A}_i \cap \bar{A}_j \cap \bar{A}_k$ as the event that ball $n+1$ comes last out of balls i, j, k, and $n+1$, which means $P(\bar{A}_i \cap \bar{A}_j \cap \bar{A}_k) = \frac{1}{4}$. In general, the probability that ball $n+1$ comes as the last of any $k+1$ specific balls is $\frac{1}{k+1}$, and there are $\binom{n}{k}$ collections of $k+1$

specific balls that include ball $n+1$. By Theorem 9, then, the probability that ball $n+1$ is the first ball chosen is given by

$$1 - \sum_{k=1}^{n} \binom{n}{k} \frac{(-1)^{k+1}}{k+1} = \sum_{k=0}^{n} \binom{n}{k} \frac{(-1)^k}{k+1}.$$

$\square$

Our second example uses the same ideas as in the probabilistic proof of Identity 24 but in a more complicated way.

Identity 46.

$$\sum_{k=0}^{n} \binom{n}{k} \frac{(-1)^k}{(k+1)^2} = \frac{H_{n+1}}{n+1}.$$

Proof. Let's think about the left side of Identity 46. The key factor is $\frac{1}{(k+1)^2}$. With the probabilistic proof of Identity 24 in mind, this expression implies a scenario in which we are drawing balls independently from two different jars. To distinguish them, let's give them different colors. Then $\frac{1}{(k+1)^2}$ could be the probability that red ball $n+1$ comes as the last of $k+1$ specific balls from the red jar and that blue ball $n+1$ comes as the last of $k+1$ specific balls from the blue jar. We only have one factor of $\binom{n}{k}$, though—not two. This means that we need some interpretation of the process of drawing balls from jars that preserves choosing the red balls independently of the blue balls but, when it comes time to analyze ball order, has us considering only one set of n items rather than two. One such interpretation is to have the ball-drawing process *pair* each red ball with a blue ball. For example, the first red ball drawn would be paired with the first blue ball drawn, the second red ball drawn would be paired with the second blue ball drawn, and so forth. Thus one instance of the selection process could produce the sequence $(R1, B5), (R2, B3), (R5, B4), (R4, B1), (R3, B2)$, where Ri and Bj denote the ith red and jth blue balls, respectively. This perspective allows us to think of $\binom{n}{k}$ as the number of ways to select k pairs of balls out of the first n pairs chosen from the jars.

Interpreting the expression $\frac{1}{(k+1)^2}$ as the probability that red ball $n+1$ is chosen last of $k+1$ red balls and that blue ball $n+1$ is chosen last of $k+1$ blue balls, while it would be correct, isn't quite as helpful now since we are choosing *pairs* of balls. Instead, we could get at the same idea but in a way more faithful to the pairs interpretation by taking $\frac{1}{(k+1)^2}$ as the probability that the last pair chosen has a higher-numbered red ball and a higher-numbered blue ball than do any of k other specific pairs. This interpretation implicitly defines $\bar{A}_i$ as the event that the ith pair has a smaller-numbered red ball and a smaller-numbered blue ball than does the $n+1$ pair. This in turn gives us A_i as the event that either the $n+1$ pair's red ball is smaller-numbered than the ith pair's red ball or the $n+1$ pair's blue ball is smaller-numbered than the ith pair's blue ball. Using inclusion-exclusion, then, we finally have a probabilistic interpretation of the left side of Identity 46: It's $P(\bigcap_{i=1}^{n} A_i)$, the probability that, for each i, $1 \leq i \leq n$, pair $n + 1$ contains either a smaller-numbered red ball than does pair i or a smaller-numbered blue ball than does pair i.

For example, the sequence $(R1, B5), (R2, B3), (R5, B4), (R4, B1), (R3, B2)$ would be part of the event $\bigcap_{i=1}^{n} A_i$ (with $n = 4$). For the first pair, $B2$ is smaller than $B5$; for the second pair, $B2$ is smaller than $B3$; for the third pair, $R3$ is smaller than $R5$ and $B2$ is smaller than $B4$; and for the fourth pair, $R3$ is smaller than $R4$.

Now for the right side. Expanding the harmonic number, we have

$$\frac{1}{n + 1} \sum_{k=1}^{n+1} \frac{1}{k},$$

so conditioning on something makes sense here. The fraction $\frac{1}{k}$ is the probability that a specific ball is the smallest-numbered out of k, so perhaps it makes sense to condition on the number of the red ball in the $n + 1$ pair. So, suppose the red ball in pair $n + 1$ has number k. There is a $\frac{1}{n+1}$ probability that this will happen, regardless of the value of k. Now, we need either the red ball in the last pair or the blue ball in the last pair to have a smaller number than in any other pairs. If the red ball has value k, then there are $k - 1$ pairs for which the red ball is not smaller. For those pairs, we need the blue ball to be smaller. Not knowing the value of the blue ball, the probability that the blue ball in pair $n + 1$ has a smaller number than the blue balls in all of these $k - 1$ pairs is $1/k$. Summing up over all possible values of k, we have

$$\frac{1}{n + 1} \sum_{k=1}^{n+1} \frac{1}{k} = \frac{H_{n+1}}{n + 1}$$

as the probability that, for each i, $1 \leq i \leq n$, pair $n + 1$ has either a smaller-numbered red ball or a smaller-numbered blue ball than does pair i.

□

Identities 24 and 46 share the following feature: When $k = 0$ the summand in the identity is 1. However, there are binomial identities that can be proved probabilistically for which this is not the case. Here is a variant of Theorem 9 that does not force this restriction.

Corollary 2. *Given events* $B, A_1, A_2, \ldots, A_n$,

$$P\left(B \cap \left(\bigcap_{i=1}^{n} A_i\right)\right) = P(B) - \sum_{i=1}^{n} P(B \cap \bar{A}_i) + \sum_{1 \leq i < j \leq n} P(B \cap \bar{A}_i \cap \bar{A}_j)$$
$$- \sum_{1 \leq i < j < k \leq n} P(B \cap \bar{A}_i \cap \bar{A}_j \cap \bar{A}_k)$$
$$+ \cdots + (-1)^n P(B \cap \bar{A}_1 \cap \cdots \cap \bar{A}_n).$$

Proof. See Exercise 32.   □

For our third example using the probabilistic version of inclusion-exclusion, let's take a look at another variation on Identity 24, a probabilistic proof of Identity 41. This one uses Corollary 2.

Identity 41.

$$\sum_{k=0}^{n} \binom{n}{k} \frac{(-1)^k}{(k+1)(k+2)} = \frac{1}{n+2}.$$

Proof. The right side is simple enough that we can propose an interpretation like that of Identity 24: Suppose balls numbered 1 through $n+2$ are placed in a jar. We select the balls, one-by-one, and remove them from the jar. What is the probability that ball $n+2$ is the first ball chosen? Clearly, this is $\frac{1}{n+2}$.

For the left side, the first thing to notice is that it has a feature that is not present in either of the two previous examples: The $k = 0$ term for the left side does not equal 1. Instead, it equals $\frac{1}{2}$. This matters because the previous two examples use the second equation in Theorem 9, and both of them have the $k = 0$ term correspond to the probability of the entire sample space (which is of course 1). This is where Corollary 2 comes in: If we use it we can have the $k = 0$ term correspond to $P(B)$. We just need to find an event B that has probability $\frac{1}{2}$.

For the left side, we need $B \cap \left(\bigcap_{i=1}^{n} A_i\right)$ to be the event that ball $n+2$ is the first ball chosen. If we use an interpretation like that in the proof of Identity 24 we get that A_i is the event that ball $n+2$ is drawn before ball i. This leaves B to be the event that ball $n+2$ is drawn before ball $n+1$. We do have $P(B) = \frac{1}{2}$, so things are good there.

Let's now look at the right side of Corollary 2. The event $\bar{A}_i$ is the event that ball i comes before ball $n+2$. Intersecting k specific $\bar{A}_i$'s with B yields the event that ball $n+1$ is drawn last of $k+2$ specific balls, and, given that, ball $n+2$ is drawn second-to-last. The first event has probability $\frac{1}{k+2}$, and, given that, the second has probability $\frac{1}{k+1}$. There are $\binom{n}{k}$ ways to choose k of the $\bar{A}_i$'s, so Corollary 2 tells us that the probability that ball $n+2$ is the first ball drawn is also

$$\sum_{k=0}^{n} \binom{n}{k} \frac{(-1)^k}{(k+1)(k+2)}.$$

□

5.4   The Beta Integral Revisited

We end this chapter with a probabilistic derivation of the formula for the binomial coefficient in terms of the beta integral. This is the only part of this chapter that requires knowledge of continuous probability. (In addition, nothing later in the text builds off this section.)

Identity 106.

$$\binom{n}{k}^{-1} = (n+1) \int_0^1 x^k (1-x)^{n-k} dx.$$

Proof. Suppose we have random variables $X_1, X_2, \ldots, X_{n+1}$ that are independent and identically distributed uniformly on the interval $[0,1]$. What is the probability that the value of X_{n+1} is the $k+1$ smallest of the $n+1$ values?

Since the random variables are independent and identically distributed, each of the $n+1$ values is equally likely to be the $k+1$ largest. Thus the answer is $\frac{1}{n+1}$.

We could also answer the question by conditioning on the value of X_{n+1}. If $X_{n+1} = x$, then X_{n+1} is the $k+1$ smallest if exactly k of $X_1, \ldots, X_n$ are smaller than x and exactly $n-k$ are larger. Since each X_i has a uniform $[0,1]$ distribution, the probability that any one particular X_i is smaller than x is exactly x, independently of the others. The number of $X_1, \ldots, X_n$ that have a value smaller than x thus has the binomial (x, p) distribution. The probability that exactly k of the X_i's have a value smaller than x is therefore $\binom{n}{k} x^k (1-x)^{n-k}$. Integrating over all possible values of X_{n+1}, we have the probability that X_{n+1} is the $k+1$ smallest is also $\int_0^1 \binom{n}{k} x^k (1-x)^{n-k} dx$.

Therefore,

$$\binom{n}{k}^{-1} = (n+1) \int_0^1 x^k (1-x)^{n-k} dx.$$

□

5.5 Exercises

1. Suppose we have a coin with success probability p, and we flip it n times. Show that the *most likely* number of successes is $\lfloor (n+1)p \rfloor$, where $\lfloor x \rfloor$ is the largest integer less than or equal to x. (Hint: Use the binomial distribution, and investigate which values of k have $P(X = k)/P(X = k-1) \geq 1$.)

2. Suppose we have a coin with success probability p, and we flip it n times. Show that the probability of obtaining an even number of flips is given by $\frac{1}{2}(1 + (1-2p)^n)$.

3. Suppose we have m red balls and n blue balls in a jar, and we select them one by one, replacing each ball in the jar before we select another one.

Let X be the number of selections it takes to draw our first red ball. Find $P(X = k)$.

4. A mathematician carries a matchbox in his left pocket and a matchbox in his right pocket. When he wishes to light his pipe, he is equally likely to choose a match from either his left pocket or his right. Suppose he reaches into a pocket and discovers that the matchbox in that pocket is empty. If each matchbox initially contains n matches, show that the probability that there are k matches remaining in the other pocket is $\binom{2n-k}{n} \left(\frac{1}{2}\right)^{2n-k}$. (This is a famous problem, known as the *Banach match problem*.)

5. Suppose the mathematician in Exercise 4 is left-handed and chooses that pocket with probability p and his right pocket with probability $1 - p$. Now when he reaches into a pocket and finds an empty matchbox, what is the probability that there are k matches left in the other pocket's matchbox?

6. Find a probabilistic proof of the following.

Identity 135. *If $0 < p \leq 1$, r is a positive integer, and n is a nonnegative integer, then*

$$\sum_{k=n+1}^{\infty} \binom{k-1}{r-1} p^r (1-p)^{k-r} = \sum_{k=0}^{r-1} \binom{n}{k} p^k (1-p)^{n-k}.$$

7. Prove the following.

Identity 129. *If $0 \leq x < 1$, then*

$$\sum_{n=k}^{\infty} \binom{n}{k} x^n = \frac{x^k}{(1-x)^{k+1}}.$$

8. Use Identity 130 to prove Identity 89 in the case $x > 0$, $y > 0$:

Identity 89.

$$\sum_{k=0}^{n} \binom{n}{k} k x^{k-1} y^{n-k} = n(x+y)^{n-1}.$$

9. Use Identity 131 to prove Identity 92 in the case $x > 0$, $y > 0$:

Identity 92.

$$\sum_{k=0}^{n} \binom{n}{k} k^2 x^{k-1} y^{n-k} = n(x+y)^{n-2}(nx+y).$$

The next several problems do not involve binomial coefficients; instead, they give practice using the indicator variable technique.

10. Suppose each of n students writes his name on a piece of paper and drops it into a hat. After all the names are in the hat each student pulls out a slip of paper from the hat. Assuming each student is equally likely to pull out any of the slips of paper, what's the expected number of students who draw their own names? (This question is related to derangements; see Chapter 3 for more on them.)

11. Suppose there are m men and n women sitting in a row at a concert. If any of the $(m + n)!$ seating arrangements is equally likely, what is the expected number of people sitting to the immediate right of a person of the opposite gender?

12. Suppose there are m men and n women sitting around a circular table. If any of the seating arrangements is equally likely, what is the expected number of people sitting to the immediate right of a person of the opposite gender?

13. Suppose there are m men and n women sitting in a row at a concert. If any of the seating arrangements is equally likely, what is the expected number of people who have a person of the opposite gender sitting next to them?

14. Suppose there are m men and n women sitting around a circular table. If any of the seating arrangements is equally likely, what is the expected number of people who have a person of the opposite gender sitting next to them?

15. Each box produced by a breakfast cereal company contains a card featuring a famous mathematician. There are n total mathematicians featured in this promotion. If each box is equally likely to contain any of the n possible cards, what is the expected number of boxes of cereal you would have to buy in order to obtain a complete set of the cards? (Hint: Use an approach like that in Proof 1 of Identity 133, the expected value of a negative binomial random variable. If B is the total number of boxes of cereal required, let B_k be the number of additional boxes needed to get the kth newest card after the $k-1$ newest card has already been obtained. In fact, B is a variant of a negative binomial random variable in which the success probabilities change as the number of successes already observed increases.)

16. Use an expected value argument to prove the following.

Identity 136.

$$\sum_{k=0}^{m} \binom{m}{k} \binom{n}{r-k} k^2 = \binom{m+n}{r} \frac{rm}{m+n} \left(\frac{(r-1)(m-1)}{m+n-1} + 1 \right).$$

17. Use Theorem 6 to show that, if X is geometric (p), then $E(X) = 1/p$.

18. Use Theorem 6 to show that, if X is geometric (p), then $E(X^2) = \dfrac{2-p}{p^2}$.

19. Prove the following.

   Identity 137. *If $0 \le x < 1$, then*

   $$\sum_{n=k}^{\infty} \binom{n}{k} nx^n = \frac{(k+x)x^k}{(1-x)^{k+2}}.$$

20. Prove the following.

   Identity 138. *If $0 \le x < 1$, then*

   $$\sum_{n=k}^{\infty} \binom{n}{k} n^2 x^n = \frac{(k^2 + 3kx + x + x^2)x^k}{(1-x)^{k+3}}.$$

21. Prove Identity 134 using a recursive argument similar to the one we used in Proof 2 of Identity 133:

   Identity 134. *If $0 < p \le 1$ and r is a positive integer, then*

   $$\sum_{k=r}^{\infty} \binom{k-1}{r-1} p^r (1-p)^{k-r} k^2 = \frac{r}{p}\left(\frac{r+1}{p} - 1\right).$$

22. The *mth factorial moment* is given by $E(X^{\underline{m}})$. Factorial moments for discrete distributions often have simpler expressions than do the corresponding raw moments $E(X^m)$. (The reason for this has to do with the usefulness of factorial powers in finite difference calculus. See, for example, Chapter 2 of *Concrete Mathematics* [32], or Identities 177 and 204 in Chapter 7.) This exercise uses indicator variables to calculate the factorial moments for the binomial distribution. Let X be a binomial (n, p) random variable. Then X is the number of successes that occur in n independent Bernoulli trials.

   (a) Suppose we are interested in the number of *pairs* of successes that occur in n Bernoulli trials. Argue that this is given by

   $$\binom{X}{2} = \sum_{j=1}^{n} \sum_{k=j+1}^{n} X_j X_k,$$

   where X_k is the indicator variable determining whether we have success on trial k.

   (b) Use the result of part (a) to show that $E(X(X-1)) = n(n-1)p^2$, yielding the following generalization of Identity 35:

Identity 139. *If* $0 \leq p \leq 1$, *then*

$$\sum_{k=1}^{n} \binom{n}{k} p^k (1-p)^{n-k} k(k-1) = n(n-1)p^2.$$

(c) Generalize part (b) to $E(X^{\underline{m}}) = n^{\underline{m}} p^m$, yielding the following generalization of Identity 36:

Identity 140. *If* $0 \leq p \leq 1$, *then*

$$\sum_{k=1}^{n} \binom{n}{k} p^k (1-p)^{n-k} k^{\underline{m}} = n^{\underline{m}} p^m.$$

(d) Use the result of part (c) to show that $E(X^3) = np + 3n(n-1)p^2 + n(n-1)(n-2)p^3$. (See also Identity 144.)

23. The technique in Exercise **22** can be applied to the hypergeometric distribution as well. Let X be a hypergeometric (m, n, r) random variable.

   (a) Calculate $E\binom{X}{2}$, and use it to prove the following.

   Identity 141.

   $$\sum_{k=0}^{n} \binom{m}{k} \binom{n}{r-k} k(k-1) = \binom{m+n}{r} \frac{r(r-1)m(m-1)}{(m+n)(m+n-1)}.$$

   (b) Calculate $E\binom{X}{p}$, and use it to prove the following.

   Identity 142.

   $$\sum_{k=0}^{n} \binom{m}{k} \binom{n}{r-k} k^{\underline{p}} = \binom{m+n}{r} \frac{r^{\underline{p}} m^{\underline{p}}}{(m+n)^{\underline{p}}}.$$

24. For the negative binomial distribution, the rising factorial moments are easier to obtain than the falling factorial moments. Use the absorption identity (Identity 6) to prove the following.

   Identity 143. *If* $0 < p \leq 1$ *and* r *is a positive integer, then*

   $$\sum_{k=r}^{\infty} \binom{k-1}{r-1} p^r (1-p)^{k-r} k^{\overline{m}} = \frac{r^{\overline{m}}}{p^m}.$$

25. Prove the following recursive relationship for the moments of the binomial distribution.

   Theorem 10. *Suppose* X *is binomial* (n, p). *Then, for* $q \geq 1$,

   $$E[X^q] = npE[(Y+1)^{q-1}],$$

   where Y *is binomial* $(n-1, p)$.

26. Use Theorem 10 to prove the following identities giving the first, second, and third moments of the binomial distribution.

(a)

Identity 130. *If* $0 \leq p \leq 1$, *then*

$$\sum_{k=0}^{n} \binom{n}{k} p^k (1-p)^{n-k} k = np.$$

(b)

Identity 131. *If* $0 \leq p \leq 1$, *then*

$$\sum_{k=0}^{n} \binom{n}{k} p^k (1-p)^{n-k} k^2 = n^2 p^2 + np(1-p).$$

(c) (A generalization of Identity 110)

Identity 144. *If* $0 \leq p \leq 1$, *then*

$$\sum_{k=0}^{n} \binom{n}{k} p^k (1-p)^{n-k} k^3 = np(1 + 3(n-1)p + (n-1)(n-2)p^2).$$

27. Prove the following recursive relationship for the moments of the negative binomial distribution.

Theorem 11. *Suppose X is negative binomial (r, p). Then, for $q \geq 1$,*

$$E[X^q] = \frac{r}{p} E[(Y-1)^{q-1}],$$

where Y is negative binomial $(r+1, p)$.

28. Use Theorem 11 to prove the following identities giving the first, second, and third moments of the negative binomial distribution.

(a)

Identity 133. *If* $0 < p \leq 1$ *and r is a positive integer, then*

$$\sum_{k=r}^{\infty} \binom{k-1}{r-1} p^r (1-p)^{k-r} k = \frac{r}{p}.$$

(b)

Identity 134. *If* $0 < p \leq 1$ *and r is a positive integer, then*

$$\sum_{k=r}^{\infty} \binom{k-1}{r-1} p^r (1-p)^{k-r} k^2 = \frac{r}{p} \left(\frac{r+1}{p} - 1 \right).$$

(c)

Identity 145. *If $0 < p \le 1$ and r is a positive integer, then*

$$\sum_{k=r}^{\infty} \binom{k-1}{r-1} p^r (1-p)^{k-r} k^3$$

$$= \frac{r}{p} \left(1 - 3\left(\frac{r+1}{p}\right) + \left(\frac{r+1}{p}\right)\left(\frac{r+2}{p}\right) \right).$$

29. Prove the following recursive relationship for the moments of the hyper-geometric distribution.

   Theorem 12. *Suppose X has a hypergeometric (m, n, r) distribution. Then, for $q \ge 1$,*

   $$E[X^q] = \frac{rm}{m+n} E[(Y+1)^{q-1}],$$

   where Y is hypergeometric $(m-1, n, r-1)$.

30. Use Theorem 12 to prove the following identities giving the first, second, and third moments of the hypergeometric distribution.

   (a)

   Identity 132.

   $$\sum_{k=0}^{m} \binom{m}{k}\binom{n}{r-k} k^2 = \frac{rm}{m+n}.$$

   (b)

   Identity 136.

   $$\sum_{k=0}^{m} \binom{m}{k}\binom{n}{r-k} k^2 = \binom{m+n}{r} \frac{rm}{m+n} \left(\frac{(r-1)(m-1)}{m+n-1} + 1 \right).$$

   (c)

   Identity 146.

   $$\sum_{k=0}^{m} \binom{m}{k}\binom{n}{r-k} k^3 = \binom{m+n}{r} \frac{rm}{m+n} A(m, n, r), \text{ where}$$

   $$A(m, n, r) = 1 + 3\frac{(r-1)(m-1)}{m+n-1} + \frac{(r-1)(r-2)(m-1)(m-2)}{(m+n-1)(m+n-2)}.$$

31. Use induction on n to prove the probabilistic version of the principle of inclusion-exclusion:

Theorem 9.

$$P\left(\bigcup_{i=1}^{n} A_i\right) = \sum_{i=1}^{n} P(A_i) - \sum_{1 \le i < j \le n} P(A_i \cap A_j)$$
$$+ \sum_{1 \le i < j < k \le n} P(A_i \cap A_j \cap A_k)$$
$$- \cdots + (-1)^{n-1} P(A_1 \cap \cdots \cap A_n).$$

32. Use the conditional probability formula $P(E \cap F) = P(E)P(F|E)$ to prove Corollary 2:

Corollary 2. *Given events* $B, A_1, A_2, \ldots, A_n$,

$$P\left(B \cap \left(\bigcap_{i=1}^{n} A_i\right)\right) = P(B) - \sum_{i=1}^{n} P(B \cap \bar{A}_i) + \sum_{1 \le i < j \le n} P(B \cap \bar{A}_i \cap \bar{A}_j)$$
$$- \sum_{1 \le i < j < k \le n} P(B \cap \bar{A}_i \cap \bar{A}_j \cap \bar{A}_k)$$
$$+ \cdots + (-1)^n P(B \cap \bar{A}_1 \cap \cdots \cap \bar{A}_n).$$

33. Exercise 10 asks you to calculate the expected number of people who get their own names when names are drawn at random from a hat. We can find a different expression for this using Identity 68, which, together with the answer in Exercise 10, gives us a double-sum binomial identity. Which identity is that?

34. Find a probabilistic proof of the following.

Identity 53.

$$\sum_{k=0}^{n} \binom{n}{k} \frac{(-1)^k}{k+2} = \frac{1}{(n+1)(n+2)}.$$

35. Find a probabilistic proof of the following generalization of Identity 24:

Identity 109.

$$\sum_{k=0}^{n} \binom{n}{k} \frac{(-1)^k}{k+m} = \frac{(m-1)! \, n!}{(m+n)!}.$$

36. Find a probabilistic proof of the following generalization of Identity 41.

Identity 147.

$$\sum_{k=0}^{n} \binom{n}{k} \frac{(m-1)! \, (-1)^k}{(k+1)(k+2) \cdots (k+m)} = \sum_{k=0}^{n} \binom{n}{k} \frac{(m-1)! \, k! \, (-1)^k}{(k+m)!} = \frac{1}{n+m}.$$

(Identity 147 is a variant of Identity 42.)

37. Find a probabilistic proof of the following.

Identity 148.

$$\sum_{k=0}^{n} \binom{n}{k} \frac{(-1)^k}{2k+1} = \frac{(2^n n!)^2}{(2n+1)!}.$$

5.6 Notes

For more information on basic probability concepts and calculations see standard texts such as Ross [59] and Grimmett and Stirzaker [34]. Ross's book in particular has a large number of examples that use indicator variables to calculate probabilities.

Peterson [51] uses properties of the exponential distribution to give a different probability-based proof of Identity 24, as well as a generalization. Spivey [70] gives a balls-and-jars proof of Identity 24 and the generalization in Peterson.

6

Generating Functions

In this chapter we discuss the technique of generating functions, an extremely powerful tool that has applications in a large number of areas. For example, generating functions can be used to find the solutions to counting problems, to solve recurrences, and to prove a wide variety of identities. In their classic text *Concrete Mathematics*, Graham, Knuth, and Patashnik call generating functions "the most important idea in this whole book." [32, p. 196]

Formally, an ordinary generating function for the sequence $(a_k) = (a_0, a_1, a_2, \ldots)$ is a power series in which a_k is the coefficient of the term x^k. This may not seem like much, but it effectively amounts to an embedding of the sequence inside a larger mathematical structure—one that has many properties that are already understood. Manipulating the larger structure in various ways allows you to extract different properties of the embedded sequence, such as identities involving the sequence. There are other kinds of generating functions besides ordinary ones, too; different ones are useful for different kinds of tasks.

The first section of this chapter walks us through the idea of a generating function: how it might naturally arise and be used to solve a counting problem. The second section looks at a variety of ways in which ordinary generating functions can be used to prove binomial identities. In the third and final section we consider exponential generating functions, in which $a_k/k!$ is the coefficient of x^k. It turns out that binomial coefficients appear naturally in the process of manipulating exponential generating functions, and we'll use this connection to prove more binomial identities.

6.1   The Idea of a Generating Function

In Exercise 2 of Chapter 3 we ask the reader to show that $\binom{n+m-1}{m-1}$ counts the number of solutions in nonnegative integers to the equation $x_1 + x_2 + \cdots + x_m = n$. The hint in the solutions says to use a stars-and-bars argument of the kind discussed in Chapter 3.

This problem can be solved another way using generating functions. Let's change the x_i's to z_i's and then ask the same question in a different fashion: How many ways are there to take x^n and express it as $x^{z_1} x^{z_2} \cdots x^{z_m}$, where

$z_i \geq 0$? Well, for each z_i, we're choosing from $0, 1, 2, \ldots, n$. At this stage there's no benefit in forcing the $z_i \leq n$ restriction, so let's just say that we're choosing from $0, 1, 2, \ldots$. Since these are the exponents for a particular x, we're effectively choosing for the ith factor x^0 or x^1 or x^2, etc. "Or" is often represented mathematically as "+", so we can also represent our set of choices for the ith factor as $x^0 + x^1 + x^2 + \cdots$. (This last is a bit ad hoc, but it turns out to be useful.) We have m of these factors, though, each of whose choices can be represented as $x^0 + x^1 + x^2 + \cdots$. Since we must choose a value for x^{z_1} and x^{z_2} and so forth, and (continuing with our theme of translating English into math) "and" is generally represented by multiplication, we can represent our entire set of choices by $(x^0 + x^1 + x^2 + \cdots)^m$.

Now, what does this buy us? Well, any given term in the product $(x^0 + x^1 + x^2 + \cdots)^m$ will be of the form $x^{z_1} x^{z_2} \cdots x^{z_m}$, where x^{z_i} is chosen from the ith factor of $(x^0 + x^1 + x^2 + \cdots)$. For example, if $m = 6$, the choice $z_1 = 5, z_2 = 4, z_3 = 3, z_4 = 2, z_5 = 1, z_6 = 5$ selects $x^5 x^4 x^3 x^2 x^1 x^5$ as the term, and, if $n = 20$, it represents one solution to $x_1 + x_2 + \cdots + x_m = n$. Thus each term in the product $(x^0 + x^1 + x^2 + \cdots)^m$ has a one-to-one correspondence to a particular selection of the x^{z_i}'s. This means that if we multiply out $(x^0 + x^1 + x^2 + \cdots)^m$ and collect like terms, then (and this is the key idea here) *the coefficient of x^n in $(x^0 + x^1 + x^2 + \cdots)^m$ is the number of solutions in nonnegative integers to $x_1 + x_2 + \cdots + x_m = n$*. The function $(x^0 + x^1 + x^2 + \cdots)^m$ is called a *generating function* for the number of solutions to the equation $x_1 + x_2 + \cdots + x_m = n$, with $x_i \geq 0$, since the answer is the coefficient of x^n in that function.

Of course, we still need a way to extract the coefficient of x^n from $(x^0 + x^1 + x^2 + \cdots)^m$. In other words, if we could express $(x^0 + x^1 + x^2 + \cdots)^m$ as a power series,

$$(x^0 + x^1 + x^2 + \cdots)^m = \sum_{k=0}^{\infty} a_k x^k,$$

then a_n is the answer we seek.

Before we proceed further with this problem, let's step back and take a look at the big picture. In general, an *ordinary generating function* for a sequence $(a_0, a_1, \ldots)$ is a function $f(x)$ satisfying

$$f(x) = \sum_{k=0}^{\infty} a_k x^k.$$

(There are several different kinds of generating functions. The only two we consider are ordinary and exponential generating functions, as those are the two most useful for proving binomial coefficient identities. When we use the term "generating function" without a qualifier, we are referring to an ordinary generating function.) We have seen several generating functions already, although we have not called them that. The most important is Identity 18, the binomial series that we saw in Chapter 2,

$$(x+1)^n = \sum_{k=0}^{\infty} \binom{n}{k} x^k.$$

From our current perspective, Identity 18 says that the generating function for the sequence given by $a_k = \binom{n}{k}$ is $(x+1)^n$.

Let's also take a look at two important properties of generating functions.

1. We frequently ignore the question of convergence of the infinite series defining the generating function. When we work with generating functions we view them as formal power series, and so the resulting manipulations are valid even when the series do not technically converge. (For some of the background theory, see, for example, Doubilet, Rota, and Stanley [24].) Of course, for values of x for which the infinite series does converge, the generating function yields an identity.

2. A generating function uniquely defines a particular sequence. This means that if two generating functions are equal, they must define the same sequence. This property will often be used to generate binomial identities.

Now, back to our problem: Our first step in finding a_n in the expression

$$(x^0 + x^1 + x^2 + \cdots)^m = \sum_{k=0}^{\infty} a_k x^k$$

is to remember the formula for the sum of a geometric series:

$$\frac{1}{1-x} = \sum_{k=0}^{\infty} x^k.$$

This means that

$$(x^0 + x^1 + x^2 + \cdots)^m = \left(\frac{1}{1-x}\right)^m = (1-x)^{-m},$$

and so now we need to find a power series for this latter expression. Fortunately, we just saw the power series we need: the binomial series! We proved in Chapter 2 that it is valid for negative exponents as well. Applying Identity 18 and Identity 19, $\binom{-n}{k} = (-1)^k \binom{n+k-1}{k}$, we have

$$(1-x)^{-m} = \sum_{k=0}^{\infty} \binom{-m}{k}(-x)^k = \sum_{k=0}^{\infty} \binom{m+k-1}{k}(-1)^k(-1)^k x^k$$

$$= \sum_{k=0}^{\infty} \binom{m+k-1}{k} x^k.$$

The nth term in the power series is $\binom{m+n-1}{n} = \binom{n+m-1}{m-1}$, which is what we wanted to show.

6.2   Ordinary Generating Functions in Action

Let's now take a look at some of the different ways generating functions can be used to prove binomial identities.

For our first example, let's give a more careful proof of the identity we just derived, an identity known as the *negative binomial series*.

Identity 149 (Negative Binomial Series). *For integers $n \geq 0$,*

$$\frac{1}{(1-x)^{n+1}} = \sum_{k=0}^{\infty} \binom{n+k}{n} x^k.$$

Proof. The proof (perhaps unsurprisingly, given what we saw in the previous section), starts with the formula for a geometric series:

$$\frac{1}{1-x} = \sum_{k=0}^{\infty} x^k.$$

If we differentiate both sides of this expression we get

$$\frac{1}{(1-x)^2} = \sum_{k=0}^{\infty} k x^{k-1} = \sum_{k=1}^{\infty} k x^{k-1}.$$

Continuing in this vein, if we differentiate n times we get

$$\frac{n!}{(1-x)^{n+1}} = \sum_{k=0}^{\infty} k^{\underline{n}} x^{k-n} = \sum_{k=n}^{\infty} k^{\underline{n}} x^{k-n}.$$

Shifting the indexing on the last sum and moving the $n!$ expression over we obtain

$$\frac{1}{(1-x)^{n+1}} = \sum_{k=0}^{\infty} \frac{(k+n)^{\underline{n}}}{n!} x^k = \sum_{k=0}^{\infty} \binom{n+k}{n} x^k.$$

$\square$

Of course, in the previous section we had $n = m - 1$.

There are at least three points worth noting about Identity 149 and our proof of it. First, while Identity 149 is a special case of Identity 18, it can also be viewed as a counterpart to it: It sums over a column of entries in Pascal's triangle rather than a row. Second, this example illustrates the use of calculus to produce new generating functions from known ones. (Recall that both differentiation and integration are valid operations on power series.) Finally, since the geometric series formula converges for $|x| < 1$, Identity 149 also gives us a different way to prove that the negative binomial probabilities actually do sum to 1 (see Exercise 7).

For our second example, let's use generating functions to solve a similar problem to the one in the previous section. Exercise 3 in Chapter 3 asks us the following:

3. Show that $\binom{n-1}{m-1}$ counts the number of solutions in *positive* integers to the equation $x_1 + x_2 + \cdots + x_m = n$.

Proof. The logic we need is similar to that in the problem with the nonnegative integer restriction. The only difference is that we cannot choose 0 for the value of one of the x_i's. This translates into disallowing x^0 as an option in our choices for x^{z_i}, and so our set of choices for x^{z_i} can be expressed as $x^1 + x^2 + \cdots$. This is true for all m factors, and so the generating function we are left with is $(x^1 + x^2 + \cdots)^m$.

Applying a little algebra to this expression, we can rewrite $x^1 + x^2 + \cdots$ as $x(x^0 + x^1 + \cdots) = x(1-x)^{-1}$, and thus our generating function is (with a little help from Identity 149)

$$(x^1 + x^2 + \cdots)^m = \frac{x^m}{(1-x)^m} = x^m \sum_{k=0}^{\infty} \binom{m+k-1}{k} x^k$$

$$= \sum_{k=0}^{\infty} \binom{m+k-1}{k} x^{m+k}.$$

Now, we want the coefficient of x^n in this expression. When $n = m+k$, the index k is given by $k = n-m$. Thus the coefficient of x^n is $\binom{n-1}{n-m} = \binom{n-1}{m-1}$. □

The central binomial coefficients $\binom{2n}{n}$ comprise one the most important subsets of binomial coefficients. However, thus far we have only seen a few identities featuring them—because proving identities with the central binomial coefficients is, in general, harder than for the entire class of binomial coefficients. Generating functions, though, make proving many central binomial coefficient identities quite easy. We'll look at several of these for the rest of this section, including generalizations of most of the central binomial coefficient identities we saw in Chapter 4.

First we need the actual generating function for the central binomial coefficients.

Identity 150.

$$\frac{1}{\sqrt{1-4x}} = \sum_{n=0}^{\infty} \binom{2n}{n} x^n.$$

Proof. We start with Identity 30:

$$\binom{-1/2}{n} = \left(-\frac{1}{4}\right)^n \binom{2n}{n}.$$

Then we can apply the ever-useful binomial series to obtain

$$\sum_{n=0}^{\infty}\binom{2n}{n}x^n = \sum_{n=0}^{\infty}\binom{-1/2}{n}(-4x)^n = (1-4x)^{-1/2},$$

a perhaps surprisingly easy derivation.   □

The proof of Identity 108 in Chapter 4 uses the beta integral to find the sum of the reciprocals of the central binomial coefficients. Generating functions can produce this sum, too, as well as a generalization. The derivation has some similarities to our proof of Identity 108. In that proof the extra factor of $2n+1$ from the beta integral (Identity 106) was one of the complications in the derivation. Let's move that to the left side to make our work easier.

Identity 151.

$$\frac{4}{\sqrt{4x-x^2}}\arcsin\left(\frac{\sqrt{x}}{2}\right) = \sum_{n=0}^{\infty}\frac{x^n}{\binom{2n}{n}(2n+1)}.$$

Proof. Applying the beta integral,

$$\binom{n}{k}^{-1} = (n+1)\int_0^1 y^k(1-y)^{n-k}dy,$$

we have

$$\sum_{n=0}^{\infty}\frac{x^n}{\binom{2n}{n}(2n+1)} = \sum_{n=0}^{\infty}\int_0^1 y^n(1-y)^n x^n\,dy = \int_0^1 \sum_{n=0}^{\infty}(xy(1-y))^n\,dy$$

$$= \int_0^1 \frac{1}{1-xy(1-y)}\,dy,$$

where we use the geometric series formula in the last step. Now, apply the substitution $t = \sqrt{x}(2y-1)$. It's not obvious at this stage that this will be helpful, but, as we shall see, it simplifies the integration. We get

$$1 - xy(1-y) = 1 - x\left(\frac{t+\sqrt{x}}{2\sqrt{x}}\right)\left(\frac{\sqrt{x}-t}{2\sqrt{x}}\right) = 1 - \frac{x-t^2}{4} = \frac{4-x+t^2}{4}.$$

Thus we have

$$\int_0^1 \frac{1}{1-xy(1-y)}\,dy = \int_{-\sqrt{x}}^{\sqrt{x}}\frac{4}{2\sqrt{x}(t^2+4-x)}\,dt = \frac{2}{\sqrt{x}}\int_{-\sqrt{x}}^{\sqrt{x}}\frac{1}{t^2+4-x}\,dt$$

$$= \frac{2}{\sqrt{x(4-x)}}\arctan\left(\frac{t}{\sqrt{4-x}}\right)\Bigg|_{-\sqrt{x}}^{\sqrt{x}}$$

$$= \frac{2}{\sqrt{4x-x^2}}\left(\arctan\left(\frac{\sqrt{x}}{\sqrt{4-x}}\right) - \arctan\left(\frac{-\sqrt{x}}{\sqrt{4-x}}\right)\right)$$

$$= \frac{4}{\sqrt{4x - x^2}} \arctan\left(\frac{\sqrt{x}}{\sqrt{4 - x}}\right)$$

$$= \frac{4}{\sqrt{4x - x^2}} \arcsin\left(\frac{\sqrt{x}}{2}\right).$$

The second-to-last step follows because arctangent is an odd function. For the last step, a triangle that has sides of $\sqrt{x}$ and $\sqrt{4 - x}$ has a hypotenuse of 2. Thus the formula can be expressed more simply through the use of the arcsine function. □

With a variable switch, Identity 151 has a simpler representation as

Identity 152.

$$\frac{\arcsin x}{x\sqrt{1 - x^2}} = \sum_{n=0}^{\infty} \frac{4^n x^{2n}}{\binom{2n}{n}(2n + 1)}.$$

Proof. Replace x with $4x^2$ in Identity 151. □

Finally, a little manipulation with Identity 152 gives us what we originally wanted, a generating function for the reciprocals of the central binomial coefficients:

Identity 153.

$$\frac{1}{1 - x^2} + \frac{x \arcsin x}{(1 - x^2)^{3/2}} = \sum_{n=0}^{\infty} \frac{4^n x^{2n}}{\binom{2n}{n}}.$$

Proof. We need to get rid of the extra factor of $2n + 1$ in the denominator of Identity 152. One way to do that is to multiply both sides of Identity 152 by x and then differentiate. The left side of Identity 153 is just the derivative of $(\arcsin x)/\sqrt{1 - x^2}$. □

For the straight-up generating function for $\binom{2n}{n}^{-1}$ with x instead of x^2 and no extra factor of 4^n, see Exercise 15. We can use Identity 153, though, to evaluate sums such as Identity 108, the sum of the reciprocals of the central binomial coefficients.

Identity 108.

$$\sum_{n=0}^{\infty} \frac{1}{\binom{2n}{n}} = \frac{4}{3} + \frac{2\pi\sqrt{3}}{27}.$$

Proof. Tracking the convergence set through the last few identities we see that in Identity 151 we need $|xy(1 - y)| < 1$ to guarantee convergence. Since $0 \leq y \leq 1$, $y(1 - y) \leq 1/4$. Moreover, since $y(1 - y) = 1/4$ only when $y = 1/2$, we can claim $y(1 - y) < 1/4$ without affecting the value of the integral in which $y(1 - y)$ appears. Thus in Identity 151 we have convergence when $|x| \leq 4$. We replace x with $4x^2$ in Identity 152, which reduces our convergence set to $|x| \leq 1$. The differentiation in Identity 153 only preserves convergence on the interior, so we have convergence when $|x| < 1$ for this identity. Letting $x = 1/2$ in Identity 153 yields

$$\sum_{n=0}^{\infty} \frac{1}{\binom{2n}{n}} = \frac{1}{1 - 1/4} + \frac{\arcsin(1/2)}{2(1 - 1/4)^{3/2}} = \frac{4}{3} + \frac{\pi}{12(3/4)^{3/2}} = \frac{4}{3} + \frac{2\pi\sqrt{3}}{27}.$$

□

We just argued that we have convergence when $|x| < 4$ for Identity 151. This includes negative values for x. However, the generating function in this identity includes an $\arcsin(\sqrt{x}/2)$ factor, which means that for $-4 < x < 0$ we have a real sum expressed in terms of imaginary numbers! In Exercise 15 of Chapter 9 we explore what it means to evaluate the inverse sine of an imaginary number as a prelude to deriving the formula for the alternating version of Identity 108.

Identities 151, 152, and 153 illustrate an idea that is often the case with generating functions: If you find deriving the generating function for a particular sequence to be difficult, try finding the generating function for a similar sequence first. Then manipulate the generating function you have into the generating function you want.

Let's now take a look at one of the most important generating function properties: the convolution property. This property is particularly useful for proving binomial coefficient identities involving sums, including, as it so happens, certain sums featuring the central binomial coefficients.

The *convolution* of two sequences (a_n) and (b_n) is the sequence given by

$$c_n = \sum_{k=0}^{n} a_k b_{n-k}.$$

The convolution property for generating functions states that if $f_a(x)$ is the generating function for sequence (a_n), and $f_b(x)$ is the generating function for sequence (b_n), then the product $f_a(x)f_b(x)$ is the generating function for the convolution of (a_n) and (b_n).

Theorem 13 (Convolution Property). *If* $f_a(x) = \sum_{n=0}^{\infty} a_n x^n$ *and* $f_b(x) = \sum_{n=0}^{\infty} b_n x^n$ *then*

$$f_a(x)f_b(x) = \sum_{n=0}^{\infty} \left(\sum_{k=0}^{n} a_k b_{n-k} \right) x^n.$$

Proof. The product $f_a(x)f_b(x)$ looks like this:

$$f_a(x)f_b(x) = \left(a_0 + a_1 x + a_2 x^2 + \cdots \right) \left(b_0 + b_1 x + b_2 x^2 + \cdots \right).$$

We cannot begin the evaluation of this expression by multiplying a_0 through all of $f_b(x)$, as we would with two polynomials: We would never finish! Instead, to multiply two infinite series we have to consider what the x^n term would be for each n, starting with $n = 0$. This means we have to think about which terms in $f_a(x)$ and $f_b(x)$ would be multiplied to give an x^n term in their product. For example, the only way to achieve an x^0 term is with $a_0 b_0$, so that must be the constant term in $f_a(x)f_b(x)$. There are two ways to obtain an x^1 term: $a_0(b_1 x)$ and $(a_1 x)b_0$. Thus the x^1 term in $f_a(x)f_b(x)$ must be $(a_0 b_1 + a_1 b_0)x$. Following this logic, the x^2 term in $f_a(x)f_b(x)$ is $(a_0 b_2 + a_1 b_1 + a_2 b_0)x^2$, and the x^3 term is $(a_0 b_3 + a_1 b_2 + a_2 b_1 + a_3 b_0)x^3$.

In general, the x^n term in the product $f_a(x)f_b(x)$ is the sum of all terms of the form $a_j x^j b_k x^k$ in which $j + k = n$. Thus the coefficient of x^n in $f_a(x)f_b(x)$ is

$$a_0 b_n + a_1 b_{n-1} + \cdots + a_n b_0 = \sum_{k=0}^{n} a_k b_{n-k}.$$

$\square$

Let's look at some examples to see how the convolution property can be used to prove binomial sum identities. Our first example is Vandermonde's identity, which we proved combinatorially in Chapter 3. That proof was only valid for m and n being nonnegative integers. This one uses Newton's binomial series (Identity 18), though, and thus is valid for real m and n. (We also give a proof of Vandermonde's identity in Chapter 4, which is valid for real m and n as well.)

Identity 57. *For real m and n,*

$$\sum_{k=0}^{r} \binom{n}{k}\binom{m}{r-k} = \binom{n+m}{r}.$$

Proof. By letting $a_k = \binom{n}{k}$ and $b_k = \binom{m}{k}$, we can view the left side of Identity 57 as the convolution of the sequences (a_k) and (b_k). To use the convolution property, we need the generating functions of $\binom{n}{k}$ and $\binom{m}{k}$. These are given by the binomial series, Identity 18:

$$(x+1)^n = \sum_{k=0}^{\infty} \binom{n}{k} x^k.$$

Now, applying the convolution property, we have

$$\sum_{r=0}^{\infty}\left(\sum_{k=0}^{r} \binom{n}{k}\binom{m}{r-k}\right) x^r = (x+1)^n (x+1)^m = (x+1)^{n+m}$$

$$= \sum_{r=0}^{\infty} \binom{n+m}{r} x^r.$$

Since the corresponding coefficients of the two generating functions must be equal, the identity follows. $\square$

Also in Chapter 3, we used a fairly complicated involution argument to prove the following identity. There's a much easier proof with generating functions.

Identity 65.

$$\sum_{k=0}^{n} \binom{n}{k}^2 (-1)^k = (-1)^{n/2} \binom{n}{n/2} [n \text{ is even}].$$

Proof. Since $\binom{n}{n-k} = \binom{n}{k}$, the left side is the convolution of the sequences $a_k = \binom{n}{k}(-1)^k$ and $b_k = \binom{n}{k}$. By Identity 18,

$$(1-x)^n = \sum_{k=0}^{\infty} \binom{n}{k}(-x)^k = \sum_{k=0}^{\infty} \binom{n}{k}(-1)^k x^k,$$

which means that the generating function for a_k is $(1-x)^n$. Similarly, the generating function for b_k is $(1+x)^n$. The left side of Identity 65 is thus the coefficient of x^n in the generating function for the convolution of a_k and b_k, which is

$$(1-x)^n (1+x)^n = (1-x^2)^n = \sum_{k=0}^{\infty} \binom{n}{k}(-1)^k x^{2k}.$$

If n is odd, the coefficient of x^n in this sum is 0. If n is even, the coefficient of x^n is $(-1)^{n/2}\binom{n}{n/2}$. $\qquad\square$

Let's now take a look at how convolution can be used to prove two identities featuring sums involving the central binomial coefficients. First, the convolution of the central binomial coefficients with themselves has a nice formula:

Identity 154.

$$\sum_{k=0}^{n} \binom{2k}{k}\binom{2n-2k}{n-k} = 4^n.$$

Proof. Since the left side is the convolution of the sequence $\binom{2k}{k}$ with itself, the convolution property and Identity 150 say that the left side has generating function $(1-4x)^{-1/2}(1-4x)^{-1/2} = (1-4x)^{-1}$. Applying the geometric series formula, we have

$$\frac{1}{1-4x} = \sum_{n=0}^{\infty}(4x)^n = \sum_{n=0}^{\infty} 4^n x^n.$$

Since $(1-4x)^{-1}$ generates the sequence 4^n, the identity is proved. $\qquad\square$

We can combine ideas from the previous two identities to find the alternating convolution of the central binomial coefficients.

Identity 155.

$$\sum_{k=0}^{n} \binom{2k}{k} \binom{2n-2k}{n-k} (-1)^k = 2^n \binom{n}{n/2} [n \text{ is even}].$$

Proof. The left side is the convolution of the sequence $(-1)^n \binom{2n}{n}$ with the sequence $\binom{2n}{n}$. Via Identity 150, we have

$$\sum_{n=0}^{\infty} \binom{2n}{n} (-1)^n x^n = \sum_{n=0}^{\infty} \binom{2n}{n} (-x)^n = (1+4x)^{-1/2}.$$

By the convolution property, then, the left side has generating function

$$(1+4x)^{-1/2}(1-4x)^{-1/2} = (1-16x^2)^{-1/2} = \sum_{n=0}^{\infty} \binom{-1/2}{n} (-16x^2)^n$$

$$= \sum_{n=0}^{\infty} \binom{-1/2}{n} (-16)^n x^{2n}.$$

Applying Identity 30, $\binom{-1/2}{n} = (-1/4)^n \binom{2n}{n}$, we obtain

$$\sum_{n=0}^{\infty} \binom{-1/2}{n} (-16)^n x^{2n} = \sum_{n=0}^{\infty} \binom{2n}{n} \left(-\frac{1}{4}\right)^n (-16)^n x^{2n} = \sum_{n=0}^{\infty} \binom{2n}{n} 4^n x^{2n}.$$

If n is odd, the coefficient of x^n is 0. If n is even, the coefficient of x^n is $2^n \binom{n}{n/2}$. □

The generating functions for the binomial series (Identity 18)

$$(1+x)^m = \sum_{k=0}^{m} \binom{m}{k} x^k$$

and the negative binomial series (Identity 149)

$$\frac{1}{(1-x)^{m+1}} = \sum_{k=0}^{\infty} \binom{m+k}{m} x^k$$

are so close to being reciprocals that at this point that we might be itching to multiply them together. For our last example with convolutions, let's do that but modify the negative binomial series so that the generating functions are actually reciprocals. This yields the following identity.

Identity 156.

$$\sum_{k=0}^{n} \binom{m}{k} \binom{m+n-k-1}{m-1} (-1)^{n-k} = [n = 0].$$

Proof. Via Identity 149, we have

$$\frac{1}{(1+x)^m} = \sum_{k=0}^{\infty} \binom{m+k-1}{m-1}(-1)^k x^k.$$

Thus the convolution of the sequences $\binom{m}{k}$ and $\binom{m+k-1}{m-1}(-1)^k$ has generating function $(1+x)^m(1+x)^{-m} = 1$. The number 1 by itself is an odd-looking generating function, but it is a legitimate generating function: We have

$$1 = \sum_{n=0}^{\infty} a_n x^n,$$

where

$$a_n = [n = 0] = \begin{cases} 1, & n = 0; \\ 0, & n \neq 0. \end{cases}$$

Thus 1 as a generating function generates the sequence $1, 0, 0, 0, \dots$. Therefore, the convolution of $\binom{m}{k}$ and $\binom{m+k-1}{m-1}(-1)^k$ is the sequence $1, 0, 0, 0, \dots$. □

6.3   Exponential Generating Functions

Probably the second-most common types of generating functions are the exponential generating functions. An *exponential generating function* for a sequence $(a_0, a_1, \dots)$ is a function $f(x)$ satisfying

$$f(x) = \sum_{n=0}^{\infty} a_n \frac{x^n}{n!}.$$

The extra factor of $n!$ in the denominator turns out to be a useful addition for proving binomial identities. The reason is the following convolution property for exponential generating functions.

Theorem 14 (Convolution Property for Exponential Generating Functions).
If $f_a(x) = \sum_{n=0}^{\infty} a_n \frac{x^n}{n!}$ and $f_b(x) = \sum_{n=0}^{\infty} b_n \frac{x^n}{n!}$ then

$$f_a(x)f_b(x) = \sum_{n=0}^{\infty} \left(\sum_{k=0}^{n} \binom{n}{k} a_k b_{n-k} \right) \frac{x^n}{n!}.$$

Proof. The proof is similar to that for Theorem 13. The x^n term in the product $f_a(x)f_b(x)$ is the sum of all terms of the form $\frac{a_j}{j!}x^j \frac{b_k}{k!}x^k$ in which $j + k = n$. Thus the coefficient of x^n in $f_a(x)f_b(x)$ is

$$\frac{a_0}{0!}\frac{b_n}{n!} + \frac{a_1}{1!}\frac{b_{n-1}}{(n-1)!} + \dots + \frac{a_n}{n!}\frac{b_0}{0!} = \sum_{k=0}^{n} \frac{a_k}{k!}\frac{b_{n-k}}{(n-k)!} = \sum_{k=0}^{n}\binom{n}{k}\frac{a_k b_{n-k}}{n!}.$$

□

If we define the *binomial convolution* of the sequences (a_n) and (b_n) to be the sequence given by

$$c_n = \sum_{k=0}^{n} \binom{n}{k} a_k b_{n-k}$$

then Theorem 14 tells us that the product of the exponential generating functions for (a_n) and (b_n) generates their binomial convolution (c_n). This can be used to produce a variety of binomial sum identities.

Before we give some examples using exponential generating functions, we should mention the basic exponential generating function—that for the sequence of 1's:

$$e^x = \sum_{n=0}^{\infty} \frac{x^n}{n!}.$$

This, of course, is more commonly known as the Maclaurin series for e^x.

For our first example, let's give yet another proof of the row sum of the binomial coefficients.

Identity 4.

$$\sum_{k=0}^{n} \binom{n}{k} = 2^n.$$

Proof. The sum is the binomial convolution of the sequences given by $a_n = 1$ and $b_n = 1$. Since

$$f_a(x)f_b(x) = e^x e^x = e^{2x} = \sum_{n=0}^{\infty} \frac{(2x)^n}{n!} = \sum_{n=0}^{\infty} 2^n \frac{x^n}{n!},$$

Theorem 14 tells us that the sum is 2^n. □

Before we do more examples, let's prove a few basic properties of exponential generating functions—some of which we will use and some of which are just good to know.

Theorem 15. *If $f_a(x)$ is the exponential generating function for the sequence (a_n), then*

1. *For a constant c, the function $f_a(cx)$ is the exponential generating function for the sequence $(c^n a_n)$.*

2. *The function $f'_a(x)$ is the exponential generating function for the sequence $(a_{n+1}) = (a_1, a_2, \ldots)$.*

3. *The function $\int_0^x f_a(t)\, dt$ is the exponential generating function for the sequence $(0, a_0, a_1, \ldots)$.*

4. *The function* $xf_a(x)$ *is the exponential generating function for the sequence given by* $(na_{n-1}) = (0, a_0, 2a_1, 3a_2, \ldots)$.

5. *The function* $(f_a(x) - a_0)/x$ *is the exponential generating function for the sequence* $(a_{n+1}/(n+1)) = (a_1, a_2/2, a_3/3, \ldots)$.

Proof. We have

1. $f_a(cx) = \displaystyle\sum_{n=0}^{\infty} a_n \frac{(cx)^n}{n!} = \sum_{n=0}^{\infty} c^n a_n \frac{x^n}{n!}$.

2. $f_a'(x) = \displaystyle\sum_{n=0}^{\infty} n a_n \frac{x^{n-1}}{n!} = \sum_{n=1}^{\infty} a_n \frac{x^{n-1}}{(n-1)!} = \sum_{n=0}^{\infty} a_{n+1} \frac{x^n}{n!}$.

3. $\displaystyle\int_0^x f_a(t)\, dt = \int_0^x \sum_{n=0}^{\infty} a_n \frac{t^n}{n!}\, dt = \sum_{n=0}^{\infty} a_n \frac{t^{n+1}}{(n+1)!}\Bigg|_0^x = \sum_{n=0}^{\infty} a_n \frac{x^{n+1}}{(n+1)!} =$

$\displaystyle\sum_{n=1}^{\infty} a_{n-1} \frac{x^n}{n!}$.

4. $xf_a(x) = \displaystyle\sum_{n=0}^{\infty} a_n \frac{x^{n+1}}{n!} = \sum_{n=0}^{\infty} (n+1) a_n \frac{x^{n+1}}{(n+1)!} = \sum_{n=1}^{\infty} n a_{n-1} \frac{x^n}{n!} =$

$\displaystyle\sum_{n=0}^{\infty} n a_{n-1} \frac{x^n}{n!}$.

5. $\dfrac{f_a(x) - a_0}{x} = \displaystyle\sum_{n=0}^{\infty} a_n \frac{x^{n-1}}{n!} - \frac{a_0}{x} = \sum_{n=1}^{\infty} a_n \frac{x^{n-1}}{n!} = \sum_{n=0}^{\infty} a_{n+1} \frac{x^n}{(n+1)!} =$

$\displaystyle\sum_{n=0}^{\infty} \frac{a_{n+1}}{n+1} \frac{x^n}{n!}$.

$\square$

The first property of Theorem 15 says that substituting cx for x multiplies the sequence by a factor of c^n. The second and third properties in Theorem 15 tell us that differentiation and integration of an exponential generating function shift the corresponding sequence backward and forward, respectively. The fourth property says that multiplying an exponential generating function by x shifts the sequence forward and multiplies it by a factor of n. The fifth property says that, after subtracting off the first element of the sequence, dividing a generating function by x multiplies the sequence by n and then shifts it backward. The last four properties can be iterated; thus, for example, $f_a^{(m)}(x)$ generates the sequence $(a_{n+m}) = (a_m, a_{m+1}, \ldots)$.

The fourth property of Theorem 15 yields another quick proof of Identity 21.

Identity 21.

$$\sum_{k=0}^{n} \binom{n}{k} k = n2^{n-1}.$$

Proof. The left side is the binomial convolution of the sequences given by $a_n = n$ and $b_n = 1$. The sequence (a_n) can be thought of as taking the sequence of 1's, shifting it forward (which only puts a 0 in front of it), and multiplying it by n. The fourth property of Theorem 15 thus yields $f_a(x) = xe^x$. By the convolution property, $f_a(x)f_b(x) = xe^{2x}$. Since e^{2x} generates the sequence $(2^n)_{n=0}^{\infty}$, Property 4 of Theorem 15 tells us that xe^{2x} generates the sequence obtained by shifting 2^n forward and multiplying it by n: $(n2^{n-1})_{n=0}^{\infty}$. $\square$

Similarly, Property 5 of Theorem 15 gives another short proof of Identity 23:

Identity 23.

$$\sum_{k=0}^{n} \binom{n}{k} \frac{1}{k+1} = \frac{2^{n+1}-1}{n+1}.$$

Proof. The left side is the binomial convolution of the sequences given by $a_n = 1/(n+1)$ and $b_n = 1$. By the fifth property of Theorem 15, the exponential generating function of (a_n) is $(e^x - 1)/x$. The convolution property (Theorem 14) says that the left side is generated by

$$\frac{e^{2x} - e^x}{x} = \frac{e^{2x} - 1}{x} - \frac{e^x - 1}{x}.$$

By Property 5 of Theorem 15, the function $(e^{2x} - 1)/x$ generates the sequence $(2^{n+1}/(n+1))_{n=0}^{\infty}$, and we already know that $(e^x - 1)/x$ generates the sequence $(1/(n+1))_{n=0}^{\infty}$. Subtracting these (valid by Theorem 17—see Exercise 1), we have that the two sides of the identity are equal. $\square$

Given a sequence (a_n), the sequence (b_n), with $b_n = \sum_{k=0}^{n} \binom{n}{k} a_k$, is called the *binomial transform* of the sequence (a_n). Similarly, the sequence (c_n), with $c_n = \sum_{k=0}^{n} \binom{n}{k} a_k (-1)^k$, is called the *alternating binomial transform* of the sequence (a_n). While this is the first time we have used these terms, many of the identities we have presented in this book have represented the binomial transform or alternating binomial transform of some sequence. The next theorem shows that the exponential generating functions for a sequence, its binomial transform, and its alternating binomial transform are closely related.

Theorem 16. *Let (a_n), (b_n), and (c_n) be sequences such that*

$$b_n = \sum_{k=0}^{n} \binom{n}{k} a_k \text{ and } c_n = \sum_{k=0}^{n} \binom{n}{k} a_k (-1)^k.$$

Then their respective exponential generating functions $f_a(x)$, $f_b(x)$, and $f_c(x)$ satisfy

$$f_b(x) = e^x f_a(x),$$
$$f_c(x) = e^x f_a(-x), \ and$$
$$f_b(x) = e^{2x} f_c(-x).$$

In other words, to get the exponential generating function of the binomial transform, just multiply the exponential generating function of the original sequence by e^x. To get the exponential generating function of the alternating binomial transform, swap x with $-x$ in the exponential generating function of the original sequence and then multiply by e^x.

Proof. The sequence (b_n) is the binomial convolution of (a_n) and the sequence of 1's. Since the latter has exponential generating function e^x, we have $f_b(x) = e^x f_a(x)$ by Theorem 14.

By the first property of Theorem 15, the sequence $(-1)^n a_n$ has exponential generating function $f_a(-x)$. Since (c_n) is the binomial convolution of $(-1)^n a_n$ and the sequence of 1's, we have $f_c(x) = e^x f_a(-x)$ by Theorem 14.

Finally, $f_a(-x) = e^{-x} f_c(x)$, which means that $f_a(x) = e^x f_c(-x)$. Thus we have $f_b(x) = e^x f_a(x) = e^x(e^x f_c(-x)) = e^{2x} f_c(-x)$. □

Let's look at three illustrations of the use of Theorem 16. The first is a very short proof of the binomial inversion formula that we saw in Chapter 2.

Theorem 1 (Binomial Inversion).

$$b_n = \sum_{k=0}^{n} \binom{n}{k} (-1)^k a_k \iff a_n = \sum_{k=0}^{n} \binom{n}{k} (-1)^k b_k.$$

Proof. By Theorem 16, if (b_n) is the alternating binomial transform of (a_n), then $f_b(x) = e^x f_a(-x)$. This means that $f_a(x) = e^x f_b(-x)$, and so (a_n) is the alternating binomial transform of (b_n). The argument in the other direction is similar. □

Another application of Theorem 16 is to prove the following double binomial sum identity, which is another way of expressing the fact that the alternating binomial transform is its own inverse.

Identity 157. *For any sequence (a_n),*

$$\sum_{k=0}^{n} \sum_{j=0}^{n} \binom{n}{k} \binom{k}{j} (-1)^{j+k} a_j = a_n.$$

Proof. By Theorem 16, if $f_a(x)$ is the exponential generating function for (a_n), then $e^x f_a(-x)$ generates the sequence given by

$$b_n = \sum_{j=0}^{n} \binom{n}{j} (-1)^j a_j.$$

Applying Theorem 16 again, $e^x e^{-x} f_a(x) = f_a(x)$ generates the sequence given by

$$c_n = \sum_{k=0}^{n} \binom{n}{k} \left(\sum_{j=0}^{k} \binom{k}{j} (-1)^j a_j \right) (-1)^k = \sum_{k=0}^{n} \sum_{j=0}^{n} \binom{n}{k} \binom{k}{j} (-1)^{j+k} a_j.$$

By definition, though, $f_a(x)$ also generates the sequence (a_n). $\square$

For our last example we find the exponential generating function of the derangement numbers D_n (where D_n is the number of permutations on n elements with no fixed points), and we use this to prove the explicit formula for D_n in Identity 68 (first proved in Chapter 3).

Identity 158.

$$\frac{e^{-x}}{1-x} = \sum_{n=0}^{\infty} D_n \frac{x^n}{n!}.$$

Proof. Identity 88 says that

$$\sum_{k=0}^{n} \binom{n}{k} D_{n-k} = n!.$$

Reindexing the sum and using the fact that $\binom{n}{k} = \binom{n}{n-k}$ (Identity 5) we can rewrite Identity 88 as $\sum_{k=0}^{n} \binom{n}{k} D_k = n!$. Thus the binomial transform of the sequence of derangement numbers is the factorial sequence $(n!)_{n=0}^{\infty}$. If $f_D(x)$ and $f_n(x)$ denote, respectively, their exponential generating functions, we have, by Theorem 16,

$$f_n(x) = e^x f_D(x).$$

Since $f_n(x) = 1/(1-x)$ (see Exercise 28), this means that

$$f_D(x) = \frac{e^{-x}}{1-x}.$$

$\square$

Now we can use exponential generating functions to prove

Identity 68.

$$D_n = \sum_{k=0}^{n} \binom{n}{k} (n-k)! (-1)^k = n! \sum_{k=0}^{n} \frac{(-1)^k}{k!}.$$

Proof. By Identity 158 and Theorem 14, D_n is the binomial convolution of the sequences generated by e^{-x} and $1/(1-x)$. The former is the sequence given by $a_n = (-1)^n$, and the latter is, by Exercise 28, the sequence given by $b_n = n!$. The identity follows. $\square$

6.4 Exercises

1. Prove the following linearity properties for generating functions.

    Theorem 17 (Linearity of Generating Functions).

    (a) *Suppose $f_a(x)$ and $f_b(x)$ are, respectively, the ordinary generating functions for the sequences (a_n) and (b_n), and suppose c and d are constants. Then the ordinary generating function for the sequence $(ca_n + db_n)$ is $cf_a(x) + df_b(x)$.*

    (b) *Suppose $f_a(x)$ and $f_b(x)$ are, respectively, the exponential generating functions for the sequences (a_n) and (b_n), and suppose c and d are constants. Then the exponential generating function for the sequence $(ca_n + db_n)$ is $cf_a(x) + df_b(x)$.*

2. In the proof of Identity 149 we use the geometric series formula

$$\frac{1}{1-x} = \sum_{n=0}^{\infty} x^n.$$

    What does this tell us about the sequence whose generating function is $(1-x)^{-1}$?

3. Prove that

$$\frac{x}{(1-x)^2} = \sum_{n=0}^{\infty} nx^n,$$

    thereby giving us the formula for the ordinary generating function of the sequence of nonnegative integers.

4. Theorem 15 contains several basic properties of exponential generating functions. Prove the following theorem, which gives a corresponding set of properties of ordinary generating functions.

    Theorem 18. *If $f_a(x)$ is the ordinary generating function for the sequence (a_n), then*

    (a) *For a constant c, the function $f_a(cx)$ is the ordinary generating function for the sequence $(c^n a_n)$.*

(b) *The function $f'_a(x)$ is the ordinary generating function for the sequence $((n+1)a_{n+1}) = (a_1, 2a_2, 3a_3, \ldots)$.*

(c) *The function $\int_0^x f_a(t)\, dt$ is the ordinary generating function for the sequence $(0, a_0, a_1/2, a_2/3, \ldots, a_{n-1}/n, \ldots)$.*

(d) *For $m \geq 0$ the function $x^m f_a(x)$ is the ordinary generating function for the sequence given by $(a_{n-m}) = (0, 0, \ldots, 0, a_0, a_1, a_2, \ldots)$, where there are m leading 0's.*

5. Prove the following. (Compare with Exercise 27.)

(a) The ordinary generating function of $(an+b)_{n=0}^{\infty}$, where a and b are constants, is
$$\frac{ax}{(1-x)^2} + \frac{b}{1-x}.$$

(b) The ordinary generating function of $(n^2)_{n=0}^{\infty}$ is $\dfrac{x(1+x)}{(1-x)^3}$.

(c) The ordinary generating function of $(5^n)_{n=0}^{\infty}$ is $\dfrac{1}{1-5x}$.

6. Starting with Identity 149, prove

Identity 129. *For integers $n \geq 0$,*
$$\frac{x^n}{(1-x)^{n+1}} = \sum_{k=0}^{\infty} \binom{k}{n} x^k.$$

7. Use Identity 149 to give a short proof that the negative binomial distribution probabilities sum to 1:

Identity 128. *If $0 < p \leq 1$ and r is a positive integer, then*
$$\sum_{k=r}^{\infty} \binom{k-1}{r-1} p^r (1-p)^{k-r} = 1.$$

8. How many solutions are there to the equation $x_1 + x_2 + \cdots + x_m = n$, where $x_i \geq 2$ for each i?

9. How many solutions are there to the equation $x_1 + x_2 + \cdots + x_m = n$, where $x_i \geq r$ for each i?

10. Let a_n denote the number of ways to make change for n cents using only pennies, nickels, dimes, quarters, half-dollars, and silver dollars. Find the ordinary generating function for (a_n).

11. Prove

Identity 159.
$$\frac{1 - \sqrt{1-4x}}{2x} = \sum_{n=0}^{\infty} \binom{2n}{n} \frac{x^n}{n+1}.$$

12. Prove

Identity 160.

$$\frac{(1-4x)^{-1/2}-1}{2x} = \sum_{n=0}^{\infty} \binom{2n+1}{n} x^n.$$

13. Prove the arcsine series:

Identity 161.

$$\arcsin x = \sum_{n=0}^{\infty} \binom{2n}{n} \frac{x^{2n+1}}{(2n+1)4^n}.$$

14. Prove

Identity 162.

$$2\ln\left(\frac{1-\sqrt{1-4x}}{2x}\right) = \sum_{n=1}^{\infty} \binom{2n}{n} \frac{x^n}{n}.$$

15. Prove

Identity 163.

$$\frac{4}{4-x} + \frac{\sqrt{x}\arcsin(\sqrt{x}/2)}{2(1-x/4)^{3/2}} = \sum_{n=0}^{\infty} \frac{x^n}{\binom{2n}{n}}.$$

16. Use a generating function argument to prove

Identity 164.

$$\sum_{k=0}^{n} \binom{n}{k}^2 = \binom{2n}{n}.$$

17. Use a generating function argument to prove

Identity 165.

$$\sum_{k=0}^{n} \binom{m}{k}\binom{m+n-k}{m}(-1)^k = 1.$$

18. Use a generating function argument to prove

Identity 61.

$$\sum_{k=0}^{n} \binom{k}{m}\binom{n-k}{r} = \binom{n+1}{m+r+1}.$$

19. Prove

Identity 166.

$$\sum_{k=0}^{n} \binom{2k}{k}\binom{2n-2k}{n-k}\frac{1}{2k+1} = \frac{16^n}{\binom{2n}{n}(2n+1)}.$$

20. The relationship between the ordinary generating function of a sequence (a_n) and that of its binomial transform (b_n), with $b_n = \sum_{k=0}^{n}\binom{n}{k}a_k$, is more complicated than that for exponential generating functions. Multiplying $(x+1)^n$, the generating function for $\binom{n}{k}$, by $f_a(x)$ via the convolution formula (Theorem 13) does not work to obtain $f_b(x)$ because the n in $(x+1)^n$ is fixed, whereas $f_b(x)$ requires n to vary. Instead, use the definition of $f_b(x)$,

$$f_b(x) = \sum_{n=0}^{\infty}\left(\sum_{k=0}^{n}\binom{n}{k}a_k\right)x^n,$$

and swap the order of summation to derive the correct relationship

$$f_b(x) = \frac{1}{1-x}f_a\left(\frac{x}{1-x}\right).$$

21. Prove the following exponential generating function for column m in Pascal's triangle.

Identity 167.

$$e^x\frac{x^m}{m!} = \sum_{n=m}^{\infty}\binom{n}{m}\frac{x^n}{n!}.$$

22. For a doubly-indexed sequence like the binomial coefficients it is possible to sum over both variables to create a *double generating function*. Prove the following double generating function for the binomial coefficients.

Identity 168.

$$e^{xy+y} = \sum_{n=0}^{\infty}\sum_{k=0}^{\infty}\binom{n}{k}x^k\frac{y^n}{n!}.$$

23. Prove that the exponential generating function of the sequence $(1,0,1,0,1,0,\ldots)$ is given by

$$\cosh x = \frac{e^x + e^{-x}}{2}.$$

Also, prove that the exponential generating function of the sequence $(0,1,0,1,0,1,\ldots)$ is given by

$$\sinh x = \frac{e^x - e^{-x}}{2}.$$

24. Generalizing Exercise 23, prove that if $f_a(x)$ is the ordinary or exponential generating function of the sequence (a_n), then the following hold.

(a) The corresponding ordinary or exponential generating function of the sequence $(a_0, 0, a_2, 0, a_4, 0, \ldots)$ is

$$\frac{f_a(x) + f_a(-x)}{2}.$$

(b) The corresponding ordinary or exponential generating function of the sequence $(0, a_1, 0, a_3, 0, a_5, \ldots)$ is

$$\frac{f_a(x) - f_a(-x)}{2}.$$

25. Use Exercise 24 to prove

Identity 16.

$$\sum_k \binom{n}{2k} x^{2k} y^{n-2k} = \frac{1}{2} \left[(x+y)^n + (y-x)^n \right].$$

26. Prove that if $f_a(x)$ is the exponential generating function of the sequence (a_n) and $m \geq 0$ then $x^m f_a(x)$ is the exponential generating function of the sequence $(n^{\underline{m}} a_{n-m})$. Then use this property to prove

Identity 36.

$$\sum_{k=0}^n \binom{n}{k} k^{\underline{m}} = n^{\underline{m}} 2^{n-m}.$$

27. Prove the following. (Compare with Exercises 3 and 5.)

(a) The exponential generating function of $(n)_{n=0}^\infty$ is xe^x.

(b) The exponential generating function of $(an+b)_{n=0}^\infty$, where a and b are constants, is $axe^x + be^x$.

(c) The exponential generating function of $(n^2)_{n=0}^\infty$ is $(x + x^2)e^x$.

(d) The exponential generating function of $(5^n)_{n=0}^\infty$ is e^{5x}.

28. Prove the following.

(a) The exponential generating function of $(n!)_{n=0}^\infty$ is $1/(1-x)$.

(b) The exponential generating function of $((n+1)!)_{n=0}^\infty$ is $1/(1-x)^2$.

(c) The exponential generating function of the sequence given by

$$a_n = (n-1)![n \geq 1]$$

is $-\ln(1-x)$.

29. Exercise 28 asks you to prove that the exponential generating function of the sequence given by $(n+1)!$ is $1/(1-x)^2$. This means that $(n+1)!$ must be the binomial convolution of the sequence generated by $1/(1-x)$ with itself. Verify this fact by calculating this binomial convolution.

30. Let $B^m(a_n)$ denote k successive applications of the binomial transform to the sequence (a_n). In other words, $B^m(a_n) = B(B(\cdots B(a_n)\cdots))$, where B occurs m times. Use exponential generating functions to prove

Identity 169.

$$B^m(a_n) = \sum_{k=0}^{n} \binom{n}{k} m^{n-k} a_k.$$

31. Use a generating function argument to prove

Identity 24.

$$\sum_{k=0}^{n} \binom{n}{k} \frac{(-1)^k}{k+1} = \frac{1}{n+1}.$$

32. Use a generating function argument to prove

Identity 53.

$$\sum_{k=0}^{n} \binom{n}{k} \frac{(-1)^k}{k+2} = \frac{1}{(n+1)(n+2)}.$$

6.5 Notes

The classic text on generating functions is Wilf's *Generatingfunctionology* [79]. Chapter 7 of *Concrete Mathematics* [32] is also a good resource. See, too, Chapter 4 of Quaintance and Gould [55], as well as short papers by Gould [31], Haukkanen [36], and Prodinger [53] on generating functions and binomial transforms. (Exercise 20 appears in Gould [31], for example.)

For an algebraic proof of Identity 169 and results on representing inverses of the binomial transform operator, see Spivey and Steil [71].

Several identities in this chapter entail sums involving the central binomial coefficients or their reciprocals. For more identities concerning the central binomial coefficients, see Lehmer [42] and Sprugnoli [74]. Sved [77] and De Angelis [3] have combinatorial proofs of Identity 154; both are more complicated than any of the combinatorial proofs presented in Chapter 3. Spivey [67] gives a combinatorial/probabilistic proof of Identity 155.

Another technique for using formal power series to evaluate binomial sums is that of *Riordan arrays*. See Sprugnoli [73].

7

Recurrence Relations and Finite Differences

In this chapter we look at recurrence relations as a tool for proving binomial identities. We'll pay particular attention to an important special case of recurrence relations: finite differences.

A *recurrence relation* is an equation that defines a sequence in terms of preceding terms and some initial or boundary conditions. The most famous recurrence relation is the one defining the Fibonacci numbers.

Identity 170.

$$F_n = F_{n-1} + F_{n-2}, \; n \geq 2, \quad F_0 = 0, \; F_1 = 1.$$

Identity 170 says that the nth Fibonacci number is the sum of the two previous Fibonacci numbers. Thus we can generate $F_2 = 1$, $F_3 = 2$, $F_4 = 3$, $F_5 = 5$, $F_6 = 8$, and so forth. Obtaining an explicit formula for the nth Fibonacci number takes some more effort, although we will work through this derivation later in the chapter. (Identity 170 is more accurately a *definition* than an identity at this point, but we'll call it an identity. Eventually we will see some other properties of the Fibonacci numbers that are equivalent to Identity 170 so that any of them could be taken as the definition of the Fibonacci numbers. Also, we can define Fibonacci numbers for negative indices by dropping the $n \geq 2$ restriction in Identity 170. See Chapter 8 for more details.)

We have already seen some other recurrence relations in this book. For example, in our proof of the equivalence of the four definitions of the binomial coefficients in Chapter 1 we saw the recurrence

$$G(n) = (x+1)G(n-1), \quad G(0) = 1.$$

This is a simpler recurrence than the one for the Fibonacci numbers, and we were easily able to solve it to find an explicit formula for $G(n)$: $G(n) = (x+1)^n$. We also saw some recurrences in Chapter 5 that we used to find moments of binomial, negative binomial, and hypergeometric distributions.

The most important recurrence in any discussion of the binomial coefficients, though, is the very first identity we saw in this book:

Identity 1 (Pascal's Recurrence). *For integers n and k,*

$$\binom{n}{k} = \binom{n-1}{k} + \binom{n-1}{k-1}, \quad \binom{n}{0} = 1, \quad \binom{0}{k} = [k = 0].$$

Pascal's recurrence is a much more complicated recurrence than the ones for F_n or $G(n)$, as it involves two variables. Although of course we already know the solution, we did not find that solution through a direct manipulation of the recurrence relation like we did to find a formula for $G(n)$. In fact, solving two-variable recurrence relations is quite a bit more difficult than solving single-variable recurrence relations. The relationship between them is analogous to that between ordinary differential equations and partial differential equations. With single-variable recurrence relations and ordinary differential equations, there is only one independent variable. With two-variable recurrence relations and partial differential equations, though, there are two independent variables.

The first section of this chapter focuses on single-variable recurrence relations. We prove a few identities involving these kinds of recurrence relations, using clever substitutions, generating functions, and the method of undetermined coefficients.

In the second section we focus on finite differences. These are equations in which $\Delta a_n = a_{n+1} - a_n = f(n)$, where $f(n)$ is some function of n. An equation involving a finite difference is thus really just an important special case of a recurrence relation. We'll see that the binomial coefficients arise naturally when taking repeated finite differences of a sequence. We'll investigate this relation and show how it can be used to prove binomial identities. In addition, we will see how the binomial transform of a sequence is related to the binomial transform of the finite difference of that sequence. We'll look at applications and variations of this relationship as well.

The third and last section focuses on two-variable recurrence relations. In this section we prove directly that, when $n \geq k \geq 0$, the solution to the recurrence relation in Identity 1 is $\frac{n!}{k!\,(n-k)!}$.

7.1 Single-Variable Recurrence Relations

Before we talk about solving recurrence relations, we need to spend some time on a more basic question: How do you even construct a recurrence relation for some problem you're trying to solve? As an illustration, we'll derive two recurrence relations for the derangement numbers. Then we'll use the second of these to prove Identity 68, an explicit expression for the nth derangement number as an alternating binomial sum that we saw in Chapter 3.

Remember that a *derangement* is a permutation on n elements with no fixed points and that D_n is the number of derangements on n elements. Besides Identity 68, Chapter 3 contains another formula involving D_n in the exercises. In Chapter 6 we found the exponential generating function for D_n.

The following, due to Euler, is considered the basic recurrence relation for the derangement numbers.

Identity 171. *For $n \geq 2$,*

$$D_n = (n-1)(D_{n-1} + D_{n-2}), \quad D_0 = 1, \; D_1 = 0.$$

Proof. First, there can be no derangements on one element, so $D_1 = 0$. The question of defining D_0 is a bit trickier. There is normally considered to be one permutation on zero elements, the empty permutation. Since there are clearly no elements mapped to themselves by the empty permutation, we have an argument for $D_0 = 1$. This agrees with the $n = 0$ case of the explicit formula for D_n given in Identity 68. Thus it makes sense to define $D_0 = 1$.

For the recurrence itself, let's interpret D_n as the number of ways n students can select their names from a hat without any student drawing her own name. First, there are $n-1$ possible names the first student could draw. Suppose she chooses the name of student i. Let students 1 and i swap names. The problem now breaks into cases.

1. If student i ends up with his own name then the group of students except for 1 and i constitute a derangement of $n-2$ elements, with D_{n-2} possibilities for how this could happen.

2. If student i does not end up with his own name then the group of students except for 1 constitute a derangement of $n-1$ elements, with D_{n-1} possibilities for how this could happen.

All together, then, $D_n = (n-1)(D_{n-1} + D_{n-2})$. □

Combinatorial arguments like the one in the proof of Identity 171 are often used to establish recurrence relations.

We can use Identity 171 to prove a second, simpler recurrence relation for the derangement numbers. This relation in turn can easily be used to derive the explicit formula for D_n as well as its exponential generating function (Identity 158).

Identity 172. *For $n \geq 1$,*

$$D_n = nD_{n-1} + (-1)^n, \quad D_0 = 1.$$

Proof. Subtract nD_{n-1} from both sides of Identity 171 to obtain $D_n - nD_{n-1} = -(D_{n-1} - (n-1)D_{n-2})$. Defining $A_n = D_n - nD_{n-1}$, this recurrence turns into $A_n = -A_{n-1}$. Also, we have $A_1 = D_1 - D_0 = 0 - 1 = -1$. Since A_n changes sign every time n increases by one, the solution to the A_n recurrence is $A_n = (-1)^n$. Thus $D_n - nD_{n-1} = (-1)^n$, completing the proof. □

As promised, Identity 172 gives us a quick proof of the sum formula for D_n.

Identity 68.

$$D_n = n! \sum_{k=0}^{n} \frac{(-1)^k}{k!} = \sum_{k=0}^{n} \binom{n}{k}(n-k)!(-1)^k.$$

Proof. Divide both sides of Identity 172 by $n!$ to get

$$\frac{D_n}{n!} = \frac{D_{n-1}}{(n-1)!} + \frac{(-1)^n}{n!}.$$

Letting $B_n = D_n/n!$, the recurrence becomes $B_n = B_{n-1} + (-1)^n/n!$. Since $B_0 = D_0/0! = 1$, unrolling the recurrence gives us $B_1 = 1 + (-1)/1!$, $B_2 = 1 + (-1)/1! + (-1)^2/2!$, and so forth, so that

$$B_n = \sum_{k=0}^{n} \frac{(-1)^k}{k!}.$$

Since $D_n = n!B_n$, the identity follows.   □

The proofs of Identity 172 and 68 presented here both illustrate a useful idea with recurrences: Turn the recurrence into a different recurrence that is easy to solve.

Also as promised, we can find the exponential generating function for the derangement numbers fairly easily from Identity 172.

Identity 158.

$$\frac{e^{-x}}{1-x} = \sum_{n=0}^{\infty} D_n \frac{x^n}{n!}.$$

Proof. Let $f_D(x)$ denote the exponential generating function for the derangement numbers. Multiply both sides of Identity 172 by $x^n/n!$ and sum over n from 1 to infinity to obtain

$$\sum_{n=1}^{\infty} D_n \frac{x^n}{n!} = \sum_{n=1}^{\infty} nD_{n-1} \frac{x^n}{n!} + \sum_{n=1}^{\infty} (-1)^n \frac{x^n}{n!}$$

$$\implies \sum_{n=0}^{\infty} D_n \frac{x^n}{n!} - D_0 = x \sum_{n=1}^{\infty} D_{n-1} \frac{x^{n-1}}{(n-1)!} + \sum_{n=0}^{\infty} (-1)^n \frac{x^n}{n!} - 1$$

$$\implies f_D(x) - 1 = xf_D(x) + e^{-x} - 1$$

$$\implies f_D(x) = \frac{e^{-x}}{1-x}.$$

                                                                                       □

The idea we use in the proof of Identity 158 is one that can often be useful for finding a generating function for a sequence, given a recurrence relation for that sequence. In fact, generating functions are often used to solve recurrences, since one can use the generating function to extract the nth term from the sequence using techniques discussed in Chapter 6. (We even did this in Chapter 6 to give a derivation of Identity 68 from Identity 158.)

Let's take a closer look at this idea by considering another example. First, we'll construct a recurrence relation for the solution to a problem, and then we'll solve the recurrence using generating functions.

Here's the problem: Suppose we draw n straight lines on a piece of paper so that every pair of lines intersect (but no three lines intersect at a common point). Into how many regions do these n lines divide the plane?

First, let's consider the problem for small values of n. Let a_n denote the number of regions we're after. If $n = 0$ there is clearly just one region, and so $a_0 = 1$. With $n = 1$ we get $a_1 = 2$. A second line gives us $a_2 = 4$. The third line cuts through the first two lines. Before, between, and after the intersection points define three regions that the third line cuts. The third line splits each of these three regions in two, and thus it adds three new regions. We get $a_3 = a_2 + 3$. Similarly, the fourth line cuts through the first three lines. This means that the fourth line splits four regions in two, creating four new regions. Thus $a_4 = a_3 + 4$. In general, the nth line splits n regions in two, creating n new regions. Therefore, we have $a_n = a_{n-1} + n$, with $a_0 = 1$.

Now, let's apply the generating function idea. In the derangements example we used an exponential generating function, but here we use an ordinary one. Multiply both sides of the recurrence by x^n and sum from 1 to infinity to obtain

$$\sum_{n=1}^{\infty} a_n x^n = \sum_{n=1}^{\infty} a_{n-1} x^n + \sum_{n=1}^{\infty} n x^n.$$

Since the generating function for the sequence $(n)_{n=0}^{\infty}$ is $x/(1-x)^2$ (see Exercise 3 in Chapter 6), we have

$$\sum_{n=0}^{\infty} a_n x^n - a_0 = x \sum_{n=1}^{\infty} a_{n-1} x^{n-1} + \sum_{n=0}^{\infty} n x^n$$

$$\implies f_a(x) - a_0 = x \sum_{n=0}^{\infty} a_n x^n + \frac{x}{(1-x)^2}$$

$$\implies f_a(x) = 1 + x f_a(x) + \frac{x}{(1-x)^2}$$

$$\implies f_a(x) = \frac{1}{1-x} + \frac{x}{(1-x)^3}.$$

We know that $1/(1-x)$ generates the sequence of 1's. According to Identity 149, $1/(1-x)^3$ generates the sequence with nth term $\binom{n+2}{2}$. Thus, by Theorem 18, $x/(1-x)^3$ generates the sequence with nth term $\binom{n+1}{2}$. Therefore, $a_n = 1 + \binom{n+1}{2}$.

We could, of course, have unrolled the recurrence to find the solution. (See Exercise 1.) However, this example illustrates the use of generating functions to solve recurrences.

We could use the same idea to solve the Fibonacci recurrence $F_n = F_{n-1} + F_{n-2}$, with $F_0 = 0$, $F_1 = 1$ (Identity 170). In fact, Exercise 4 asks you to do just that. However, the Fibonacci recurrence can also be solved using a common method for solving homogeneous linear ordinary differential equations with constant coefficients: the method of undetermined coefficients.

As a recap of how this method works, suppose you have the differential equation $y'' + 5y' + 6 = 0$. The method of undetermined coefficients says to posit a solution of the form $y = e^{rx}$. Substituting this into the differential equation yields $0 = r^2 e^{rx} + 5r e^{rx} + 6 e^{rx} = (r^2 + 5r + 6)e^{rx}$. Since e^{rx} cannot equal 0, we solve the *characteristic equation* $0 = r^2 + 5r + 6 = (r + 3)(r + 2)$ to find two solutions: $y_1 = e^{-3x}$ and $y_2 = e^{-2x}$. Since these two solutions are linearly independent, the theory then says that all solutions to $y'' + 5y' + 6 = 0$ are of the form $y = Ae^{-3x} + Be^{-2x}$. Substituting the initial conditions into this equation produces the values of A and B. (For more on this method, see any standard differential equations text such as Blanchard, Devaney, and Hall [11].)

It turns out that the same method works for homogeneous linear ordinary *difference equations* (i.e., recurrence relations) with constant coefficients, such as the Fibonacci recurrence. However, instead of positing a solution of the form $y = e^{rx}$, we posit one of the form $y = r^n$. Let's use this method to prove the following explicit formula for the Fibonacci numbers, known as *Binet's formula*.

French mathematician **Jacques Philippe Marie Binet** (1786–1856) lived through tumultuous times in the history of his country: the French Revolution, Napoleon's reign and fall, the Bourbon restoration, the Revolutions of 1830 and 1848, and the beginning of the Second Empire under Louis-Napoléon. A devout Catholic and strong backer of the monarchy, his support for Charles X cost him his position at the l'École Polytechnique after the Revolution of 1830. While the explicit formula for the Fibonacci numbers given here was named after him, he was not the first to find it (that honor goes to Abraham de Moivre a hundred years earlier). Binet does have a strong mathematical claim to fame, though: He discovered the rule for matrix multiplication.

Identity 173 (Binet's Formula).

$$F_n = \frac{\phi^n - \psi^n}{\sqrt{5}},$$

where $\phi = (1 + \sqrt{5})/2$ *(the golden ratio), and* $\psi = (1 - \sqrt{5})/2$.

Proof. Assuming a solution of the form $F_n = r^n$, substitute into Identity 170, $F_n = F_{n-1} + F_{n-2}$. We obtain $r^n = r^{n-1} + r^{n-2}$. Moving all terms to the left side and factoring, we have $r^{n-2}(r^2 - r - 1) = 0$. Since r^{n-2} cannot equal 0 unless $r = 0$ (and $F_n = 0$ clearly does not fit the initial condition $F_1 = 1$), we need to solve the characteristic equation $r^2 - r - 1 = 0$ to find values for r that work. Applying the quadratic formula, we get

$$r = \frac{1 \pm \sqrt{5}}{2}.$$

As with differential equations, this means the general solution to the Fibonacci recurrence is

$$F_n = A\left(\frac{1 + \sqrt{5}}{2}\right)^n + B\left(\frac{1 - \sqrt{5}}{2}\right)^n.$$

Substituting the initial conditions into this equation, we obtain the system

$$0 = A + B,$$

$$1 = \frac{1 + \sqrt{5}}{2}A + \frac{1 - \sqrt{5}}{2}B.$$

Solving yields $A = 1/\sqrt{5}$, $B = -1/\sqrt{5}$. □

Since $|\psi| < 1$, an interesting consequence of Binet's formula is that $F_n \approx \phi^n/\sqrt{5}$ for all but small values of n.

We have now seen three methods for solving single-variable recurrence relations: (1) unrolling the recurrence, (2) generating functions, and (3) undetermined coefficients.

As we shall prove in Chapter 8, the Fibonacci numbers can also be expressed in terms of binomial coefficients.

7.2 Finite Differences

The *finite difference* of a sequence (a_n) is the sequence given by $(\Delta a_n)_{n=0}^{\infty} = (a_{n+1} - a_n)_{n=0}^{\infty}$. We have already seen finite differences; we used them in our proof of Identity 25 in Chapter 2. The relationship between binomial sums and finite differences is a rich one, and we will see many examples of this in this section.

Taking finite differences may seem like an unusual operation at first, but they are really just derivatives on the integers rather than the real numbers. In fact, the finite difference of $f(x)$ is the difference quotient $(f(x + h) - f(x))/h$ when $h = 1$. Thus, when f is defined over the reals, the finite difference of $f(x)$ is an approximation to the derivative of $f(x)$.

An equation containing a sequence and its finite difference, such as $a_n = \Delta a_n$, is a *difference equation* (mentioned in the previous section) and is analogous to a differential equation. One can take higher-order finite differences, too, just as one can take higher-order derivatives. Thus, for example, $\Delta^2 a_n = \Delta(a_{n+1} - a_n) = \Delta a_{n+1} - \Delta a_n = (a_{n+2} - a_{n+1}) - (a_{n+1} - a_n) = a_{n+2} - 2a_{n+1} + a_n$. There is also the notion of an *antidifference*: The sequence (b_n) is the antidifference of the sequence (a_n) if $\Delta b_n = a_n$. An antidifference is really a sum, though, since $\Delta b_n = a_n$ implies

$$\sum_{k=0}^{n-1} a_k = \sum_{k=0}^{n-1} (b_{k+1} - b_k) = b_n - b_0$$

(thanks to telescoping), a discrete version of the Fundamental Theorem of Calculus. Thus

$$b_n = b_0 + \sum_{k=0}^{n-1} a_k. \tag{7.1}$$

There is an entire calculus built around finite differences; the notes contain some references.

A difference equation, from another viewpoint, is also a recurrence relation. For example, $a_n = \Delta a_n$ can be expressed as $a_n = a_{n+1} - a_n$, or $a_{n+1} = 2a_n$. So finite differences are really just another lens through which we can view recurrence relations. Sometimes the finite difference lens is more illuminating, and sometimes the usual recurrence relation lens is more illuminating.

Taking higher-order finite differences of some common sequences can lead to interesting results. One way to view these is with difference triangles. To construct the *difference triangle* for a particular sequence (a_n), start by placing the sequence on the left diagonal of the triangle. Then let every other entry in the triangle be the number to its left minus the number diagonally above it and to the left. The result is a sort of subtractive version of Pascal's triangle for a given sequence. If we count diagonals starting with zero, then the zeroth left diagonal is the original sequence. The first left diagonal then consists of the sequence of first-order finite differences of (a_n), the second left diagonal is the sequence of second-order finite differences of (a_n), and so forth. For the zeroth right diagonal of the difference triangle, it is not too hard to see that this is the sequence $(\Delta^n a_0) = (\Delta^0 a_0, \Delta^1 a_0, \Delta^2 a_0, \ldots)$. (Similarly, the first right diagonal is the sequence $(\Delta^n a_1)$, and so on.)

As an example, let's look at the difference triangle for the factorial numbers (where $a_n = n!$). The first several factorial numbers are 1, 1, 2, 6, 24, 120, and 720, and so these are placed along the left diagonal of the triangle. The resulting difference triangle is given in Figure 7.1.

Looking at the sequence on the right diagonal of the triangle in Figure 7.1, we see 1, 0, 1, 2, 9, 44, and 265. These are the first several derangement numbers! What's going on here? This seems to imply some sort of relationship between the factorials and the derangement numbers, but what is this relationship? The answer, perhaps unsurprisingly, turns out to involve the binomial coefficients.

				1				
			1		0			
		2		1		1		
	6		4		3		2	
24		18		14		11		9
120		96		78		64		53	44
720		600		504		426		362	309	265

FIGURE 7.1
Difference triangle for the factorial numbers

To understand the relationship, let's take a look at how the smaller derangement numbers are produced from the smaller factorials via the difference triangle. We have $D_0 = 1 = 0!$, of course. Then $D_1 = 0$, which is $1 - 1 = 1! - 0!$. The number D_2 is formed from $1 = 1 - 0 = (2! - 1!) - (1! - 0!) = 2! - 2(1!) + 0!$. The next derangement number D_3 comes from $2 = 3 - 1 = (4 - 1) - (1 - 0) = ((3! - 2!) - (2! - 1!)) - ((2! - 1!) - (1! - 0!)) = 3! - 3(2!) + 3(1!) - 0!$. Ignoring the alternating signs, the coefficients 1, 3, 3, and 1 on the factorials constitute the third row of Pascal's triangle. Looking at the previous examples, the corresponding coefficients for D_2 are 1, 2, and 1; those for D_1 are 1 and 1; and that for D_0 is 1. These are the second, first, and zeroth rows of Pascal's triangle, respectively. With the alternating signs it appears that this triangle is expressing the relationship $D_n = \sum_{k=0}^{n} \binom{n}{k}(n-k)!(-1)^k$, or, switching indices, $D_n = \sum_{k=0}^{n} \binom{n}{k}k!(-1)^{n-k}$. We have already seen this formula; it is Identity 68.

Generalizing from this example, it seems that the sequence appearing on the right-hand side of the difference triangle for (a_n) is $(-1)^n$ times the alternating binomial transform of (a_n). More formally, we can express this relationship as $\sum_{k=0}^{n} \binom{n}{k}a_k(-1)^{n-k} = \Delta^n a_0$. Since we can place any number in the (a_n) sequence at the top of the triangle, though, not just the zeroth one, we can formulate this relationship slightly more generally as the following.

Identity 174.

$$\sum_{k=0}^{n} \binom{n}{k} a_{m+k}(-1)^{n-k} = \Delta^n a_m.$$

Proof. The cleanest proof of this identity uses the *shift operator* E. Formally, we define $Ea_m = a_{m+1}$, so that the shift operator just shifts the sequence by 1 unit. This means that, as operators, we have $\Delta = E - 1$. Thus $\Delta^n = (E - 1)^n$. Expanding the latter via the binomial theorem, we have

$$\sum_{k=0}^{n} \binom{n}{k} E^k (-1)^{n-k} = \Delta^n.$$

Again, this is an operator equation. Applying both sides to the sequence beginning with a_m, we have $E^k a_m = a_{m+k}$, and thus the identity holds. □

Identity 174 can also be proved by induction; see Exercise 12.

There is a lot going on with Identity 174. We will spend the rest of this section examining some of the consequences of this identity, answering questions raised by it, and, of course, proving other binomial identities with it.

One straightforward consequence of Identity 174 is that the alternating binomial transform of a sequence is (up to sign) just the sequence of successive finite differences of the original sequence. More specifically, we have the following.

Identity 175. *If*

$$c_n = \sum_{k=0}^{n} \binom{n}{k} a_k (-1)^k,$$

then

$$(-1)^n c_n = \Delta^n a_0.$$

Proof. Multiply both sides of the equation defining c_n by $(-1)^n$, and substitute with Identity 174. □

Identity 175 gives a new interpretation for the alternating binomial transform identities we have seen thus far. For example, from this perspective Identity 24,

$$\sum_{k=0}^{n} \binom{n}{k} \frac{(-1)^k}{k+1} = \frac{1}{n+1},$$

rewritten as

$$\sum_{k=0}^{n} \binom{n}{k} \frac{(-1)^{n-k}}{k+1} = \frac{(-1)^n}{n+1},$$

tells us that the sequence appearing on the right side of the difference triangle for the reciprocals of the natural numbers is just $((-1)^n/(n+1))_{n=0}^{\infty}$, the alternating reciprocals of the natural numbers. See the exercises for more examples.

We can also use Identity 174 to prove binomial identities involving sequences that have easy-to-calculate high-order finite differences. One such sequence, thanks to Pascal's recurrence, is the binomial coefficients themselves, as we see in the next identity.

Identity 176. *For integers m and r,*

$$\sum_{k=0}^{n} \binom{n}{k} \binom{m+k}{r} (-1)^{n-k} = \binom{m}{r-n}.$$

Proof. If the upper index is considered to be the variable, we have $\Delta\binom{m}{r} = \binom{m+1}{r} - \binom{m}{r} = \binom{m}{r-1}$, by Identity 1. Thus $\Delta^n\binom{m}{r} = \binom{m}{r-n}$. Applying Identity 174 completes the proof. □

For a generalization of Identity 176, see Exercise 18.

Another sequence that has simple higher-order finite differences is the sequence given by $a_m = m^{\underline{r}}$ of falling powers (where r is fixed). Before we apply Identity 174 to this sequence, though, we need both to define negative falling powers and prove the finite difference formula for falling powers. First, we have, for $n \geq 1$,

$$x^{\underline{-n}} = \frac{1}{(x+1)(x+2)\cdots(x+n)}.$$

Note the offset by 1 when compared to positive falling powers.

The finite difference formula for falling powers, Identity 177, is the finite difference counterpart to the differentiation rule for powers: $\frac{d}{dx}x^n = nx^{n-1}$.

Identity 177. *For $x \in \mathbb{R}$ and integers n,*

$$\Delta x^{\underline{n}} = nx^{\underline{n-1}},$$

where the finite difference is taken with respect to x.

Proof. Let $n \geq 0$. Then we have

$$\Delta x^{\underline{n}} = (x+1)^{\underline{n}} - x^{\underline{n}}$$
$$= (x+1)(x)(x-1)\cdots(x-n+2) - x(x-1)\cdots(x-n+1)$$
$$= x(x-1)\cdots(x-n+2)\big(x+1-(x-n+1)\big)$$
$$= nx(x-1)\cdots(x-n+2)$$
$$= nx^{\underline{n-1}}.$$

On the other hand, for $n \geq 1$, we have

$$\Delta x^{\underline{-n}} = (x+1)^{\underline{-n}} - x^{\underline{-n}}$$
$$= \frac{1}{(x+2)(x+3)\cdots(x+n+1)} - \frac{1}{(x+1)(x+2)\cdots(x+n)}$$
$$= \frac{1}{(x+2)(x+3)\cdots(x+n)}\left(\frac{1}{x+n+1} - \frac{1}{x+1}\right)$$
$$= \frac{1}{(x+2)(x+3)\cdots(x+n)}\left(\frac{x+1-(x+n+1)}{(x+n+1)(x+1)}\right)$$
$$= \frac{-n}{(x+1)(x+2)(x+3)\cdots(x+n)(x+n+1)}$$
$$= -nx^{\underline{-n-1}}.$$

$\square$

Identity 177 can be applied successively to obtain the following.

Identity 178. *For $x \in \mathbb{R}$,*

$$\Delta^m x^{\underline{n}} = n^{\underline{m}}x^{\underline{n-m}},$$

where the finite difference is taken with respect to x.

The antidifference of a falling power also has a simple formula; see Exercise 10.

We can use Identities 174 and 178 to prove binomial identities involving positive and negative falling powers. However, Identity 176 is, in disguise, the identity obtained with positive falling powers. (See Exercise 17.) The basic identity obtained with negative falling powers is the following; Exercise 19 contains a generalization.

Identity 109.

$$\sum_{k=0}^{n} \binom{n}{k} \frac{(-1)^k}{m+k} = \frac{n!}{m(m+1)\cdots(m+n)}.$$

Proof. Let $a_m = (m-1)^{\underline{-1}}$. (Remember, $(m-1)^{\underline{-1}} = 1/m$.) Then

$$\Delta^n a_m = (-1)^{\underline{n}}(m-1)^{\underline{-1-n}} = \frac{(-1)(-2)\cdots(-n)}{m(m+1)(m+2)\cdots(m+n)}$$

$$= (-1)^n \frac{n!}{m(m+1)(m+2)\cdots(m+n)}.$$

By Identity 174, we have

$$\sum_{k=0}^{n} \binom{n}{k} \frac{(-1)^{n-k}}{m+k} = (-1)^n \frac{n!}{m(m+1)\cdots(m+n)},$$

which, upon dividing both sides by $(-1)^n$, yields the identity. □

Our third example of a sequence with simple higher-order differences is the Fibonacci sequence. This is because the finite difference of the Fibonacci sequence is just a shifted version of itself. Let's establish this result, and then look at how to apply Identity 174 to it.

Identity 179. *For integers m, $\Delta^n F_m = F_{m-n}$.*

Proof. By the Fibonacci recurrence (Identity 170), $\Delta F_m = F_{m+1} - F_m = F_{m-1}$. Thus $\Delta(\Delta F_m) = F_m - F_{m-1} = F_{m-2}$, and, in general, $\Delta^n F_m = F_{m-n}$. □

By defining Fibonacci numbers for negative indices through the process of extending the Fibonacci recurrence below F_0, we do not need the restrictions $m \geq n$ or even $m \geq 0$ in Identity 179. See the discussion immediately preceding Identity 220 in Chapter 8.

Identity 179 combined with Identity 174 not only can be used to find the alternating binomial transform of the Fibonacci numbers (in the manner of the the proof of Identity 109 we just gave), but it can also be used to find the binomial transform of the Fibonacci numbers!

Identity 180. *For integers* m,

$$\sum_{k=0}^{n} \binom{n}{k} F_{m+k}(-1)^k = (-1)^n F_{m-n}.$$

Proof. By Identities 174 and 179,

$$\sum_{k=0}^{n} \binom{n}{k} F_{m+k}(-1)^{n-k} = \Delta^n F_m = F_{m-n}.$$

Multiplying by $(-1)^n$ completes the proof. □

Identity 180 gives us the alternating binomial transform of the Fibonacci numbers.

Identity 181.

$$\sum_{k=0}^{n} \binom{n}{k} F_k(-1)^k = -F_n.$$

Proof. By Identity 220, $F_{-n} = (-1)^{n+1} F_n$. Setting $m = 0$ in Identity 180 thus yields

$$\sum_{k=0}^{n} \binom{n}{k} F_k(-1)^k = (-1)^n F_{-n} = (-1)^n (-1)^{n+1} F_n = (-1)^{2n}(-1)F_n = -F_n.$$

□

As promised, we can obtain the binomial transform of the Fibonacci numbers as well. First, a generalization:

Identity 182. *For integers* m,

$$\sum_{k=0}^{n} \binom{n}{k} F_{m+k} = F_{2n+m}.$$

Proof. In Theorem 2 (see Exercise 27 of Chapter 2) we saw the following variation of binomial inversion:

$$f(n) = \sum_{k=0}^{n} \binom{n}{k} g(k) \iff g(n) = \sum_{k=0}^{n} \binom{n}{k} f(k)(-1)^{n-k}.$$

Applying this formula to the formula

$$\sum_{k=0}^{n} \binom{n}{k} F_{m+k}(-1)^{n-k} = F_{m-n}$$

found in the proof of Identity 180, we obtain

$$\sum_{k=0}^{n} \binom{n}{k} F_{m-k} = F_{m+n}.$$

Reversing the order of summation gives

$$\sum_{k=0}^{n} \binom{n}{k} F_{m-n+k} = F_{m+n}.$$

Finally, replacing m with $m+n$ yields the identity.   □

Now we have the binomial transform of the Fibonacci numbers:

Identity 183.

$$\sum_{k=0}^{n} \binom{n}{k} F_k = F_{2n}.$$

Proof. Let $m = 0$ in Identity 182.   □

Yet another use of Identity 174 is to give an interpretation of sum triangles. The *sum triangle* for a sequence (a_n) has its left diagonal consisting of (a_n), while every other number in the triangle is the sum of the number to its left and the number diagonally above it and to the left. In other words, a sum triangle is the same as a difference triangle except that the subtraction operation is replaced with addition. For example, the sum triangle for the derangement numbers is in Figure 7.2. This is, of course, the mirror image of the difference triangle for the factorial numbers.

					1						
				0		1					
			1		1		2				
		2		3		4		6			
	9		11		14		18		24		
44		53		64		78		96		120	
265	309		362		426		504		600		720

FIGURE 7.2
Sum triangle for the derangement numbers

How is the right diagonal of a sum triangle obtained from the left diagonal? The next theorem tells us.

Theorem 19. *Suppose (a_n) is the sequence on the left diagonal of a sum triangle. Then the right diagonal (b_n) is the binomial transform of (a_n); i.e.,*

$$b_n = \sum_{k=0}^{n} \binom{n}{k} a_k.$$

Proof. By Identity 174 and the fact that a sum triangle is the mirror image of a difference triangle, we know that if (a_n) forms the left diagonal of a sum triangle, and (b_n) forms its right diagonal, then (a_n) and (b_n) are related via

$$a_n = \sum_{k=0}^{n} \binom{n}{k} b_k (-1)^{n-k}.$$

Thanks to the variation of binomial inversion in Theorem 2 (see Exercise 27 of Chapter 2),

$$f(n) = \sum_{k=0}^{n} \binom{n}{k} g(k) \iff g(n) = \sum_{k=0}^{n} \binom{n}{k} f(k)(-1)^{n-k},$$

we obtain

$$b_n = \sum_{k=0}^{n} \binom{n}{k} a_k.$$

$\square$

For a combinatorial proof of Theorem 19, as well as a generalization to repeated iterations of sum triangles, see Spivey and Steil [71].

Now we know how to interpret any entry in a sum or difference triangle in terms of its left and right diagonals. To obtain a given entry in a sum triangle starting from its left triangle, you take successive sums, or equivalently, a binomial transform. To obtain the same entry starting from its right diagonal, you take successive finite differences, or, equivalently, (up to sign) an alternating binomial transform. Mixing these interpretations leads to two more binomial coefficient identities. If we take the pairing of "binomial transform from the left/finite differences from the right" we obtain Identity 184. This identity can itself be exploited to obtain more sophisticated formulas relating binomial sums and finite differences, as we shall see in the rest of the section. We give two proofs of Identity 184: One from the point of view of sum triangles, as well as a simple induction proof using Pascal's recurrence.

Identity 184. *Let (b_n) be the binomial transform of the sequence (a_n); i.e.,*

$$b_n = \sum_{k=0}^{n} \binom{n}{k} a_k.$$

Then

$$\sum_{k=0}^{m} \binom{m}{k} b_{n+k} (-1)^{m-k} = \Delta^m b_n = \sum_{k=0}^{n} \binom{n}{k} a_{k+m}.$$

The first equality in the conclusion is a restatement of Identity 174, and so the only part that needs to be shown is the second equality.

Proof 1. Construct the sum triangle for (a_n), and let b_n be its right diagonal. Let $T_{n,m}$ be the entry in row $n+m$, entry m in this triangle. Then $T_{n,m}$ is the nth number on the right diagonal of the subtriangle that has a_m at its apex. By Theorem 19, then,

$$T_{n,m} = \sum_{k=0}^{n} \binom{n}{k} a_{m+k}.$$

However, $T_{n,m}$ is also the mth number on the left diagonal of the subtriangle that has b_n at its apex. Since a sum triangle is the mirror image of a difference triangle, this means that, by Identity 174, $T_{n,m} = \Delta^m b_n$. The result follows.

□

Proof 2. We have

$$\Delta b_n = b_{n+1} - b_n$$

$$= \sum_{k=0}^{n+1} \binom{n+1}{k} a_k - \sum_{k=0}^{n} \binom{n}{k} a_k$$

$$= \sum_{k=0}^{n+1} \binom{n}{k-1} a_k, \qquad\qquad \text{by Identity 1,}$$

$$= \sum_{k=0}^{n} \binom{n}{k} a_{k+1},$$

remembering that $\binom{n}{n+1} = \binom{n}{-1} = 0$.
    Therefore,

$$\Delta(\Delta b_n) = \sum_{k=0}^{n} \binom{n}{k} a_{k+2},$$

and, in general, we have

$$\Delta^m b_n = \sum_{k=0}^{n} \binom{n}{k} a_{k+m}.$$

□

Identity 184 describes an interesting relationship between shifts, finite differences, and binomial transforms: Shifting a sequence (a_n) by m and then taking its binomial transform gives the same result as first taking the binomial transform and then calculating its mth order finite difference.

Let's put Identity 184 to work. For example, Exercise 14 in Chapter 4 asks for the binomial transform of $1/(k+2)$. Thanks to Identity 184, this is now straightforward, given that we already know the binomial transform of $1/(k+1)$.

Identity 119.

$$\sum_{k=0}^{n} \binom{n}{k} \frac{1}{k+2} = \frac{n2^{n+1} + 1}{(n+1)(n+2)}.$$

Proof. We have proved Identity 23 multiple times:

$$\sum_{k=0}^{n} \binom{n}{k} \frac{1}{k+1} = \frac{2^{n+1} - 1}{n+1}.$$

By Identity 184, then, we have

$$\sum_{k=0}^{n} \binom{n}{k} \frac{1}{k+2} = \frac{2^{n+2} - 1}{n+2} - \frac{2^{n+1} - 1}{n+1}$$

$$= \frac{(n+1)2^{n+2} - (n+1) - (n+2)2^{n+1} + (n+2)}{(n+1)(n+2)}$$

$$= \frac{n2^{n+1} + 1}{(n+1)(n+2)}.$$

$\square$

Identity 184 can be used with alternating binomial sums as well. For example, Exercise 28 in Chapter 5 asks for the alternating binomial transform of $1/(k+2)$. Since we know the alternating binomial transform of $1/(k+1)$, this one also becomes straightforward.

Identity 53.

$$\sum_{k=0}^{n} \binom{n}{k} \frac{(-1)^k}{k+2} = \frac{1}{(n+1)(n+2)}.$$

Proof. Multiplying Identity 24 by -1, we have

$$\sum_{k=0}^{n} \binom{n}{k} \frac{(-1)^{k+1}}{k+1} = -\frac{1}{n+1}.$$

By Identity 184, then, we have

$$\sum_{k=0}^{n} \binom{n}{k} \frac{(-1)^{k+2}}{k+2} = -\frac{1}{n+2} + \frac{1}{n+1}$$

$$= \frac{-(n+1) + (n+2)}{(n+1)(n+2)}$$

$$= \frac{1}{(n+1)(n+2)}.$$

Since $(-1)^{k+2} = (-1)^k(-1)^2 = (-1)^k$, we are done.

$\square$

Identities 174 and 184 are parallel identities, of a sort: They show what happens when you shift the input sequence to, respectively, an alternating binomial transform and a regular binomial transform. However, they differ in this sense: Identity 174 gives the result in terms of finite differences of the original sequence, and Identity 184 gives the result in terms of finite differences of the binomial transform of the original sequence. The next identity connects Identities 174 and 184 to produce a reciprocity result for alternating binomial sums that generalizes both Identity 175 and the binomial inversion formula (Theorem 1).

Identity 185. *If a_n and c_n are sequences such that*

$$c_n = \sum_{k=0}^{n} \binom{n}{k} a_k (-1)^k$$

then

$$\sum_{k=0}^{m} \binom{m}{k} c_{n+k} (-1)^k = (-1)^m \Delta^m c_n = (-1)^n \Delta^n a_m = \sum_{k=0}^{n} \binom{n}{k} a_{m+k} (-1)^k.$$

Proof. By Identities 174 and 184 we have

$$\sum_{k=0}^{m} \binom{m}{k} c_{n+k} (-1)^{m-k} = \Delta^m c_n = \sum_{k=0}^{n} \binom{n}{k} a_{k+m} (-1)^{k+m}$$

$$= (-1)^m \sum_{k=0}^{n} \binom{n}{k} a_{m+k} (-1)^k$$

$$= (-1)^{m+n} \sum_{k=0}^{n} \binom{n}{k} a_{m+k} (-1)^{n-k} = (-1)^{m+n} \Delta^n a_m.$$

Multiplying by $(-1)^m$ proves the result.   □

As we mentioned, Identity 185 generalizes the binomial inversion formula in Theorem 1. This formula says that

$$c_n = \sum_{k=0}^{n} \binom{n}{k} a_k (-1)^k \iff a_n = \sum_{k=0}^{n} \binom{n}{k} c_k (-1)^k.$$

Letting $m = 0$ in Identity 185 produces the left equation in Theorem 1, and letting $n = 0$, followed by substitution of n for m, produces the right side. (See Exercise 23 for the corresponding generalization of the variation of binomial inversion in Theorem 2.)

Let's take a look at Identity 184 from another angle. It tells us that if we shift the sequence (a_n) *forwards*, then that is equivalent to taking finite differences of the binomial transform of (a_n). What if we want to shift the sequence (a_n) *backwards*, though? It shouldn't be too surprising that the result is the same as taking the *antidifference* of the binomial transform of (a_n). The next identity makes this explicit.

Identity 186. *If*

$$b_n = \sum_{k=0}^{n} \binom{n}{k} a_k$$

then

$$\sum_{k=0}^{n-1} b_k = \sum_{k=1}^{n} \binom{n}{k} a_{k-1}.$$

Proof. For the purposes of this proof, define a_{-1} to be 0. Then let

$$\sum_{k=0}^{n} \binom{n}{k} a_{k-1} = \sum_{k=1}^{n} \binom{n}{k} a_{k-1} = s_n.$$

This implies $s_0 = 0$. By Identity 184, $\Delta s_n = b_n$. The identity then follows from Equation (7.1). □

For the record, Identity 186 generalizes the ideas in the proof of Identity 25 in Chapter 2. In that proof, we wanted the binomial sum $\sum_{k=1}^{n} \binom{n}{k} \frac{1}{k}$. We let $f(n)$ be the value of that sum, found an expression for the finite difference of $f(n)$, and then summed to obtain the value of $f(n)$. Identity 186 basically automates this process, as we can see by comparing this with a proof of Identity 25 that uses Identity 186.

Identity 25.

$$\sum_{k=1}^{n} \binom{n}{k} \frac{1}{k} = \sum_{k=1}^{n} \frac{2^k - 1}{k}.$$

Proof. Identity 23 tells us

$$\sum_{k=0}^{n} \binom{n}{k} \frac{1}{k+1} = \frac{2^{n+1} - 1}{n+1}.$$

By Identity 186, then, we have

$$\sum_{k=1}^{n} \binom{n}{k} \frac{1}{k} = \sum_{k=0}^{n-1} \frac{2^{k+1} - 1}{k+1} = \sum_{k=1}^{n} \frac{2^k - 1}{k}.$$

□

For a second example using Identity 186, let's prove the alternating version of Identity 25.

Identity 43.

$$\sum_{k=1}^{n} \binom{n}{k} \frac{(-1)^k}{k} = -H_n.$$

Proof. Multiplying Identity 24 by -1, we have

$$\sum_{k=0}^{n} \binom{n}{k} \frac{(-1)^{k+1}}{k+1} = -\frac{1}{n+1}.$$

By Identity 186, then, we have

$$\sum_{k=1}^{n} \binom{n}{k} \frac{(-1)^k}{k} = \sum_{k=0}^{n-1} \frac{-1}{k+1} = -H_n.$$

$\square$

Our third example with Identity 186 is not actually a binomial coefficient identity at all!

Identity 187.

$$\sum_{k=0}^{n} F_{2k} = F_{2n+1} - 1.$$

Proof. Identity 183 gives the binomial transform of the Fibonacci numbers:

$$\sum_{k=0}^{n} \binom{n}{k} F_k = F_{2n}.$$

Applying Identity 186, we obtain

$$\sum_{k=1}^{n} \binom{n}{k} F_{k-1} = \sum_{k=0}^{n-1} F_{2k}.$$

However, we already know from letting $m = -1$ in Identity 182 that

$$\sum_{k=0}^{n} \binom{n}{k} F_{k-1} = F_{2n-1}.$$

Thus

$$\sum_{k=0}^{n-1} F_{2k} = \sum_{k=0}^{n} \binom{n}{k} F_{k-1} - F_{-1} = F_{2n-1} - 1.$$

Replacing n with $n+1$ yields the identity. $\square$

Let's take a look at Identity 184 from yet another perspective. It says that if we shift a binomial transform's input sequence (a_n), then that gives the same result as taking the finite difference of the binomial transform (b_n). What if we take the finite difference of the input sequence, though? How does that affect the resulting binomial transform? The next identity answers this question.

Identity 188. *Suppose*

$$b_n = \sum_{k=0}^{n} \binom{n}{k} a_k, \; and$$

$$b_n^{\Delta} = \sum_{k=0}^{n} \binom{n}{k} \Delta a_k.$$

Then

$$b_n^{\Delta} = b_{n+1} - 2b_n, \; and$$

$$b_n = 2^n \left(a_0 + \sum_{k=1}^{n} \frac{b_{k-1}^{\Delta}}{2^k} \right).$$

Proof. For the first part, we have

$$b_n^{\Delta} = \sum_{k=0}^{n} \binom{n}{k} \Delta a_k = \sum_{k=0}^{n} \binom{n}{k} (a_{k+1} - a_k) = \Delta b_n - b_n = b_{n+1} - 2b_n,$$

where the second-to-last step follows from Identity 184.

For the second, let $B(x)$ and $B^{\Delta}(x)$ denote the generating functions of b_n and b_n^{Δ}, respectively. Also, $b_0 = \binom{0}{0} a_0 = a_0$. Now, take the expression for b_n^{Δ} we just proved, shift the indexes by 1, multiply by x^n, and sum. We have

$$\sum_{n=1}^{\infty} b_n x^n = \sum_{n=1}^{\infty} b_{n-1}^{\Delta} x^n + 2 \sum_{n=1}^{\infty} b_{n-1} x^n$$

$$\implies b_0 + \sum_{n=1}^{\infty} b_n x^n = b_0 + \sum_{n=1}^{\infty} b_{n-1}^{\Delta} x^n + 2 \sum_{n=0}^{\infty} b_n x^{n+1}$$

$$\implies B(x) = a_0 + \sum_{n=1}^{\infty} b_{n-1}^{\Delta} x^n + 2x B(x)$$

$$\implies B(x) = \frac{a_0 + \sum_{n=1}^{\infty} b_{n-1}^{\Delta} x^n}{1 - 2x}.$$

(Since

$$\sum_{n=1}^{\infty} b_{n-1}^{\Delta} x^n = \sum_{n=0}^{\infty} b_n^{\Delta} x^{n+1} = x \sum_{n=0}^{\infty} b_n^{\Delta} x^n = x B^{\Delta}(x),$$

this can be expressed more compactly as

$$B(x) = \frac{a_0 + x B^{\Delta}(x)}{1 - 2x},$$

although it is not necessary for the rest of the argument.)

We have $1/(1-2x) = \sum_{n=0}^{\infty} (2x)^n$, and so $1/(1-2x)$ generates the sequence

$(2^n)_{n=0}^{\infty}$. By Theorem 13, then, (b_n) is the convolution of $(2^n)_{n=0}^{\infty}$ and the sequence $(a_0, b_0^{\Delta}, b_1^{\Delta}, \ldots)$. Therefore,

$$b_n = 2^n a_0 + \sum_{k=1}^{n} b_{k-1}^{\Delta} 2^{n-k} = 2^n \left(a_0 + \sum_{k=1}^{n} \frac{b_{k-1}^{\Delta}}{2^k} \right).$$

□

Before we look at some examples using Identity 188, let's prove its alternating version. Then we'll prove examples using both identities in tandem.

Identity 189. *Suppose*

$$c_n = \sum_{k=0}^{n} \binom{n}{k} a_k (-1)^k, \ \text{and}$$

$$c_n^{\Delta} = \sum_{k=0}^{n} \binom{n}{k} \Delta a_k (-1)^k.$$

Then

$$c_n^{\Delta} = -c_{n+1}, \ \text{and}$$

$$c_n = -c_{n-1}^{\Delta}[n \geq 1] + a_0[n = 0] = \begin{cases} a_0, & n = 0; \\ -c_{n-1}^{\Delta}, & n \geq 1. \end{cases}$$

Proof. By Identity 184,

$$c_n^{\Delta} = \sum_{k=0}^{n} \binom{n}{k} \Delta a_k (-1)^k = \sum_{k=0}^{n} \binom{n}{k} a_{k+1} (-1)^k - \sum_{k=0}^{n} \binom{n}{k} a_k (-1)^k$$

$$= -\Delta c_n - c_n = -c_{n+1} + c_n - c_n = -c_{n+1}.$$

Shifting indices, then, $c_n = -c_{n-1}^{\Delta}$ when $n \geq 1$. In addition, $c_0 = \binom{0}{0} a_0 (-1)^0 = a_0$. □

When the finite difference of a sequence is simpler than the original sequence, or its binomial or alternating binomial sum is easy to calculate, Identities 188 and 189 can be quite powerful. For example, they give immediate proofs for the row sum and alternating row sum of the binomial coefficients. Then we can bootstrap these results to obtain more complicated binomial sum identities. (In these and subsequent identities we use the notation of Identities 188 and 189.)

Identity 4.

$$\sum_{k=0}^{n} \binom{n}{k} = 2^n.$$

Proof. Let $a_k = 1$. Then $\Delta a_k = 1 - 1 = 0$ for each k. This means $b_n^\Delta = 0$. Therefore, $b_n = 2^n(1 + 0) = 2^n$. $\square$

Identity 12.

$$\sum_{k=0}^{n} \binom{n}{k}(-1)^k = [n = 0].$$

Proof. Let $a_k = 1$. Then $\Delta a_k = 1 - 1 = 0$ for each k. This means $c_n^\Delta = 0$. By Identity 189, then, $c_n = 0 + a_0[n = 0] = 1[n = 0]$. $\square$

As mentioned, we can use the row sum and alternating row sum results to create proofs for other binomial sums. For example, if $a_k = k$, then $\Delta a_k = (k + 1) - k = 1$. This yields the next two identities.

Identity 21.

$$\sum_{k=0}^{n} \binom{n}{k}k = n2^{n-1}.$$

Proof. Let $a_k = k$. Then $\Delta a_k = 1$. Thus $b_n^\Delta = 2^n$, by Identity 4. Identity 188 then implies

$$b_n = \sum_{k=0}^{n} \binom{n}{k}k = 2^n\left(0 + \sum_{k=1}^{n} \frac{2^{k-1}}{2^k}\right) = 2^n \sum_{k=1}^{n} \frac{1}{2} = n2^{n-1}.$$

$\square$

Similarly, we have

Identity 22.

$$\sum_{k=0}^{n} \binom{n}{k}k(-1)^k = -[n = 1].$$

Proof. If $a_k = k$, then $\Delta a_k = 1$. Thus $c_n^\Delta = 1[n = 0]$, by Identity 12. Identity 189 then tells us that $c_n = -c_{n-1}^\Delta[n \geq 1] + a_0[n = 0] = -1[n - 1 = 0] = -[n = 1]$. $\square$

We can continue bootstrapping, using the finite difference formula $\Delta x^{\underline{m}} = mx^{\underline{m-1}}$ given in Identity 177. Instead of considering $m = 2, 3$, etc., separately, though, we'll just jump to the general m case.

Identity 36.

$$\sum_{k=0}^{n} \binom{n}{k}k^{\underline{m}} = n^{\underline{m}}2^{n-m}.$$

Proof. By Identities 4 and 21, the identity is true for $m = 0$ and $m = 1$. Fix $m \geq 2$, and assume that the identity is true for $k \in \{1, 2, \ldots, m-1\}$. Then, with $a_k = k^{\underline{m}}$, we have $\Delta a_k = mk^{\underline{m-1}}$. According to the induction hypothesis, then, $b_n^{\Delta} = mn^{\underline{m-1}}2^{n-(m-1)}$. Thus, by Identity 188, we have

$$\sum_{k=0}^{n}\binom{n}{k}k^{\underline{m}} = 2^n\sum_{k=1}^{n}\frac{m(k-1)^{\underline{m-1}}2^{k-1-(m-1)}}{2^k} = m2^{n-m}\sum_{k=1}^{n}(k-1)^{\underline{m-1}}$$

$$= m2^{n-m}\sum_{k=0}^{n-1}k^{\underline{m-1}} = n^{\underline{m}}2^{n-m},$$

where the last step follows from the sum formula for falling powers, Identity 204:

$$\sum_{k=0}^{n}k^{\underline{m}} = \frac{(n+1)^{\underline{m+1}}}{m+1}.$$

(See Exercise 10 for a proof.)   □

We can also use this approach to prove the identity for the alternating binomial transform of the falling power $k^{\underline{m}}$.

Identity 38.

$$\sum_{k=0}^{n}\binom{n}{k}k^{\underline{m}}(-1)^k = (-1)^m m![n = m].$$

Proof. By Identities 12 and 22, the formula is true when $m = 0$ and $m = 1$. Now, fix $m \geq 2$, and assume that the identity is true for $k \in \{1, 2, \ldots, m-1\}$. Again, $a_k = k^{\underline{m}}$ implies $\Delta a_k = mk^{\underline{m-1}}$. The induction hypothesis says that $c_n^{\Delta} = m(-1)^{m-1}(m-1)![n = m-1]$. By Identity 189, then,

$$\sum_{k=0}^{n}\binom{n}{k}k^{\underline{m}}(-1)^k = m(-1)(-1)^{m-1}(m-1)![n-1 = m-1] = (-1)^m m![n = m].$$

                                                                                    □

For our last set of examples in this section we extend the ideas in the proofs of Identities 188 and 189 to (1) aerated binomial sums and (2) binomial sums on the upper parameter.

An *aerated binomial sum* is one of the form $\sum_k \binom{n}{rk+m}a_k$, where r is a positive integer and $m \in \{0, 1, \ldots, r-1\}$. The simplest aerated binomial sums are those with $r = 2$: $\sum_k \binom{n}{2k}a_k$ and $\sum_k \binom{n}{2k+1}a_k$. (These are the two kinds of aerated binomial sums we consider in this chapter, although the ideas can be extended to sums featuring larger values of r.) By viewing sums such as these as binomial transforms of *aerated sequences*, that is, sequences such as $(a_0, 0, a_1, 0, a_2, 0, a_3, 0, \ldots)$ and $(0, a_0, 0, a_1, 0, a_2, 0, a_3, \ldots)$, respectively, we can apply the ideas behind the proof of Identity 188 to obtain aerated versions of Identity 188.

Identity 190. *Suppose*

$$e_n = \sum_{k \geq 0} \binom{n}{2k} a_k, \quad and$$

$$e_n^\Delta = \sum_{k \geq 0} \binom{n}{2k} \Delta a_k.$$

Then

$$e_n^\Delta = e_{n+2} - 2e_{n+1}, \quad and$$

$$e_n = \begin{cases} a_0, & n = 0; \\ 2^{n-1} a_0 + 2^n \sum_{k=2}^{n} \dfrac{e_{k-2}^\Delta}{2^k}, & n \geq 1. \end{cases}$$

Proof. By Identity 184, the binomial transform of the aerated sequence $(a_1, 0, a_2, 0, a_3, 0, \ldots)$ is given by $\Delta^2 e_n = e_{n+2} - 2e_{n+1} + e_n$. Thus e_n^Δ, the binomial transform of the sequence $(a_1 - a_0, 0, a_2 - a_1, 0, a_3 - a_2, 0, \ldots)$, is given by

$$e_n^\Delta = e_{n+2} - 2e_{n+1}.$$

Now, let $E(x)$ and $E^\Delta(x)$ denote the generating functions of e_n and e_n^Δ, respectively. As a reminder, $e_0 = \binom{0}{0} a_0 = a_0$, and $e_1 = \binom{1}{0} a_0 + \binom{1}{1} 0 = a_0$. Shifting the indexes for the expression for e_n^Δ that we just proved by 2, multiplying by x^n, and summing, we have

$$\sum_{n=2}^{\infty} e_n x^n = \sum_{n=2}^{\infty} e_{n-2}^\Delta x^n + 2 \sum_{n=2}^{\infty} e_{n-1} x^n$$

$$\implies e_0 + e_1 x + \sum_{n=2}^{\infty} e_n x^n = e_0 + e_1 x + \sum_{n=2}^{\infty} e_{n-2}^\Delta x^n + 2 \sum_{n=1}^{\infty} e_n x^{n+1}$$

$$\implies E(x) = a_0 + a_0 x + \sum_{n=2}^{\infty} e_{n-2}^\Delta x^n + 2x E(n) - 2e_0 x$$

$$\implies E(x) = \frac{a_0 - a_0 x + \sum_{n=2}^{\infty} e_{n-2}^\Delta x^n}{1 - 2x}.$$

(As with the result from Identity 184, this can be expressed more compactly:

$$E(x) = \frac{a_0 - a_0 x + x^2 E^\Delta(x)}{1 - 2x}.)$$

By Theorem 13, e_n is the convolution of 2^n and the sequence $a_0, -a_0, e_0^\Delta, e_1^\Delta, \ldots$. This means that if $n = 0$ then $e_n = a_0$. Otherwise,

$$e_n = 2^n a_0 - 2^{n-1} a_0 + \sum_{k=2}^{n} e_{k-2}^\Delta 2^{n-k} = 2^{n-1} a_0 + \sum_{k=2}^{n} \frac{e_{k-2}^\Delta}{2^k}.$$

$\square$

We have an almost-identical result for the aerated binomial transform with odd indices:

Identity 191. *Suppose*

$$f_n = \sum_{k \geq 0} \binom{n}{2k+1} a_k, \text{ and}$$

$$f_n^\Delta = \sum_{k \geq 0} \binom{n}{2k+1} \Delta a_k.$$

Then

$$f_n^\Delta = f_{n+2} - 2f_{n+1}, \text{ and}$$

$$f_n = \begin{cases} 0, & n = 0; \\ 2^{n-1}a_0 + 2^n \sum_{k=2}^{n} \dfrac{f_{k-2}^\Delta}{2^k}, & n \geq 1. \end{cases}$$

Proof. See Exercise 29. □

With Identities 190 and 191 in view, the following two identities are almost immediate.

Identity 13.

$$\sum_{k \geq 0} \binom{n}{2k} = 2^{n-1} + \frac{1}{2}[n = 0].$$

Proof. Take the sequence $a_k = 1$ in Identity 190. □

Identity 14.

$$\sum_{k \geq 0} \binom{n}{2k+1} = 2^{n-1}[n \geq 1].$$

Proof. Take the sequence $a_k = 1$ in Identity 191. □

As we did with Identities 188 and 189, we can bootstrap these results to obtain more complicated identities. We get the following.

Identity 192.

$$\sum_{k \geq 0} \binom{n}{2k} k = n2^{n-3}[n \geq 2].$$

Proof. Let $a_k = k$; then $\Delta a_k = 1$. By Identity 13, then, $e_n^\Delta = 2^{n-1} + \frac{1}{2}[n = 0]$. We have that $e_0 = e_1 = a_0 = 0$. For $n \geq 2$, Identity 190 gives us

$$\sum_{k \geq 0} \binom{n}{2k} k = 2^n \sum_{k=2}^{n} \frac{2^{k-3} + \frac{1}{2}[k - 2 = 0]}{2^k} = 2^n \sum_{k=2}^{n} \left(\frac{1}{8} + \frac{[k = 2]}{2^{k+1}} \right)$$

$$= 2^n \left(\frac{n-1}{8} + \frac{1}{8} \right) = n2^{n-3}.$$

□

Similarly, we have

Identity 193.

$$\sum_{k \geq 0} \binom{n}{2k+1} k = (n-2)2^{n-3} [n \geq 2].$$

Proof. See Exercise 30. □

Generalizing Identity 192 to the case where $a_k = k^{\underline{m}}$ is a bit trickier. In fact, the proof of Identity 194 is the most complicated in this chapter.

Identity 194. *For $m \geq 1$,*

$$\sum_{k \geq 0} \binom{n}{2k} k^{\underline{m}} = n(n-m-1)^{\underline{m-1}} 2^{n-2m-1} [n \geq m+1].$$

Proof. By Identity 192, the formula is true for $m = 1$. Fix $m \geq 2$, and assume the identity is true for $1, 2, \ldots, m-1$. If $a_k = k^{\underline{m}}$, then $\Delta a_k = mk^{\underline{m-1}}$. By the induction hypothesis, $e_n^{\Delta} = mn(n-m)^{\underline{m-2}} 2^{n-2m+1} [n \geq m]$. Therefore, by Identity 190,

$$\sum_{k \geq 0} \binom{n}{2k} k^{\underline{m}} = m2^n \sum_{k=2}^{n} \frac{(k-2)(k-2-m)^{\underline{m-2}} 2^{k-2-2m+1}}{2^k} [k \geq m+2]$$

$$= m2^{n-2m-1} \sum_{k=m+2}^{n} (k-2)(k-2-m)^{\underline{m-2}}$$

$$= m2^{n-2m-1} \sum_{k=0}^{n-m-2} (k+m)k^{\underline{m-2}}.$$

At this point we run into a difficulty: We do not know to evaluate the sum directly. Unlike the corresponding stage in the proof of Identity 36, the sum is not in a form where we can use Identity 204 to simplify it. A little manipulation gets us there, though.

We have that $k^{\underline{m-2}} = k(k-1) \cdots (k-m+3)$. This means that $(k+1)k(k-1) \cdots (k-m+3) = (k+1)^{\underline{m-1}}$. Thus we can write rewrite our sum and then apply Identity 204.

$$m2^{n-2m-1} \sum_{k=0}^{n-m-2} (k+m)k^{\underline{m-2}}$$

$$= m2^{n-2m-1} \sum_{k=0}^{n-m-2} \left((k+1)k^{\underline{m-2}} + (m-1)k^{\underline{m-2}} \right)$$

$$= m2^{n-2m-1} \sum_{k=0}^{n-m-2} (k+1)^{\underline{m-1}} + m(m-1)2^{n-2m-1} \sum_{k=0}^{n-m-2} k^{\underline{m-2}}$$

$$= m2^{n-2m-1} \frac{(n-m)^{\underline{m}}}{m} [n \geq m+2]$$

$$+ m(m-1)2^{n-2m-1} \frac{(n-m-1)^{\underline{m-1}}}{m-1} [n \geq m+2]$$

$$= 2^{n-2m-1} \left((n-m)^{\underline{m}} + m(n-m-1)^{\underline{m-1}} \right) [n \geq m+2]$$

$$= 2^{n-2m-1} \left((n-m)(n-m-1)^{\underline{m-1}} + m(n-m-1)^{\underline{m-1}} \right) [n \geq m+2]$$

$$= n(n-m-1)^{\underline{m-1}} 2^{n-2m-1} [n \geq m+2]$$

$$= n(n-m-1)^{\underline{m-1}} 2^{n-2m-1} [n \geq m+1],$$

where the last step follows because the expression before the Iverson bracket evaluates to 0 when $n = m+1$. □

(Actually, summation by parts is an even easier way to evaluate the troublesome sum in Identity 194; see Exercise 31.)

The corresponding formula for the odd-index binomial coefficients is given by the following.

Identity 195. *For $m \geq 1$,*

$$\sum_{k \geq 0} \binom{n}{2k+1} k^{\underline{m}} = (n-m-1)^{\underline{m}} 2^{n-2m-1} [n \geq m+1].$$

Proof. See Exercise 32. □

For a final example with aerated sums, let's take a look at $\sum_{k \geq 0} \binom{n}{2k} x^k$, which will give us the even aerated version of the binomial theorem. The proof of Identity 196 uses not only Identity 190 but also a technique from Section 7.1.

Identity 196.

$$\sum_{k \geq 0} \binom{n}{2k} x^k = \frac{1}{2} \left((1 + \sqrt{x})^n + (1 - \sqrt{x})^n \right).$$

Proof. If $a_k = x^k$, then $\Delta a_k = x^{k+1} - x^k = x^k(x-1)$. Thus

$$e_n^{\Delta} = \sum_{k \geq 0} \binom{n}{2k} x^k (x-1) = (x-1)e_n.$$

Applying Theorem 190, then, we obtain the recurrence $(x-1)e_n = e_{n+2} - 2e_{n+1}$, or $e_{n+2} - 2e_{n+1} - (x-1)e_n = 0$, with $e_0 = \binom{0}{0} x^0 = 1$ and $e_1 = \binom{1}{0} x^0 = 1$. Assuming a solution of the form $e_n = r^n$ (as in the proof of Identity 173 in

Section 7.1), we obtain the characteristic equation $r^2 - 2r - (x-1) = 0$. Applying the quadratic formula yields

$$r = 1 \pm \sqrt{x}.$$

Thus

$$e_n = A(1 + \sqrt{x}) + B(1 - \sqrt{x})$$

for some A and B. With the initial conditions $e_0 = e_1 = 1$, we obtain the system

$$1 = A + B$$
$$1 = A(1 + \sqrt{x}) + B(1 - \sqrt{x}).$$

The solution is $A = B = 1/2$, giving us the result. □

For our last set of examples on finite differences, we extend the ideas behind Identity 188 to upper binomial sums, producing Identity 197. In a manner similar to that with Identity 188 and its alternating and aerated versions, we use Identity 197 to build up to an upper binomial transform for *rising* (rather than falling) factorial powers.

Identity 197.

$$\sum_{k=0}^{n} \binom{k}{m} a_k = \binom{n+1}{m+1} a_n - \sum_{k=1}^{n} \binom{k}{m+1} \Delta a_{k-1}, \quad m \geq 0;$$

$$\sum_{k=0}^{n} \binom{k}{m} \Delta a_k = \binom{n+1}{m} a_{n+1} - \sum_{k=0}^{n} \binom{k}{m-1} a_{k+1}, \quad m \geq 1.$$

Proof. The first identity is just a shifted and relabeled version of the second, so it suffices to prove the second. We have

$$\sum_{k=0}^{n} \binom{k}{m} \Delta a_k = \sum_{k=0}^{n} \binom{k}{m} a_{k+1} - \sum_{k=0}^{n} \binom{k}{m} a_k$$

$$= \sum_{k=1}^{n+1} \binom{k-1}{m} a_k - \sum_{k=0}^{n+1} \binom{k}{m} a_k + \binom{n+1}{m} a_{n+1}$$

$$= -\sum_{k=1}^{n+1} \left(\binom{k}{m} - \binom{k-1}{m} \right) a_k - \binom{0}{m} a_0 + \binom{n+1}{m} a_{n+1}$$

$$= -\sum_{k=1}^{n+1} \binom{k-1}{m-1} a_k + \binom{n+1}{m} a_{n+1}$$

$$= -\sum_{k=0}^{n} \binom{k}{m-1} a_{k+1} + \binom{n+1}{m} a_{n+1}.$$

□

Let's look at some examples of the use of Identity 197, in the spirit of many of the examples we have seen in this section.

Identity 58.

$$\sum_{k=0}^{n} \binom{k}{m} = \binom{n+1}{m+1}.$$

Proof. Take $a_k = 1$ in the first part of Identity 197. □

Identity 73.

$$\sum_{k=0}^{n} \binom{k}{m} k = n \binom{n+1}{m+1} - \binom{n+1}{m+2}.$$

Proof. Taking $a_k = k$ in the first part of Identity 197, we have

$$\sum_{k=0}^{n} \binom{k}{m} k = n \binom{n+1}{m+1} - \sum_{k=1}^{n} \binom{k}{m+1} = n \binom{n+1}{m+1} - \binom{n+1}{m+2},$$

where the second step applies Identity 58, together with the fact that $\binom{0}{m+1} = 0$ for $m \geq 0$. □

Identity 198.

$$\sum_{k=0}^{n} \binom{k}{m} k(k+1) = n(n+1) \binom{n+1}{m+1} - 2n \binom{n+1}{m+2} + 2 \binom{n+1}{m+3}.$$

Proof. Let $a_k = k(k+1) = k^{\overline{2}} = (k+1)^{\underline{2}}$. Then $\Delta a_{k-1} = 2k^{\underline{1}} = 2k$, by Identity 177. Applying Identity 197, we have

$$\sum_{k=0}^{n} \binom{k}{m} k(k+1) = n(n+1) \binom{n+1}{m+1} - 2 \sum_{k=1}^{n} \binom{k}{m+1} k$$

$$= n(n+1) \binom{n+1}{m+1} - 2n \binom{n+1}{m+2} + 2 \binom{n+1}{m+3},$$

where we use Identity 73 in the second step. □

We choose a_k to be $k(k+1)$ in the previous identity because it makes $\Delta a_{k-1} = k$ and thus the proof simpler. By manipulating Identities 58, 73, and 198, one can find $\sum_{k=0}^{n} \binom{k}{m} f(k)$ for any quadratic function $f(k)$.

If we were following the pattern we had been following thus far in the chapter we would then build up to an identity for $\sum_{k=0}^{n} \binom{k}{m} k^{\underline{r}}$. However, as we can see from the previous identity, it makes more sense to choose the rising power $k^{\overline{r}}$ when $r = 2$, rather than the falling power $k^{\underline{r}}$. This turns out to be true in general; the rising power does lead to the simpler binomial coefficient identity. The reason is twofold: 1) Identity 197 gives $\sum_{k=0}^{n} \binom{k}{m} a_k$ in terms of Δa_{k-1}; thus we need a shift by one every time we move to the finite difference. 2) The following theorem shows that rising powers have the shift property that makes applying Identity 197 easy.

Identity 199. *For* $x \in \mathbb{R}$,

$$\Delta(x-1)^{\overline{n}} = nx^{\overline{n-1}},$$

where the finite difference is taken with respect to x.

Proof. We have, by Identity 31,

$$(x-1)^{\overline{n}} = (x-1)x(x+1)(x+2)\cdots(x-1+n-1) = (x+n-2)^{\underline{n}}.$$

By Identity 177, then,

$$\begin{aligned}
\Delta(x-1)^{\overline{n}} &= n(x+n-2)^{\underline{n-1}} \\
&= n(x+n-2)(x+n-1)\cdots(x+n-1-n+1) \\
&= nx^{\overline{n-1}}.
\end{aligned}$$

$\square$

We are now ready to jump to the general r case.

Identity 200.

$$\sum_{k=0}^{n}\binom{k}{m}k^{\overline{r}} = \sum_{k=0}^{r}\binom{n+1}{m+1+k}\frac{r!}{(r-k)!}n^{\overline{r-k}}(-1)^k.$$

Proof. We have already shown that the identity is true in the cases $r = 0, 1, 2$. Fix r, and assume now that it is true for $1, 2, \ldots, r-1$. Then, by Identity 199 and Identity 197, we have

$$\begin{aligned}
\sum_{k=0}^{n}\binom{k}{m}k^{\overline{r}} &= n^{\overline{r}}\binom{n+1}{m+1} - r\sum_{k=1}^{n}\binom{k}{m+1}k^{\overline{r-1}} \\
&= n^{\overline{r}}\binom{n+1}{m+1} - r\sum_{k=0}^{r-1}\binom{n+1}{m+2+k}\frac{(r-1)!}{(r-1-k)!}n^{\overline{r-1-k}}(-1)^k \\
&= n^{\overline{r}}\binom{n+1}{m+1} + \sum_{k=0}^{r-1}\binom{n+1}{m+2+k}\frac{r!}{(r-1-k)!}n^{\overline{r-1-k}}(-1)^{k+1} \\
&= n^{\overline{r}}\binom{n+1}{m+1} + \sum_{k=1}^{r}\binom{n+1}{m+1+k}\frac{r!}{(r-k)!}n^{\overline{r-k}}(-1)^k \\
&= \sum_{k=0}^{r}\binom{n+1}{m+1+k}\frac{r!}{(r-k)!}n^{\overline{r-k}}(-1)^k.
\end{aligned}$$

$\square$

7.3   Two-Variable Recurrence Relations

Just as single-variable recurrence relations are sometimes called "difference equations," recurrence relations with two or more variables are sometimes called *partial difference equations*. As is the case with partial differential equations versus ordinary differential equations, two-variable recurrence relations are much harder to solve than single-variable recurrence relations. Consequently, in this section we focus on only two topics, both of which were promised in Chapter 1: (1) Deriving the factorial definition of the binomial coefficients from their recursive definition; and (2) Giving a formula for the row sum for a wide class of numbers that satisfy two-variable recurrence relations (a class that includes the binomial coefficients). In Chapter 8 we introduce the Stirling and Lah numbers, numbers that are related to the binomial coefficients and that satisfy other two-variable recurrence relations.

In Chapter 1 we promised a direct proof that the recursive definition of the binomial coefficients, Identity 1,

$$\binom{n}{k} = \binom{n-1}{k} + \binom{n-1}{k-1}, \quad \binom{n}{0} = 1, \quad \binom{0}{k} = [k=0],$$

implies, in the case where $n \geq k \geq 0$, the factorial definition, Identity 2:

$$\binom{n}{k} = \frac{n!}{k!\,(n-k)!}.$$

We now have the tools to complete this proof.

Let's first go general and consider two-variable recurrence relations of the form

$$\left| \begin{matrix} n \\ k \end{matrix} \right| = f_1(n,k) \left| \begin{matrix} n-1 \\ k \end{matrix} \right| + f_2(n,k) \left| \begin{matrix} n-1 \\ k-1 \end{matrix} \right| + [n=k=0], \quad \text{for } n,k \geq 0, \quad (7.2)$$

where $\left| \begin{smallmatrix} n \\ k \end{smallmatrix} \right|$ is understood to be 0 for $n < 0$ or $k < 0$. These boundary conditions, together with the recurrence itself, force the only nonzero elements to be those for which $n \geq k \geq 0$. The binomial recurrence in Identity 1 is the case $f_1(n,k) = f_2(n,k) = 1$, which is probably the simplest nontrivial version of Equation (7.2). (Strictly speaking, as we saw in Chapter 2, $\binom{n}{k}$ can be nonzero when n is negative. However, in this section we are only interested in $\binom{n}{k}$ when $n,k \geq 0$, and so this assumption will not cause us any problems.) We will see other examples of Equation (7.2) in Chapter 8.

Speaking of simplest, a natural question to ask is the following: What are the simplest functions f_1 and f_2 such that $\left| \begin{smallmatrix} n \\ k \end{smallmatrix} \right|$ is just the triangle of 1's? In other words, what are the simplest functions f_1 and f_2 such that Equation (7.2) produces the triangle in Figure 7.3?

Let's start with the boundaries of the triangle. Since $\left| \begin{smallmatrix} n \\ -1 \end{smallmatrix} \right| = 0$, when generating the left diagonal $\left| \begin{smallmatrix} n \\ 0 \end{smallmatrix} \right| = 1$ the recurrence simplifies to $\left| \begin{smallmatrix} n \\ 0 \end{smallmatrix} \right| = \left| \begin{smallmatrix} n-1 \\ 0 \end{smallmatrix} \right|$. This

$$
\begin{array}{ccccccccccccc}
& & & & & & 1 & & & & & & \\
& & & & & 1 & & 1 & & & & & \\
& & & & 1 & & 1 & & 1 & & & & \\
& & & 1 & & 1 & & 1 & & 1 & & & \\
& & 1 & & 1 & & 1 & & 1 & & 1 & & \\
& 1 & & 1 & & 1 & & 1 & & 1 & & 1 & \\
1 & & 1 & & 1 & & 1 & & 1 & & 1 & & 1
\end{array}
$$

FIGURE 7.3
Triangle of all 1's

implies that $f_1(n, 0) = 1$. Similarly, since $\left|{n-1 \atop n}\right| = 0$, when considering the right diagonal $\left|{n \atop n}\right| = 1$ the recurrence simplifies to $\left|{n \atop n}\right| = \left|{n-1 \atop n-1}\right|$; thus $f_2(n, n) = 1$. Off of the boundaries, substituting 1's into $\left|{n \atop k}\right|$, $\left|{n-1 \atop k}\right|$, and $\left|{n-1 \atop k-1}\right|$, we get that $f_1(n, k) + f_2(n, k) = 1$. This rules out constant solutions for f_1 and f_2. The simplest solutions that depend on n and k and that yield $f_2(n, n) = 1$ are probably $f_2(n, k) = n/k$ and $f_2(n, k) = k/n$. The former leads to $f_1(n, k) = 1 - n/k$, which is undefined when $k = 0$ and thus is ruled out. However, the latter leads to $f_1(n, k) = 1 - k/n$, which satisfies the other boundary condition $f_1(n, 0) = 1$. Putting all of this together gives us the following result:

Theorem 20. *If $f_1(n, k) = 1 - k/n$ and $f_2(n, k) = k/n$, then the solution to Equation* (7.2) *is $\left|{n \atop k}\right| = 1[n \geq k \geq 0]$.*

With Theorem 20 we have a place to begin constructing solutions to Equation (7.2). The following property is quite helpful.

Theorem 21. *Suppose*

$$
\left|{n \atop k}\right| = f_1(n, k)\left|{n-1 \atop k}\right| + f_2(n, k)\left|{n-1 \atop k-1}\right| + [n = k = 0]
$$

and

$$
\left\|{n \atop k}\right\| = h(n)g_1(n-k)f_1(n, k)\left\|{n-1 \atop k}\right\| + h(n)g_2(k)f_2(n, k)\left\|{n-1 \atop k-1}\right\| + [n = k = 0].
$$

Then $\left|{n \atop k}\right|$ and $\left\|{n \atop k}\right\|$ are related by

$$
\left\|{n \atop k}\right\| = \left(\prod_{j=1}^{n} h(j)\right)\left(\prod_{j=1}^{n-k} g_1(j)\right)\left(\prod_{j=1}^{k} g_2(j)\right)\left|{n \atop k}\right|. \tag{7.3}
$$

Proof. The theorem is clearly true in the case $n = k = 0$. Otherwise,

$$
h(n)g_1(n-k)f_1(n, k)\left(\prod_{j=1}^{n-1} h(j)\right)\left(\prod_{j=1}^{n-1-k} g_1(j)\right)\left(\prod_{j=1}^{k} g_2(j)\right)\left|{n-1 \atop k}\right|
$$

$$+ h(n)g_2(k)f_2(n,k)\left(\prod_{j=1}^{n-1} h(j)\right)\left(\prod_{j=1}^{n-k} g_1(j)\right)\left(\prod_{j=1}^{k-1} g_2(j)\right)\left|\begin{matrix} n-1 \\ k-1 \end{matrix}\right|$$

$$= \left(\prod_{j=1}^{n} h(j)\right)\left(\prod_{j=1}^{n-k} g_1(j)\right)\left(\prod_{j=1}^{k} g_2(j)\right) f_1(n,k)\left|\begin{matrix} n-1 \\ k \end{matrix}\right|$$

$$+ \left(\prod_{j=1}^{n} h(j)\right)\left(\prod_{j=1}^{n-k} g_1(j)\right)\left(\prod_{j=1}^{k} g_2(j)\right) f_2(n,k)\left|\begin{matrix} n-1 \\ k-1 \end{matrix}\right|$$

$$= \left(\prod_{j=1}^{n} h(j)\right)\left(\prod_{j=1}^{n-k} g_1(j)\right)\left(\prod_{j=1}^{k} g_2(j)\right)\left|\begin{matrix} n \\ k \end{matrix}\right|.$$

Since the right-hand side of Equation (7.3) satisfies the recurrence for $\left\|\begin{matrix} n \\ k \end{matrix}\right\|$, the theorem holds.   □

Theorem 21 tells us three things:

1. If we multiply f_1 and f_2 by the same function $h(n)$, the solution to the recurrence is multiplied by a factor of $\prod_{j=1}^{n} h(j)$.

2. If we multiply f_1 by the function $g_1(n-k)$, the solution to the recurrence is multiplied by a factor of $\prod_{j=1}^{n-k} g_1(j)$.

3. If we multiply f_2 by the function $g_2(k)$, the solution to the recurrence is multiplied by a factor of $\prod_{j=1}^{k} g_2(j)$.

With Theorems 20 and 21, it becomes quite easy to prove that the factorial definition of the binomial coefficients follows from the binomial recurrence.

Identity 2. *For $n \geq k \geq 0$,*

$$\binom{n}{k} = \frac{n!}{k!\,(n-k)!}.$$

Proof. For the purposes of this proof we are assuming that $\binom{n}{k}$, for integers $n \geq 0$, $k \geq 0$, is defined by Identity 1 in the form of Equation (7.2):

$$\binom{n}{k} = \binom{n-1}{k} + \binom{n-1}{k-1} + [n = k = 0],$$

where $\binom{n}{k}$ is taken to be 0 if $n < 0$ or $k < 0$.

Thanks to Theorem 20, we know that if $f_1(n,k) = 1 - \frac{k}{n}$ and $f_2(n,k) = \frac{k}{n}$ in Equation (7.2) then the solution is $\left|\begin{matrix} n \\ k \end{matrix}\right| = 1$ for $n \geq k \geq 0$. Now, let's find functions h, g_1, and g_2 such that $h(n)g_1(n-k)f_1(n,k) = h(n)g_2(k)f_2(n,k) = 1$. We need to clear the denominators of f_1 and f_2; the simplest way to do that is to have $h(n) = n$. This leaves $g_1(n-k)(n-k) = 1$ and $g_2(k)k = 1$. Therefore, $g_1(n-k) = \frac{1}{n-k}$ and $g_2(k) = \frac{1}{k}$. By Theorem 21, then, for $n \geq k \geq 0$,

$$\binom{n}{k} = \left(\prod_{j=1}^{n} j\right)\left(\prod_{j=1}^{n-k} \frac{1}{j}\right)\left(\prod_{j=1}^{k} \frac{1}{j}\right) = \frac{n!}{(n-k)!\,k!}.$$

$\square$

Our second example with two-variable recurrence relations features a proof of the following identity, mentioned in Chapter 1:

Identity 201. *If $\left|{n \atop k}\right|$ satisfies a two-variable recurrence relation of the form*

$$\left|{n \atop k}\right| = (\alpha(n-1)+\beta k+\gamma)\left|{n-1 \atop k}\right| + (\alpha'(n-1)+\beta'(k-1)+\gamma)\left|{n-1 \atop k-1}\right| + [n = k = 0],$$

for $n, k \geq 0$, where $\left|{n \atop k}\right|$ is taken to be 0 if $n < 0$ or $k < 0$, and $\beta + \beta' = 0$, then the row sum for $\left|{n \atop k}\right|$ is given by

$$\sum_{k=0}^{n}\left|{n \atop k}\right| = \prod_{i=0}^{n-1}((\alpha+\alpha')i + \gamma + \gamma').$$

The recurrence relation in Identity 201 is an important special case of Equation (7.2), as several famous triangles of numbers satisfy the recurrence for different values of the parameters. In addition to the binomial coefficients, these include the Lah numbers and both kinds of Stirling numbers, as we shall see in Chapter 8.

Proof. First, the boundary conditions imply that $\left|{n \atop k}\right| = 0$ when $k < 0$ or $n < k$. Now, let $S_n = \sum_{k=0}^{n}\left|{n \atop k}\right|$. Summing both sides of the recurrence in the statement of Identity 201 and including the boundary conditions gives

$$S_n = \sum_{k=0}^{n}(\alpha(n-1)+\beta k+\gamma)\left|{n-1 \atop k}\right| + \sum_{k=0}^{n}(\alpha'(n-1)+\beta'(k-1)+\gamma')\left|{n-1 \atop k-1}\right|$$

$$= \sum_{k=0}^{n}(\alpha(n-1)+\beta k+\gamma)\left|{n-1 \atop k}\right| + \sum_{k=-1}^{n-1}(\alpha'(n-1)+\beta'k+\gamma')\left|{n-1 \atop k}\right|$$

$$= \sum_{k=0}^{n-1}((\alpha+\alpha')(n-1)+\gamma+\gamma')\left|{n-1 \atop k}\right|$$

$$= ((\alpha+\alpha')(n-1)+\gamma+\gamma')\,S_{n-1}.$$

Since $S_0 = \left|{0 \atop 0}\right| = 1$, the solution to this recurrence is

$$\sum_{k=0}^{n}\left|{n \atop k}\right| = S_n = \prod_{i=0}^{n-1}((\alpha+\alpha')i + \gamma + \gamma').$$

$\square$

As we saw in Chapter 1, Identity 201 can be used to give a quick proof for the row sum identity for Pascal's triangle. See Exercise 8 in Chapter 8 for another example of the use of Identity 201.

7.4   Exercises

1. Unroll the recurrence $a_n = a_{n-1} + n$, $a_0 = 1$, to prove that n lines, each of which intersect the others without any three lines intersecting at a common point, divide the plane into $1 + \binom{n+1}{2}$ regions.

2. Solve the following second-order recurrence relations via the method of undetermined coefficients.

    (a) $a_n = 7a_{n-1} - 12a_{n-2}$, valid for $n \geq 2$, with $a_0 = 0, a_1 = 1$.

    (b) $a_n = 6a_{n-1} + 16a_{n-2}$, valid for $n \geq 2$, with $a_0 = 5, a_1 = 10$.

3. When using the method of undetermined coefficients on a second-order recurrence relation, sometimes you get a single, repeated root r instead of two distinct roots r_1 and r_2. In that case the solution is of the form $a_n = Ar^n + Bnr^n$, where A and B are the undetermined coefficients that can be found by using the initial conditions. (This is similar to the situation when solving a differential equation using undetermined coefficients and obtaining a repeated root.) Use this method to solve the following recurrence relation:

$$a_n = 6a_{n-1} - 9a_{n-2}, \text{ valid for } n \geq 2, \text{ with } a_0 = 2, a_1 = 3.$$

4. Work through the following steps to prove Binet's formula via generating functions.

    (a) Use an approach like the one we applied to the problem of using lines to divide the plane to obtain the generating function for the Fibonacci numbers:

    Identity 202.

    $$\frac{x}{1 - x - x^2} = \sum_{n=0}^{\infty} F_n x^n.$$

    (b) Use partial fractions decomposition on $-x/(x^2 + x - 1)$ to obtain

    $$\frac{-x}{x^2 + x - 1} = -\frac{\phi/\sqrt{5}}{x + \phi} + \frac{\psi/\sqrt{5}}{x + \psi},$$

    with $\phi = (1 + \sqrt{5})/2$, $\psi = (1 - \sqrt{5})/2$.

(c) Show that $\phi = -1/\psi$, and use this to obtain

$$\frac{-x}{x^2 + x - 1} = \frac{1}{\sqrt{5}} \left(\frac{1}{1 - \phi x} - \frac{1}{1 - \psi x} \right).$$

(d) Finally, apply the geometric series formula $\sum_{n=0}^{\infty} r^n = \frac{1}{1-r}$ to find an infinite series expression for $-x/(x^2 + x - 1)$ whose nth term is $(\phi^n - \psi^n)/\sqrt{5}$, thus proving Binet's formula:

Identity 173.

$$F_n = \frac{\phi^n - \psi^n}{\sqrt{5}}.$$

5. Use ordinary generating functions to solve the following recurrences. (You will need several results on ordinary generating functions from Chapter 6, including some from early exercises in that chapter.)

   (a) $a_n = 3a_{n-1}$, valid for $n \geq 1$, with $a_0 = 5$.

   (b) $a_n = a_{n-1} + 3$, valid for $n \geq 1$, with $a_0 = 1$.

   (c) $a_n = 4a_{n-1} + 3$, valid for $n \geq 1$, with $a_0 = 1$.

   (d) $a_n = a_{n-1} + 3^n$, valid for $n \geq 1$, with $a_0 = 4$.

   (e) $a_n = a_{n-1} + 2n$, valid for $n \geq 1$, with $a_0 = 0$.

   (f) $a_n = 2a_{n-1} + n2^n$, valid for $n \geq 1$, with $a_0 = 2$.

6. We can use recurrence relations and generating functions to find formulas for sums, too. Let's step through this approach to find a formula for the first n odd numbers: $1 + 3 + 5 + \cdots + 2n - 1$.

   (a) First, let $S_n = 1 + 3 + 5 + \cdots + 2n - 1 = \sum_{k=1}^{n} (2k - 1)$.

   (b) Then, find a recurrence relation satisfied by S_n. Since S_n is a sum, this is easy: $S_n = S_{n-1} + 2n - 1$, valid for $n \geq 1$, with $S_0 = 0$.

   (c) Next, show that the ordinary generating function for $(S_n)_{n=0}^{\infty}$ is

$$f_S(x) = \frac{2x}{(1 - x)^3} - \frac{1}{(1 - x)^2}.$$

   (d) Finally, use $f_S(x)$ to obtain the formula $S_n = n^2$.

7. Use the ideas in Exercise 6 to prove the following, the formula for the sum of squares. (Also, compare this result with Identity 74.)

Identity 203.

$$\sum_{k=0}^{n} k^2 = \binom{n+2}{3} + \binom{n+1}{3} = \frac{n(n+1)(2n+1)}{6}.$$

8. Because the exponential generating function for a sequence (a_n) features the term $a_n x^n / n!$ in its infinite series expansion, when solving a recurrence relation that includes a factorial expression it's often a good idea to use exponential generating functions instead of ordinary ones. Use exponential generating functions to solve the following:

$$a_n = n a_{n-1} + n!, \text{ valid for } n \geq 1, \text{ with } a_0 = 0.$$

9. Prove that the finite difference operator Δ is linear; i.e., prove that, for sequences (a_n) and (b_n) and constants c and d,

$$\Delta(c\, a_n + d\, b_n) = c\, \Delta a_n + d\, \Delta b_n.$$

10. Use Identity 58,

$$\sum_{k=0}^{n} \binom{k}{m} = \binom{n+1}{m+1},$$

to give a proof of the following formula for the falling power sum when $m \geq 0$:

Identity 204.

$$\sum_{k=0}^{n} k^{\underline{m}} = \frac{(n+1)^{\underline{m+1}}}{m+1}.$$

Note the similarity of Identity 204 with the integral of the power function:

$$\int_0^n x^m \, dx = \frac{n^{m+1}}{m+1}.$$

These formulas deepen the analogy between the usual calculus and the discrete calculus of finite differences that we discussed in the chapter.

11. Another important formula in the calculus of finite differences is *summation by parts* (analogous, of course, to integration by parts). Prove the following version of this formula.

Identity 205 (Summation by Parts).

$$\sum_{k=0}^{n} a_k \Delta b_k = a_{n+1} b_{n+1} - a_0 b_0 - \sum_{k=0}^{n} \Delta a_k b_{k+1}.$$

12. Use induction to give a different proof of the formula for the nth finite difference of a sequence:

Identity 174.

$$\Delta^n a_m = \sum_{k=0}^{n} \binom{n}{k} a_{m+k} (-1)^{n-k}.$$

13. Let $m \in \mathbb{R}$, and let $a_n = m^n$. Prove that the sequence given by $(\Delta^0 a_0, \Delta^1 a_0, \Delta^2 a_0, \ldots)$ on the right side of the difference triangle for (a_n) is the sequence $((m-1)^n)_{n=0}^{\infty}$.

14. Find the sequence (a_n) that has the nonnegative integers $(n)_{n=0}^{\infty}$ as the right diagonal of its difference triangle.

15. Suppose $f(x)$ is a polynomial of degree m, and let (a_n) be the sequence with $a_n = f(n)$. Prove that if $n > m$, then the sequence $(\Delta^n a_0, \Delta^n a_1, \Delta^n a_2, \ldots)$ is the zero sequence. (Note this is not asking about the sequence $(\Delta^0 a_0, \Delta^1 a_0, \Delta^2 a_0, \ldots)$. In terms of the difference triangle, this question is asking about the numbers on the nth left diagonal, not the right diagonal. Incidentally, this property is used in data analysis to determine whether a polynomial of a certain degree is a good model for a data set. See, for example, a modeling text such as Giordano, Fox, and Horton [28].)

16. Our proof of Identity 177 proceeds by expanding $x^{\underline{n}}$. Give another proof of Identity 177 in the case $n \geq 0$ by using Pascal's recurrence (Identity 1).

17. In Exercise 16 the argument goes in both directions, so that Pascal's recurrence is a consequence of the finite difference formula for falling powers in Identity 177. Since our proof of Identity 176 uses Pascal's recurrence, this means that Identity 176 can also be proved via Identity 178. In fact, Identity 178 is the application of the higher-order difference formula (Identity 174) to nonnegative falling powers. Use Identity 178 to give a direct proof of Identity 176.

18. Starting from Identity 176, prove its generalization

Identity 206.

$$\sum_{k=-s}^{n-s} \binom{n}{k+s} \binom{m+k}{r} (-1)^k = (-1)^{n+s} \binom{m-s}{r-n}.$$

19. Our finite difference proof of Identity 109 uses -1 as the falling power. With $a_m = (m-1)^{\underline{-r-1}}$, prove the following, which generalizes both Identity 109 and Identity 147:

Identity 207. *For $m \geq 1$,*

$$\sum_{k=0}^{n} \binom{n}{k} \frac{(-1)^k}{(m+k)(m+1+k)\cdots(m+r+k)} = \frac{(m-1)!\,(r+n)!}{(m+r+n)!\,r!}$$
$$= \frac{1}{m\,r!\binom{m+r+n}{r+n}}.$$

In view of Exercise 17, Identities 176 and 207 are really just the consequences of Identity 178 for positive and negative powers, respectively.

20. Use Identity 184 to give yet another proof of Identity 109:

    Identity 109.

    $$\sum_{k=0}^{n} \binom{n}{k} \frac{(-1)^k}{m+k} = \frac{n!}{m(m+1)\cdots(m+n)}.$$

21. Prove

    Identity 208.

    $$\sum_{k=0}^{n} \binom{n}{k}\binom{k+1}{m} = 2^{n-m}\left(\binom{n+1}{m} + \binom{n}{m-1}\right).$$

22. Prove, where D_n is the nth derangement number,

    Identity 209.

    $$D_n + D_{n+1} = \sum_{k=0}^{n} \binom{n}{k}(n+1-k)!\,(-1)^k.$$

23. Prove the following generalization of the variation of binomial inversion given in Theorem 2.

    Identity 210. *Suppose*

    $$d_n = \sum_{k=0}^{n} \binom{n}{k} a_k (-1)^{n-k}.$$

    Then

    $$\sum_{k=0}^{n} \binom{n}{k} a_{k+m}(-1)^{n-k} = \sum_{k=0}^{m} \binom{m}{k} d_{k+n}.$$

24. Prove

    Identity 211.

    $$\sum_{k=1}^{n} \binom{n}{k} \frac{(-1)^k}{k(k+1)} = -H_{n+1} + 1.$$

25. Prove the following generalization of Identity 211:

    Identity 212.

    $$\sum_{k=1}^{n} \binom{n}{k} \frac{(-1)^k}{k(k+1)\cdots(k+m)} = -\frac{H_{n+m} - H_m}{m!}.$$

26. Use Identity 188 to prove the following. (Hint: Use partial fractions to determine which sequence has $\frac{1}{(k+1)(k+2)}$ as its finite difference.)

Identity 39.

$$\sum_{k=0}^{n} \binom{n}{k} \frac{1}{(k+1)(k+2)} = \frac{2^{n+2}-1}{(n+1)(n+2)} - \frac{1}{n+1}.$$

27. Use finite differences to prove this identity giving the binomial transform of the harmonic numbers:

Identity 213.

$$\sum_{k=0}^{n} \binom{n}{k} H_k = 2^n \left(H_n - \sum_{k=1}^{n} \frac{1}{k2^k} \right).$$

28. Use finite differences to prove the identity giving the alternating binomial transform of the harmonic numbers:

Identity 27.

$$\sum_{k=0}^{n} \binom{n}{k} H_k (-1)^k = -\frac{1}{n} [n \geq 1].$$

29. Prove

Identity 191. *Suppose*

$$f_n = \sum_{k \geq 0} \binom{n}{2k+1} a_k, \ \ and$$

$$f_n^{\Delta} = \sum_{k \geq 0} \binom{n}{2k+1} \Delta a_k.$$

Then

$$f_n^{\Delta} = f_{n+2} - 2f_{n+1}, \ \ and$$

$$f_n = \begin{cases} 0, & n = 0; \\ 2^{n-1} a_0 + 2^n \sum_{k=2}^{n} \frac{f_{k-2}^{\Delta}}{2^k}, & n \geq 1. \end{cases}$$

30. Prove

Identity 193.

$$\sum_{k \geq 0} \binom{n}{2k+1} k = (n-2)2^{n-3} [n \geq 2].$$

31. Use summation by parts (Identity 205 in Exercise 11) to evaluate

$$\sum_{k=0}^{n-m-2} (k+m)k^{\underline{m-2}}$$

in a different fashion than the evaluation in the text, and then finish the derivation of Identity 194 from there.

32. Prove

Identity 195. *For $m \geq 1$,*

$$\sum_{k\geq 0} \binom{n}{2k+1} k^{\underline{m}} = (n-m-1)^{\underline{m}} 2^{n-2m-1} \, [n \geq m+1].$$

33. Prove the even-index version of Identity 48:

Identity 214.
$$\sum_{k\geq 0} \binom{n}{2k} \frac{1}{k+1} = \frac{n2^{n+1}+2}{(n+1)(n+2)}.$$

34. Prove the following version of the binomial theorem for odd aerated binomial sums:

Identity 215.

$$\sum_{k\geq 0} \binom{n}{2k+1} x^k = \frac{1}{2\sqrt{x}} \left((1+\sqrt{x})^n - (1-\sqrt{x})^n \right).$$

35. Show that the second equation in Identity 197 is a special case of summation by parts (Identity 205 in Exercise 11).

36. Prove

Identity 216.

$$\sum_{k=0}^{n} \binom{k}{m} \binom{r+k-1}{r} = \sum_{k=0}^{r} \binom{n+1}{m+1+k} \binom{n+r-k-1}{r-k} (-1)^k.$$

37. Prove

Identity 217.

$$\sum_{k=0}^{n} \binom{k}{m}(-1)^k = \left(\frac{-1}{2}\right)^m [n \text{ is even}] + \frac{(-1)^n}{2} \sum_{k=1}^{m} \binom{n+1}{k} \left(\frac{-1}{2}\right)^{m-k}.$$

38. Solve the recurrence

$$\left|{n \atop k}\right| = (n - k)\left|{n - 1 \atop k}\right| + \left|{n - 1 \atop k - 1}\right| + [n = k = 0],$$

where $\left|{n \atop k}\right|$ is understood to be 0 for $n < 0$ or $k < 0$.

39. Solve the recurrence

$$\left|{n \atop k}\right| = (n - k)(n + k)\left|{n - 1 \atop k}\right| + k^2\left|{n - 1 \atop k - 1}\right| + [n = k = 0],$$

where $\left|{n \atop k}\right|$ is understood to be 0 for $n < 0$ or $k < 0$.

40. Solve the recurrence

$$\left|{n \atop k}\right| = (n + k)\left|{n - 1 \atop k}\right| + \left|{n - 1 \atop k - 1}\right| + [n = k = 0],$$

where $\left|{n \atop k}\right|$ is understood to be 0 for $n < 0$ or $k < 0$.

41. Suppose we repeatedly select balls from a jar that initially contains m red balls. If the ball we select is red, we replace the red ball with a blue ball. If the ball we select is blue, we place the blue ball back in the jar. Let $\left|{n \atop k}\right|$ denote the probability of drawing exactly k red balls in n trials.

    (a) Show that $\left|{n \atop k}\right|$ satisfies the recurrence

    $$\left|{n \atop k}\right| = \frac{k}{m}\left|{n - 1 \atop k}\right| + \frac{m - k + 1}{m}\left|{n - 1 \atop k - 1}\right| + [n = k = 0],$$

    for $n, k \geq 0$, where $\left|{n \atop k}\right|$ is taken to be 0 if $n < 0$ or $k < 0$.

    (b) Solve the recurrence in (a) to show that

    $$\left|{n \atop k}\right| = \left\{{n \atop k}\right\}\frac{m^{\underline{k}}}{m^n},$$

    where $\left\{{n \atop k}\right\}$ satisfies the recurrence $\left\{{n \atop k}\right\} = k\left\{{n-1 \atop k}\right\} + \left\{{n-1 \atop k-1}\right\} + [n = k = 0]$, for $n, k \geq 0$, where $\left\{{n \atop k}\right\}$ is taken to be 0 if $n < 0$ or $k < 0$. (The numbers $\left\{{n \atop k}\right\}$ are the *Stirling numbers of the second kind*, which we shall see more of in Chapter 8.)

    (c) Finally, show that the probability of drawing the kth red ball on the nth trial is

    $$\left\{{n - 1 \atop k - 1}\right\}\frac{m^{\underline{k}}}{m^n}.$$

42. Prove the following, which gives a binomial-type theorem for two-variable recurrence relations and generalizes Identity 201.

Identity 218. *If $\left|{n \atop k}\right|$ satisfies a two-variable recurrence relation of the form*

$$\left|{n \atop k}\right| = (\alpha(n-1)+\beta k+\gamma)\left|{n-1 \atop k}\right|+(\alpha'(n-1)+\beta'(k-1)+\gamma)\left|{n-1 \atop k-1}\right|+[n=k=0],$$

for $n, k \geq 0$, where $\left|{n \atop k}\right|$ is taken to be 0 if $n < 0$ or $k < 0$, and x, y are such that $\beta y + \beta' x = 0$, then

$$\sum_{k=0}^{n}\left|{n \atop k}\right|x^k y^{n-k} = \prod_{i=0}^{n-1}((\alpha y + \alpha' x)i + \gamma y + \gamma' x).$$

7.5 Notes

Chapter 5 of Charalambides' *Enumerative Combinatorics* [15] is on derangements and related problems.

Ronald Mickens's *Difference Equations* [48] contains more on the method of undetermined coefficients applied to difference equations.

The classic text on finite differences is Charles Jordan's *Calculus of Finite Differences* [39]. Sections 3.4 and 3.5 of the *Handbook of Discrete and Combinatorial Mathematics* [58] contain a good introduction to finite difference calculus. Section 2.6 of *Concrete Mathematics* [32] does as well. Section 5.2 of that text includes a nice discussion of the use of Identity 174 in proving binomial coefficient identities.

Spivey and Steil [71] contains a path-counting proof for the relationship between sum triangles and binomial transforms discussed in Section 7.2. The paper also generalizes this to representing multiple iterations of the binomial transform on a single triangle. See also Exercise 30 of Chapter 6.

Much of Section 7.2 appears in Spivey [63]. The finite difference $\Delta a_n = a_{n+1} - a_n$ that is the focus of Section 7.2 is sometimes called the *forward difference* operator. The *backwards difference*, defined by $\nabla a_n = a_n - a_{n-1}$, produces the same sequence shifted by an index. Boyadzhiev [12] contains several results relating binomial sums and variants of the backwards difference, especially the operator $n\nabla$.

The proof of Identity 2 in Section 7.3 appears in Spivey [68]. Identity 201 is due to Neuwirth [50].

Finding the general solution to two-variable recurrence relations of the form of Equation (7.2) when f_1 and f_2 are linear functions of n and k (i.e., the recurrence relation satisfied by $\left|{n \atop k}\right|$ in Identity 201) is Research Problem 6.94 in *Concrete Mathematics* [32]. Spivey [65] uses Theorem 21 and some similar results to find explicit solutions to several cases of Equation (7.2). He derives Identity 201 as well. In addition, Barbero, Salas, and Villaseñor [27] give a complete solution to the *Concrete Mathematics* problem in terms of generating functions.

8

Special Numbers

This chapter is devoted to several collections of numbers related to the binomial coefficients. Each of these sets of numbers—the Fibonacci, Stirling, Bell, Lah, Catalan, and Bernoulli numbers—has an expression featuring the binomial coefficients. Each satisfies a variety of identities involving the binomial coefficients. Each (with the exception of the Bernoulli numbers) has one or more combinatorial interpretations. In fact, because of these combinatorial interpretations this chapter functions to a large degree like a second chapter on using combinatorial arguments to prove binomial identities. In particular, when discussing Catalan numbers we introduce more sophisticated lattice path arguments than we saw in Chapter 3. We don't just use combinatorial arguments, though; we'll also see recurrence relations and generating functions in our proofs in this chapter.

Finally, in the last section we briefly discuss the On-Line Encyclopedia of Integer Sequences, a major resource for numbers like those featured in this chapter.

8.1 Fibonacci Numbers

The Fibonacci numbers, as we have seen, are defined by the recurrence relation in Identity 170: $F_n = F_{n-1} + F_{n-2}$, for $n \geq 2$, with $F_0 = 0, F_1 = 1$. The first several Fibonacci numbers are given in Table 8.1.

TABLE 8.1
First several Fibonacci numbers

n	0	1	2	3	4	5	6	7	8	9	10	11	12	13
F_n	0	1	1	2	3	5	8	13	21	34	55	89	144	233

In Chapter 7 we proved several identities for the Fibonacci numbers, such as Binet's explicit formula for the nth Fibonacci number (Identity 173), their nth finite difference (Identity 179), their binomial and alternating binomial transforms (Identities 180 through 183), the sum of the first n even index Fibonacci numbers (Identity 187), and their generating function (Identity 202). Since the Fibonacci numbers satisfy what is probably the most basic two-term recurrence relation in existence, they make a great "recurring" example in a chapter about recurrence relations. But where do these numbers come from, and why might we be interested in them?

The Fibonacci numbers have actually been with us for a long time. They were introduced by Leonardo of Pisa (also known as "Fibonacci"—hence the name) way back in 1202. The context was the following. Suppose rabbits mate starting at the age of one month, and rabbit pregnancies last a month. Suppose also that each pregnancy produces exactly one male and one female rabbit. If we start with a new pair of rabbits from birth, and none of the rabbits die, how many pairs of rabbits will there be in one year? Let's work through this problem and see how the Fibonacci numbers arise.

Denote the number of pairs of rabbits at the beginning of month n by R_n. Initially there is only the first pair of rabbits, so $R_1 = 1$. This is true at the beginning of the second month as well, so $R_2 = 1$. At this point, though, the first pair is mature enough to mate. The pregnancy lasts a month. Thus at the beginning of month three there are two pairs of rabbits, making $R_3 = 2$. The new pair of rabbits is not mature enough to mate, though, and so only the original pair reproduces during month three. This means at the beginning of month four there are three pairs of rabbits: $R_4 = 3$. Two of these three pairs are mature enough to mate, and so at the beginning of month five there are five pairs of rabbits, yielding $R_5 = 5$. At this point we can see how the Fibonacci recurrence arises in this problem: The number of pairs of rabbits at the beginning of month n (R_n) is the number of pairs of rabbits that were around at the beginning of month $n-1$ (R_{n-1}) plus the number

Leonardo of Pisa, more commonly known as **Fibonacci**, was born in the latter part of the 12th century in what is now Italy. He traveled widely in North Africa as a young man, although he eventually returned to Pisa to settle down. He subsequently wrote several books on mathematics, the most famous of which is *Liber Abaci*. While this masterpiece of calculation techniques does discuss the sequence of numbers that bears his name, its greatest contributions to mathematics are probably (1) introducing to the Western world the Hindu-Arabic number system still in use today, and (2) providing for the Western world its first comprehensive treatment of the algebra developed in the Middle East.

of those that were mature enough to mate (R_{n-2}). Following this logic, and given Table 8.1, the answer to Fibonacci's question is 144.

The rabbit problem is rather unrealistic (of course rabbits can die, and a rabbit pregnancy will not always produce exactly one male and one fe-

TABLE 8.2
Sequences of coin flips that do not contain consecutive heads

n	Sequences of n flips without consecutive heads
1	H, T
2	HT, TH, TT
3	HTH, HTT, THT, TTH, TTT
4	HTHT, HTTH, HTTT, THTH, THTT, TTHT, TTTH, TTTT

male), but in the centuries since Leonardo of Pisa first posed his problem the simple-to-construct Fibonacci numbers have found a large number of uses. These include family trees of honeybees (see Exercise 3), antecedents of X chromosomes in humans [38], and arrangements of leaves and seeds in certain plants [23]. Unfortunately, there have also been many specious claims as to the presence of the Fibonacci numbers or the golden ratio in art or nature. (See Markowsky [46], Livio [43], or Devlin [21] for discussions of some of these misconceptions.)

The Fibonacci numbers arise in more strictly mathematical contexts as well. For example, the analysis of the Euclidean algorithm for finding greatest common divisors involves the Fibonacci numbers, and one of the breakthroughs of 20th century mathematics—the solution of Hilbert's Tenth Problem by Yuri Matiyasevich in 1970—also features the Fibonacci numbers. Before we investigate the connection between the Fibonacci numbers and the binomial coefficients (of course there is one, given what this book is about!) let's take a look at another mathematically-inclined problem whose solution entails the Fibonacci numbers.

Question: Suppose we flip a fair coin n times. What is the probability that this sequence does not contain consecutive heads?

The total number of ways to flip a coin n times is 2^n, so let's focus on the numerator of our answer and count the number of ways we can flip n coins and not obtain heads twice in a row. Table 8.2 lists the possible sequences for a few small values of n. The number of allowable sequences listed are 2, 3, 5, and 8, which certainly makes it look like the Fibonacci numbers are involved. Peering a little closer at the table, we can see that the sequences in, say, row 4 are obtained in one of two ways: Either take a sequence in row 3 that begins with T and place an H in front of it, or take any sequence in row 3 and place a T in front of it. This sounds a bit like the rabbits problem, where "placing an H in front" corresponds to a pair reproducing (only some pairs do this), and "placing a T in front" corresponds to a pair simply surviving until the next month (all pairs do this). Given all this, it's perhaps not surprising that we have the following.

Theorem 22. *Suppose we flip a fair coin n times. The probability that this sequence of coin flips does not feature consecutive heads is $\dfrac{F_{n+2}}{2^n}$.*

Proof. Let h_n be the number of ways to flip a coin n times without obtaining consecutive heads. From Table 8.2 we can see that $h_1 = 2 = F_3$ and $h_2 = 3 = F_4$, so the initial conditions hold. Now, assuming $n \geq 3$, let's condition on what happens with the first flip in a sequence of n coin flips. If the first flip is a tail, then the only restriction on the rest of the flips is that they cannot feature consecutive heads. Thus there are h_{n-1} sequences of flips in which the first flip is a tail and that avoid consecutive heads. However, if the first flip is a head, the second flip must be a tail. After that, though, the remaining flips have no restriction other than that there cannot be consecutive heads. Thus there are h_{n-2} sequences of flips in which the first flip is a head and that avoid consecutive heads. All together, then, $h_n = h_{n-1} + h_{n-2}$. Since $F_{n+2} = F_{n+2-1} + F_{n+2-2}$ as well, and the initial conditions match, we have $h_n = F_{n+2}$. Since the total number of sequences of n coin flips is 2^n, the theorem follows. $\qquad\qquad\square$

Let's now see how the Fibonacci numbers relate to the binomial coefficients. Figure 8.1 contains Pascal's triangle, but expressed in right-triangle format rather than the isosceles-triangle format we saw in Figure 1.1.

n	$\binom{n}{0}$	$\binom{n}{1}$	$\binom{n}{2}$	$\binom{n}{3}$	$\binom{n}{4}$	$\binom{n}{5}$	$\binom{n}{6}$	$\binom{n}{7}$
0	1							
1	1	1						
2	1	2	1					
3	1	3	3	1				
4	1	4	6	4	1			
5	1	5	10	10	5	1		
6	1	6	15	20	15	6	1	
7	1	7	21	35	35	21	7	1

FIGURE 8.1
Rows 0 through 7 of Pascal's triangle

Looking at the sums of the up-right-to-down-left diagonals, the zeroth and first diagonals each consist solely of the number 1. The second diagonal sums to $1 + 1 = 2$. The third sums to $2 + 1 = 3$, the fourth to $1 + 3 + 1 = 5$, and the next few to $8, 13$, and 21. Comparing these numbers with those in Table 8.1, it appears that sums of the diagonals in Pascal's triangle are the Fibonacci numbers. (In the isosceles Pascal's triangle in Figure 1.1, the equivalent would be summing what are called the "shallow diagonals.") Why might this be? Let's take a look at the number 10 in the fifth row, second position of Pascal's triangle. This number, $\binom{5}{2}$, is on the seventh diagonal. Thanks to Pascal's recurrence $\binom{n}{k} = \binom{n-1}{k} + \binom{n-1}{k-1}$ (Identity 1), $10 = \binom{5}{2} = \binom{4}{2} + \binom{4}{1} = 6 + 4$. Thus $\binom{5}{2}$ is the sum of a number, $\binom{4}{2}$, on the sixth diagonal, and a number $\binom{4}{1}$, on the fifth diagonal. Moreover, remembering that $\binom{n}{-1} = \binom{n}{n+1} = 0$, each number on the seventh diagonal is the sum of a number on the sixth diagonal and a number on the fifth diagonal, and all of the numbers on the fifth and sixth diagonals are used exactly once when constructing the seventh diagonal from them. Thanks to Identity 1, this is true for the nth diagonal, too. This means that the diagonals in Pascal's triangle satisfy the Fibonacci recurrence. Let's prove this formally.

Identity 219.

$$\sum_{k=0}^{n} \binom{n-k}{k} = F_{n+1}.$$

Proof. Let S_n denote the sum of the nth up-right-to-down-left diagonal in Pascal's triangle; i.e.,

$$S_n = \sum_{k=0}^{n} \binom{n-k}{k}.$$

Clearly, $S_0 = 1 = F_1$, and $S_1 = 1 = F_2$. Then we have, thanks to Identity 1 and the fact that $\binom{n}{-1} = \binom{n}{n+1} = 0$,

$$S_n = \sum_{k=0}^{n} \binom{n-k}{k} = \sum_{k=0}^{n} \binom{n-1-k}{k} + \sum_{k=0}^{n} \binom{n-1-k}{k-1}$$

$$= \sum_{k=0}^{n-1} \binom{n-1-k}{k} + \sum_{k=-1}^{n-1} \binom{n-2-k}{k} = S_{n-1} + \sum_{k=0}^{n-2} \binom{n-2-k}{k}$$

$$= S_{n-1} + S_{n-2}.$$

We have shown that the sequence $(S_n)_{n=0}^{\infty}$ satisfies the same recurrence and the same initial conditions as the sequence $(F_{n+1})_{n=0}^{\infty}$. Therefore, $S_n = F_{n+1}$.
□

Thanks to Identity 219, every identity involving the Fibonacci numbers directly translates into an identity involving the binomial coefficients.

In Chapter 7 we mentioned, but did not prove, that by dropping the restriction $n \geq 2$ on the Fibonacci recurrence we can define Fibonacci numbers with negative indices. For example, since $F_1 = 1$ and $F_0 = 0$, the assumption that $F_n = F_{n-1} + F_{n-2}$ for $n = 1$ holds produces $1 = 0 + F_{-1}$. Thus $F_{-1} = 1$. The same logic yields $F_{-2} = -1$, $F_{-3} = 2$, and $F_{-4} = -3$. It does not take long to realize that this process yields the following formula for Fibonacci numbers with negative indices. Let's prove that now.

Identity 220. *For $n \geq 0$,*

$$F_{-n} = (-1)^{n+1} F_n.$$

Proof. The proof is by induction. We have already seen that $F_{-1} = 1 = (-1)^2 F_1$ and $F_{-2} = 1 = (-1)^3 F_2$. Now, assume the formula holds for a fixed n. Then, by the Fibonacci recurrence (Identity 170), we have

$$F_{-(n+1)} = F_{-n-1} = F_{-n+1} - F_{-n} = F_{-(n-1)} - F_{-n}$$

$$= (-1)^n F_{n-1} - (-1)^{n+1} F_n = (-1)^{n+2} F_{n-1} + (-1)^{n+2} F_n$$
$$= (-1)^{n+2}(F_{n-1} + F_n) = (-1)^{n+2} F_{n+1}.$$

By induction, then, the formula holds for all $n \geq 0$.   □

In Chapter 3 we used combinatorial arguments to prove binomial coefficient identities. We can prove Fibonacci identities with combinatorial arguments as well; we just need a combinatorial interpretation of the Fibonacci numbers. There are several of these; see the notes for references. We use the following interpretation popularized by Benjamin and Quinn [7, p. 1].

Suppose you have a rectangle of size $1 \times n$, and you wish to tile it with squares of size 1×1 and dominoes of size 1×2. How many ways are there to do this? When $n = 1$, there is only way: with a square. With $n = 2$, there are two ways: two squares, or one domino. With $n = 3$, you can tile with three squares, a domino followed by a square, or a square followed by a domino: three ways. The answer for a general n involves the Fibonacci numbers.

Theorem 23. *Let f_n denote the number of ways to tile a $1 \times n$ board with squares and dominoes. Then $f_n = F_{n+1}$.*

Proof. Since there is one way to tile a 1×0 board (i.e., with the empty tile), we have $f_0 = 1 = F_1$. We have already shown $f_1 = 1 = F_2$, $f_2 = 2 = F_3$, and $f_3 = 3 = F_4$. Now, for $n \geq 2$, assume we want to tile a $1 \times n$ board. We could use a square for the first tile, leaving us with a $1 \times (n-1)$ board left to tile. Thus there are f_{n-1} ways to tile the $1 \times n$ board if we use a square first. If we use a domino for the first tile, though, we have a $1 \times (n-2)$ board left to tile. Thus there are f_{n-2} ways to tile the original board when using a domino first. All together, then, there are $f_n = f_{n-1} + f_{n-2}$ ways to tile a $1 \times n$ board with squares and dominoes. Thus the sequence $(f_n)_{n=0}^{\infty}$ satisfies the same recurrence and initial conditions as does the sequence $(F_{n+1})_{n=0}^{\infty}$. We therefore have $f_n = F_{n+1}$.   □

Let's take a look now at three Fibonacci identities that can be proved using the tiling interpretation in Theorem 23. First we have a combinatorial proof of Identity 219.

Identity 219.

$$\sum_{k=0}^{n} \binom{n-k}{k} = F_{n+1}.$$

Proof. Thanks to Theorem 23, we know that the right side counts the number of ways to tile a $1 \times n$ board with squares and dominoes. The left side must count this as well, but conditioning on something. The obvious choice is that it conditions either on the number of squares or the number of dominoes used in the tiling. However, for some values of k it is not possible to tile a $1 \times n$ board using exactly k squares: Since each domino covers two spaces, k and n must have the same parity. Thus the "conditioning on the number of squares" interpretation does not fit. On the other hand, if a tiling uses k dominoes, then it uses $n - 2k$ squares, and thus it uses exactly $n - k$ total tiles (squares plus dominoes). This means that $\binom{n-k}{k}$ is the number of ways to position k dominoes among $n - k$ total tiles when tiling a $1 \times n$ board. Once the dominoes are in position, every other position must be occupied by a square. Summing over all values of k, this means that the left side also counts the number of ways to tile a $1 \times n$ board with squares and dominoes. $\square$

We can also prove the formula for the sum of the first n Fibonacci numbers using a combinatorial argument.

Identity 221.

$$\sum_{k=0}^{n} F_k = F_{n+2} - 1.$$

Proof. By Theorem 23, F_{n+2} is the number of ways to tile a $1 \times (n+1)$ board with squares and dominoes. With one way to tile a board using only squares, $F_{n+2} - 1$ is the number of ways to tile a $1 \times (n+1)$ board using at least one domino. (If n is even then tiling a $1 \times (n+1)$ board with only dominoes is not possible.)

For the left side, condition on the position of the last domino. If the last domino covers spaces k and $k+1$, then there is a $1 \times (k-1)$ board before it that was tiled with squares and dominoes, and spaces $k+2$ through $n+1$ are tiled with squares. There are, by Theorem 23, F_k ways to tile a $1 \times (k-1)$ board with squares and dominoes, and there is one way to place squares on spaces $k+2$ through $n+1$. The allowed values of k run from 1 to n, and since $F_0 = 0$, we have that $\sum_{k=0}^{n} F_k$ also counts the number of ways to tile a $1 \times (n+1)$ board using at least one domino. $\square$

For our third combinatorial proof of a Fibonacci identity, let's prove the formula for the binomial transform of the Fibonacci numbers that we saw in Chapter 7.

Identity 183.

$$\sum_{k=0}^{n} \binom{n}{k} F_k = F_{2n}.$$

Proof. When $n = 0$ the identity claims that $0 = 0$. Otherwise, by Theorem 23, F_{2n} is the number of ways to tile a $1 \times (2n - 1)$ board with squares and dominoes. For such a board, there must be at least n tiles and at least one square. To understand the left side, condition on the number of squares among the first n tiles. If there are k such squares, there are $\binom{n}{k}$ ways to distribute them among the first n tiles. The remaining $n - k$ tiles must be dominoes, covering a total of $2(n-k)+k = 2n-k$ tiles, meaning there are $k-1$ spaces left to be tiled. There are F_k ways to tile those spaces with squares and dominoes. Thus there are a total of $\binom{n}{k} F_k$ ways to tile a $1 \times (2n - 1)$ board when using

k squares among the first n tiles. Summing over all values of k, we obtain the left side. $\qquad\qquad\qquad\qquad\qquad\qquad\qquad\qquad\qquad\qquad\qquad\qquad\qquad$ □

8.2   Stirling Numbers

One important binomial sum that we have not yet discussed is the binomial transform of the power sum, $\sum_{k=0}^n \binom{n}{k} k^m$. The closest we have come to this sum is the binomial transform of the falling power sum, Identity 36:

$$\sum_{k=0}^n \binom{n}{k} k^{\underline{m}} = n^{\underline{m}} 2^{n-m}.$$

The falling power $k^{\underline{m}}$ is a polynomial in powers of k. For example, $k^{\underline{3}} = k(k-1)(k-2) = k^3 - 3k^2 + 2k$. This process can, not surprisingly, be reversed: k^m is a polynomial in falling powers of k. For example, $k^3 = k^{\underline{3}} + 3k^{\underline{2}} + k^{\underline{1}}$, as can be seen by the fact that $k(k-1)(k-2) + 3k(k-1) + k = k^3 - 3k^2 + 2k + 3k^2 - 3k + k = k^3$. If we could find the coefficients used to convert k^m to falling powers of k, we could use Identity 36 to obtain a formula for $\sum_{k=0}^n \binom{n}{k} k^m$. Those coefficients are the topic of this section.

Let's define $\left\{ {n \atop k} \right\}$, for $n, k \geq 0$, to be the coefficient of $x^{\underline{k}}$ when converting x^n to falling powers. More formally, $\left\{ {n \atop k} \right\}$ is defined by the following.

Identity 222.

$$x^n = \sum_{k=0}^n \left\{ {n \atop k} \right\} x^{\underline{k}}.$$

The numbers $\left\{ {n \atop k} \right\}$ are known as *Stirling numbers of the second kind*, or *Stirling subset numbers*. (As with the Fibonacci numbers, we will eventually prove equivalent identities for the Stirling numbers of the second kind, so that any of them could be taken to be their definition. Hence our use of "identity" in Identity 222.) A table of Stirling numbers for small values of n and k is shown in Figure 8.2.

With the $\left\{ {n \atop k} \right\}$ numbers, we are now ready to prove the expression for the binomial transform of the sequence of mth powers.

Identity 223.

$$\sum_{k=0}^n \binom{n}{k} k^m = \sum_{j=0}^m \left\{ {m \atop j} \right\} n^{\underline{j}} 2^{n-j} = \sum_{j=0}^m \left\{ {m \atop j} \right\} \binom{n}{j} j! \, 2^{n-j}.$$

Proof. By Identities 222 and 36, we have

$$\sum_{k=0}^n \binom{n}{k} k^m = \sum_{k=0}^n \binom{n}{k} \sum_{j=0}^m \left\{ {m \atop j} \right\} k^{\underline{j}} = \sum_{j=0}^m \left\{ {m \atop j} \right\} \sum_{k=0}^n \binom{n}{k} k^{\underline{j}}$$

n	$\left\{ {n \atop 0} \right\}$	$\left\{ {n \atop 1} \right\}$	$\left\{ {n \atop 2} \right\}$	$\left\{ {n \atop 3} \right\}$	$\left\{ {n \atop 4} \right\}$	$\left\{ {n \atop 5} \right\}$	$\left\{ {n \atop 6} \right\}$	$\left\{ {n \atop 7} \right\}$
0	1							
1	0	1						
2	0	1	1					
3	0	1	3	1				
4	0	1	7	6	1			
5	0	1	15	25	10	1		
6	0	1	31	90	65	15	1	
7	0	1	63	301	350	140	21	1

FIGURE 8.2
Rows 0 through 7 of triangle of Stirling numbers of the second kind

$$= \sum_{j=0}^{m} \left\{ {m \atop j} \right\} n^{\underline{j}} 2^{n-j} = \sum_{j=0}^{m} \left\{ {m \atop j} \right\} \binom{n}{j} j! \, 2^{n-j},$$

where the last equality follows from the generalized binomial coefficient formula $\binom{n}{j} = n^{\underline{j}}/j!$ (Identity 17). □

Identity 223 is not very satisfying at this point because we do not know anything about the Stirling numbers of the second kind beyond Identity 222 (which, of course, merely defines them to be what we need them to be) and Figure 8.2. Let's look now at some of their other properties so that may get a sense of what these numbers are all about. The first is a combinatorial interpretation which gives rise to the alternate name *Stirling subset numbers*.

Theorem 24. *The Stirling numbers of the second kind* $\left\{ {n \atop k} \right\}$ *count the number of ways that a set of size n can be partitioned into k nonempty subsets.*

Before we prove Theorem 24, let's look at an example. Suppose we partition the set $\{1, 2, 3, 4\}$ into two nonempty subsets. Here are the possibilities:

Scottish mathematician **James Stirling** (1692–1770) spent a few years in Venice as a young man. While there, he learned a trade secret of some Venetian glassmakers and had to flee to Britain to escape assassination. His most important mathematical work was *Methodus Differentialis*, which contains his famous approximation for factorials: $n! \approx \sqrt{2\pi n} \left(\frac{n}{e} \right)^{n}$. He also worked for many years as a mining engineer. The city of Glasgow once awarded him a silver teakettle in appreciation for some river surveys he made to help the city construct a series of locks.

$\{1\} \cup \{2, 3, 4\}$ $\quad$ $\{2\} \cup \{1, 3, 4\}$ $\quad$ $\{3\} \cup \{1, 2, 4\}$ $\quad$ $\{4\} \cup \{1, 2, 3\}$

$$\{1,2\} \cup \{3,4\} \qquad \{1,3\} \cup \{2,4\} \qquad \{1,4\} \cup \{2,3\}$$

Thus $\left\{^4_2\right\} = 7$, which we also see in Figure 8.2.

Proof. Let $S(n, k)$ be the number of ways that a set of size n can be partitioned into k nonempty subsets. We will show that $S(n, k) = \left\{^n_k\right\}$.

Suppose we have a set of n balls numbered 1 through n. How many ways are there to place them in jars numbered 1 through x? Well, there are x choices for ball 1, x choices for ball 2, and so forth, for a total of x^n ways to distribute the balls in the jars.

A more complicated method of assigning balls to jars would be to place the balls in groups first, and then distribute each group to a different jar. If we use k groups, there are $S(n, k)$ ways to partition the n balls into k groups. The groups have a natural ordering to them, in the sense that the group with ball 1 in it can be considered group 1, the group with the next smallest-numbered ball not in group 1 can be considered group 2, and so forth, until we have k groups. When assigning group 1, there are x choices for a jar. Since group 2 cannot be assigned to the same jar as group 1, there are $x - 1$ choices for the jar to which group 2 is assigned. This process continues until the $x - k + 1$ choices for group k. Multiplying all of these numbers together, we have that there are $S(n, k)x^{\underline{k}}$ ways to distribute the balls into k groups and then assign those k groups to different jars. Summing over all values of k, there are $\sum_{k=0}^{n} S(n, k)x^{\underline{k}}$ ways to distribute n balls into x jars.

Equating our two expressions, we have, for x a positive integer,

$$x^n = \sum_{k=0}^{n} S(n, k)x^{\underline{k}}.$$

However, both sides of this equation are nth degree polynomials in x. A polynomial is specified by $n + 1$ values, and we have proved this equation for an infinite number of values. Thus this formula must be true for all values of x, not just positive integer ones. This means that $S(n, k)$ is the coefficient of $x^{\underline{k}}$ when expanding x^n in falling powers of x. By Identity 222, though, $\left\{^n_k\right\}$ is the coefficient of $x^{\underline{k}}$ when expanding x^n in falling powers of x. Therefore, $S(n, k) = \left\{^n_k\right\}$. Thus $\left\{^n_k\right\}$ must count the number of ways that a set of size n can be partitioned into k nonempty subsets.   □

The combinatorial interpretation of the Stirling numbers of the second kind leads directly to an explicit formula for these numbers in terms of the binomial coefficients. Like the Fibonacci numbers, then, any identity featuring the Stirling numbers can be converted to one containing the binomial coefficients.

Back in Chapter 3 we proved that the number of onto functions from the set $\{1, 2, \ldots, n\}$ to itself is counted in two different ways by Identity 67:

$$\sum_{k=0}^{n} \binom{n}{k}(n-k)^n(-1)^k = n!.$$

Variations of this argument, for different-sized input and output sets, can be used to prove the identities in Exercises 37 and 38 in Chapter 3. In general, the problem of finding the number of onto functions from a set of size n to a set of size k is closely related to the problem of finding the number of ways to partition a set of size n into k nonempty subsets. In both cases, the n objects have to be assigned to k other objects in such a way that all of the k objects are used at least once. The difference between the two problems is that in the onto-functions problem, the elements in the set of size k are *distinguishable*; that is, there is an internal labeling or ordering to the k elements. In the nonempty-subsets problem, though, the k sets are *indistinguishable*; that is, there is no internal labeling or ordering to the k sets. For example, the assignment $\{n-1, n\}$ and $\{n-2\}$ is the same assignment as $\{n-2\}$ and $\{n-1, n\}$ in the nonempty-subsets problem, whereas the assignment $f(n-1) = f(n) = 1$ and $f(n-2) = 2$ is a different assignment than is $f(n-2) = 1$ and $f(n-1) = f(n) = 2$ in the onto-functions problem. This difference is only a factor of $k!$, though, as one way to create an onto function from a set of size n to a set of size k is to partition the n objects into k nonempty subsets, in $\left\{ {n \atop k} \right\}$ ways (thus grouping the n objects according to which have the same function value), and then assigning the k subsets to values from 1 to k, in $k!$ ways (thus deciding which groups go with which function values). These ideas, together with the inclusion-exclusion argument for Identity 67 that we used in Chapter 3, allow us to prove the following representation of the Stirling numbers of the second kind in terms of binomial coefficients.

Identity 224.

$$\left\{ {n \atop m} \right\} = \frac{1}{m!} \sum_{k=0}^{m} \binom{m}{k} (m-k)^n (-1)^k.$$

Proof. For $1 \le j \le m$, let A_j be the set of functions f from the set $\{1, 2, \ldots, n\}$ to the set $\{1, 2, \ldots, m\}$ for which there exists i such that $f(i) = j$. Then the intersection of the A_j's is the set of onto functions from $\{1, 2, \ldots, n\}$ to $\{1, 2, \ldots, m\}$. As we argued above, the number of such functions is $m! \left\{ {n \atop m} \right\}$, as there are $\left\{ {n \atop m} \right\}$ ways to partition the n elements in the domain into m nonempty groups, and then there are $m!$ ways to assign these m groups to elements in the codomain.

The set $\bar{A}_j$ is the set of functions for which no element in $\{1, 2, \ldots, n\}$ maps to j. Thus $|\bar{A}_j| = (m-1)^n$, since any of the n elements in the domain can be mapped to any element but j in the codomain. Similarly, there are $(m-k)^n$ functions in the intersection of any k of the $\bar{A}_j$'s because any of the n elements in the domain can be mapped to any of the $m-k$ allowed elements in the codomain. In addition, there are m^n total functions with no restrictions. By the simplified version of inclusion-exclusion given in Corollary 1, then, the number of onto functions from $\{1, 2, \ldots, n\}$ to $\{1, 2, \ldots, m\}$ is also

$$\sum_{k=0}^{m} \binom{m}{k} (m-k)^n (-1)^k.$$

Therefore,

$$m!\left\{\begin{matrix} n \\ m \end{matrix}\right\} = \sum_{k=0}^{m} \binom{m}{k}(m-k)^n(-1)^k.$$

Dividing by $m!$ completes the proof.   □

Our combinatorial interpretation of the Stirling numbers of the second kind can also be used to prove that they satisfy a two-variable recurrence relation of the form of Equation (7.2), with $f_1(n,k) = k$ and $f_2(n,k) = 1$.

Identity 225.

$$\left\{\begin{matrix} n \\ k \end{matrix}\right\} = k\left\{\begin{matrix} n-1 \\ k \end{matrix}\right\} + \left\{\begin{matrix} n-1 \\ k-1 \end{matrix}\right\} + [n=k=0], \;\; for \; n,k \geq 0,$$

where $\left\{\begin{smallmatrix} n \\ k \end{smallmatrix}\right\}$ is understood to be 0 for $n < 0$ or $k < 0$.

Proof. Thanks to Theorem 24 we know that $\left\{\begin{smallmatrix} n \\ k \end{smallmatrix}\right\}$ counts the number of ways to partition a set of size n into k nonempty subsets. When $n = 0$, we have that $\left\{\begin{smallmatrix} 0 \\ 0 \end{smallmatrix}\right\} = 1$ because there is one way to partition an empty set into zero nonempty subsets: Use the empty partition. Otherwise, $\left\{\begin{smallmatrix} 0 \\ k \end{smallmatrix}\right\} = 0$.

For $n \geq 1$, let's count the number of ways to partition the elements $\{1, 2, \ldots, n\}$ into k nonempty subsets by conditioning on where element n falls. If n is in a subset by itself, then there are $\left\{\begin{smallmatrix} n-1 \\ k-1 \end{smallmatrix}\right\}$ ways to partition the remaining $n-1$ elements into $k-1$ subsets.

Otherwise, n is in a subset with other elements. The remaining $n-1$ elements must be partitioned into k subsets, in $\left\{\begin{smallmatrix} n-1 \\ k \end{smallmatrix}\right\}$ ways. Since element n can then be in any of the k subsets, we have a total of $k\left\{\begin{smallmatrix} n-1 \\ k \end{smallmatrix}\right\}$ ways to partition n elements into k nonempty subsets if n is not in a subset by itself.

All total, then, for $n \geq 1$ there are $\left\{\begin{smallmatrix} n-1 \\ k-1 \end{smallmatrix}\right\} + k\left\{\begin{smallmatrix} n-1 \\ k \end{smallmatrix}\right\}$ ways to form a partition of n elements into k nonempty subsets by conditioning on the location of element n. By Theorem 24, this is also $\left\{\begin{smallmatrix} n \\ k \end{smallmatrix}\right\}$.   □

We have now seen the major properties of the Stirling numbers of the second kind. Let's look at two more things we can do with them before we move on to the Stirling numbers of the first kind.

At the beginning of this section we defined Stirling numbers of the second kind as the coefficients required when converting from an ordinary power function like x^n to falling factorial powers. Stirling numbers of the second kind can also be used to convert from ordinary powers to rising factorial powers $x^{\overline{n}} = x(x+1)\cdots(x+n-1)$.

Identity 226.

$$x^n = \sum_{k=0}^{n} \left\{\begin{matrix} n \\ k \end{matrix}\right\} x^{\overline{k}}(-1)^{n-k}.$$

Proof. Thanks to Identity 32, $x^{\underline{k}} = (-1)^k(-x)^{\overline{k}}$. Thus we have, by Identity 222,

$$x^n = \sum_{k=0}^{n} \left\{ {n \atop k} \right\} x^{\underline{k}} = \sum_{k=0}^{n} \left\{ {n \atop k} \right\} (-x)^{\overline{k}}(-1)^k.$$

Replacing x with $-x$, this becomes

$$\sum_{k=0}^{n} \left\{ {n \atop k} \right\} (x)^{\overline{k}}(-1)^k = (-x)^n = x^n(-1)^n.$$

Multiplying both sides by $(-1)^n$ completes the proof. $\qquad\square$

For our final example with Stirling numbers of the second kind, let's look at the power sum $\sum_{k=0}^{n} k^m$. For $m = 0$ the sum is of course $n+1$. For $m = 1$ the sum is $\binom{n+1}{2}$, as Identity 59 states. There are formulas for other values of m (see Exercise 20 in Chapter 3, for example), but we have not yet seen an expression for the general m case. However, since the Stirling numbers of the second kind can be used to convert ordinary powers to falling powers, and since there is a simple formula for the sum of consecutive falling powers (Identity 204), we now have the tools to find such an expression. This expression involves not only the Stirling numbers but also the binomial coefficients.

Identity 227.

$$\sum_{k=0}^{n} k^m = \sum_{j=0}^{m} \binom{n+1}{j+1} \left\{ {m \atop j} \right\} j!.$$

We give two proofs. The more obvious one uses Identity 222 to convert ordinary powers to falling powers. The second is a combinatorial proof.

Proof 1. By Identities 222, 17, and 58,

$$\sum_{k=0}^{n} k^m = \sum_{k=0}^{n} \sum_{j=0}^{m} \left\{ {m \atop j} \right\} k^{\underline{j}} = \sum_{j=0}^{m} \left\{ {m \atop j} \right\} j! \sum_{k=0}^{n} \binom{k}{j} = \sum_{j=0}^{m} \left\{ {m \atop j} \right\} j! \binom{n+1}{j+1}.$$

(We could also have summed the falling powers via Identity 204 and then converted to binomial coefficients.) $\qquad\square$

Proof 2. Both sides count the number of functions f from the set $\{0, 1, 2, \ldots, m\}$ to the set $\{0, 1, 2, \ldots, n\}$ such that $f(0) > f(j)$ for any $j \in \{1, 2, \ldots, m\}$.

For the left side, condition on the value of $f(0)$. If $f(0) = k$, then there are k choices for $f(1)$ (i.e., any of $0, 1, \ldots, k-1$), k choices for $f(2)$, etc., for a total of k^m functions. Then sum over all possible values of k.

For the right side, condition on the number of elements in the image of f. If the image of f has $j+1$ elements, there are $\binom{n+1}{j+1}$ ways to choose exactly which elements will comprise the image. The largest of these must be $f(0)$, and the remaining m elements in the domain must be mapped to the other j elements in the image. As we argued in the proof of Identity 224, there are $\left\{{m \atop j}\right\}j!$ onto functions from an m-set to a j-set. Then sum over all possible values of j.   □

Since there are Stirling numbers of the second kind, there must be Stirling numbers of the first kind. These are not connected quite as closely to the binomial coefficients as the Stirling numbers of the second kind, but we will spend a little time considering their most important properties. Not surprisingly, they are used to express factorial powers in terms of ordinary powers.

Identity 228. *The Stirling numbers of the first kind,* $\left[{n \atop k}\right]$, *are the coefficients required to convert from rising factorial powers to ordinary powers via*

$$x^{\overline{n}} = \sum_{k=0}^{n} \left[{n \atop k}\right] x^k.$$

(Again, while this is technically a definition, as with the Fibonacci numbers and the Stirling numbers of the second kind, we eventually derive identities that are equivalent to this definition. Thus we can call Identity 228 an "identity.")

Stirling numbers of the first kind can also be used to convert from falling factorial powers to ordinary powers; see Identity 259 in Exercise 21. In fact, some authors *define* the Stirling numbers of the first kind this way. Doing so introduces a factor of $(-1)^{n-k}$ to the values of the numbers.

The first several Stirling numbers of the first kind are given in Figure 8.3.

n	$\left[{n \atop 0}\right]$	$\left[{n \atop 1}\right]$	$\left[{n \atop 2}\right]$	$\left[{n \atop 3}\right]$	$\left[{n \atop 4}\right]$	$\left[{n \atop 5}\right]$	$\left[{n \atop 6}\right]$	$\left[{n \atop 7}\right]$
0	1							
1	0	1						
2	0	1	1					
3	0	2	3	1				
4	0	6	11	6	1			
5	0	24	50	35	10	1		
6	0	120	274	225	85	15	1	
7	0	720	1764	1624	735	175	21	1

FIGURE 8.3
Rows 0 through 7 of triangle of Stirling numbers of the first kind

As with the Stirling numbers of the second kind, a combinatorial interpretation of the Stirling numbers of the first kind allows us to prove many identities involving them.

Theorem 25. *The Stirling numbers of the first kind* $\left[{n \atop k}\right]$ *count the number of ways that a set of size n can be partitioned into k nonempty cycles.*

First of all, a *cycle* is a permutation that has no first or last element: What would otherwise be the "beginning" of the permutation is attached to its "end" to create a cycle. A cycle with n elements can be thought of as an arrangement of n people around a circular table or as n beads on a necklace. For example, the cycle $[1, 2, 3, 4]$ expresses the idea that 2 comes after 1, 3 comes after 2, 4 comes after 3, and 1 comes after 4. It is thus equivalent to the cycles $[2, 3, 4, 1]$, $[3, 4, 1, 2]$, and $[4, 1, 2, 3]$. As a further example, here are the ways in which one can partition $\{1, 2, 3, 4\}$ into two nonempty cycles:

$$[1]\,[2, 3, 4] \qquad [1]\,[2, 4, 3] \qquad [2]\,[1, 3, 4] \qquad [2]\,[1, 4, 3]$$
$$[3]\,[1, 2, 4] \qquad [3]\,[1, 4, 2] \qquad [4]\,[1, 2, 3] \qquad [4]\,[1, 3, 2]$$
$$[1, 2]\,[3, 4] \qquad [1, 3]\,[2, 4] \qquad [1, 4]\,[2, 3]$$

Thus $\left[{4 \atop 2}\right] = 11$, as we see in Figure 8.3. Note, for instance, that there is only one possible cycle containing one element and only one possible cycle consisting of two elements. Standard cycle notation also frequently places the smallest-numbered element first, as we have done.

Stirling numbers of the first kind are sometimes called *Stirling cycle numbers* because of Theorem 25. Let's prove this theorem now.

Proof. Let $Y(n, k)$ be the number of ways that a set of size n can be partitioned into k nonempty cycles.

Suppose we want to partition the numbers from 1 to n into nonempty cycles, where each cycle can be colored one of x colors, independently of the others. How many ways are there to do this?

Well, one approach is to condition on the number of cycles. There are $Y(n, k)$ ways to partition $\{1, 2, \ldots, n\}$ into k cycles. For each cycle, there are x ways to color it, for a total of x^k ways to color all k cycles. Summing up over all possible values of k, we have that the total number of ways to form colored cycles from n elements is $\sum_{k=0}^{n} Y(n, k) x^k$.

Another approach is to build the colored cycles by placing the numbers 1 through n in order into cycles. Starting with the number 1, it must "begin" a cycle (under standard cycle notation that has the smallest number in the cycle first). There are x choices for the color of the cycle that includes 1. For the number 2, it can start a new colored cycle, in x ways, or it can go after the number 1 in 1's cycle, in one way, for a total of $x + 1$ choices for the number 2. The number 3 can start a new colored cycle, in x ways, or it can go after the number 1 or the number 2 in whatever cycles they happen to be, in two ways. This gives a total of $x + 2$ choices for the number 3. Continuing this process, the number k can start a new colored cycle in x ways or go after any of the $k - 1$ numbers already placed into cycles, for a total $x + k - 1$ choices. After the number n has been placed, we will have seen a total of $x(x + 1) \cdots (x + n - 1) = x^{\overline{n}}$ choices for placing all n numbers into colored cycles.

All together, then, when x is a positive integer, we have

$$x^{\bar{n}} = \sum_{k=0}^{n} Y(n,k)x^k.$$

As in the proof of Theorem 24, though, both sides of this equation are nth degree polynomials in x. We have proved this equation for infinitely many values of x, and so this equation must be true for all real x. Thus $Y(n,k)$ is the coefficient of x^k when expressing $x^{\bar{n}}$ in powers of x. By Identity 228, then, $Y(n,k) = \begin{bmatrix} n \\ k \end{bmatrix}$, and so $\begin{bmatrix} n \\ k \end{bmatrix}$ is the number of ways to place n objects into k nonempty cycles.   □

Like the Stirling numbers of the second kind, the Stirling numbers of the first kind satisfy a two-variable recurrence relation of the form of Equation (7.2). In this case the coefficient functions are $f_1(n,k) = n - 1$ and $f_2(n,k) = 1$.

Identity 229.

$$\begin{bmatrix} n \\ k \end{bmatrix} = (n-1)\begin{bmatrix} n-1 \\ k \end{bmatrix} + \begin{bmatrix} n-1 \\ k-1 \end{bmatrix} + [n = k = 0], \;\; for \; n, k \geq 0,$$

where $\begin{bmatrix} n \\ k \end{bmatrix}$ *is understood to be* 0 *for* $n < 0$ *or* $k < 0$.

Proof. According to Theorem 25, $\begin{bmatrix} n \\ k \end{bmatrix}$ counts the number of ways to partition n elements into k nonempty cycles. There is one way to place 0 elements into zero nonempty cycles: Use the empty cycle. Thus $\begin{bmatrix} 0 \\ 0 \end{bmatrix} = 1$. Otherwise, $\begin{bmatrix} 0 \\ k \end{bmatrix} = 0$.

For $n \geq 1$, let's count $\begin{bmatrix} n \\ k \end{bmatrix}$ by conditioning on the location of element n in the cycles. If n is in a cycle by itself, then there are $\begin{bmatrix} n-1 \\ k-1 \end{bmatrix}$ ways to form the remaining $k - 1$ cycles from the remaining $n - 1$ elements.

Suppose now that n is in a cycle with other elements. There are $\begin{bmatrix} n-1 \\ k \end{bmatrix}$ ways to form k cycles from $n - 1$ elements, and then the element n can be placed after any of the $n-1$ already-placed elements, for a total of $(n-1)\begin{bmatrix} n-1 \\ k-1 \end{bmatrix}$ ways to form k cycles from n elements if n is not in a cycle by itself.

All together, then, for $n \geq 1$ there are $\begin{bmatrix} n-1 \\ k-1 \end{bmatrix} + (n-1)\begin{bmatrix} n-1 \\ k \end{bmatrix}$ ways to create k cycles from n elements. By Theorem 25, this is also $\begin{bmatrix} n \\ k \end{bmatrix}$.   □

The row sum of the Stirling numbers of the first kind has a very nice formula:

Identity 230.

$$\sum_{k=0}^{n} \begin{bmatrix} n \\ k \end{bmatrix} = n!.$$

We will give three proofs, the last two of which are combinatorial.

Proof 1. Substitute 1 for x in Identity 228. Since $1^{\bar{n}} = 1 \cdot 2 \cdots n = n!$, the result is proved.   □

Proof 2. One of the basic properties of permutations is that each one can be expressed uniquely as the decomposition of disjoint cycles [22, p. 3]. Since there are $n!$ total permutations on n elements, and $\sum_{k=0}^{n} \left[{n \atop k}\right]$ gives the number of ways to partition n elements into nonempty cycles by conditioning on the number of cycles, the result follows. $\square$

Proof 3. Since Proof 2 relies on citing a property of permutations, let's see if we can give a combinatorial proof of Identity 230 from something closer to first principles. One way to do so is by modifying the proof of Theorem 25. We have just argued that $\sum_{k=0}^{n} \left[{n \atop k}\right]$ is the number of ways to partition n elements into nonempty cycles by conditioning on the number of cycles.

We can also partition n elements into nonempty cycles by considering the numbers 1 through n in order. For the number 1, we can create a new cycle in one way. For the number 2, we can either create a new cycle with it or place it after the number 1, for two choices. The number 3 can start a new cycle or go after 1 or after 2, for a total of three choices. This process continues, so that the number k can start a new cycle or go after any of the $k-1$ numbers previously placed, for a total of k choices. For the numbers 1 through n, then, we have a total of $1 \cdot 2 \cdots n = n!$ ways to partition n elements into nonempty cycles. $\square$

See Exercise 8 for a fourth proof of Identity 230, one that utilizes Identity 201.

The alternating row sum of the Stirling numbers of the first kind also has a nice formula:

Identity 231.

$$\sum_{k=0}^{n} \left[{n \atop k}\right] (-1)^k = [n = 0] - [n = 1].$$

The quick proof of Identity 231 uses Identity 228. However, we can also give a combinatorial proof by using the sign-reversing involution technique introduced in Chapter 3.

Proof 1. Let $x = -1$ in Identity 228. If $n \geq 2$, then $(-1)^{\overline{n}} = (-1) \cdot 0 \cdot 1 \cdots n = 0$. We also have $(-1)^{\overline{1}} = -1$ and $(-1)^{\overline{0}} = 1$. $\square$

Proof 2. There is one permutation on zero elements, the empty permutation, which has even parity. There is also one permutation on one element, the identity permutation, which has odd parity. Let $n \geq 2$, and let S_n be the set of permutations on n elements. Now, suppose π is a permutation in S_n where 1 and 2 appear in the same cycle. Such a cycle looks like the following, where we follow the convention of having the smallest number in the cycle appear first:

$$[1, x_1, \ldots, x_j, 2, y_1, \ldots, y_k].$$

In this case, define $\phi(\pi)$ to be the same permutation as π except that instead of having 1 and 2 in the same cycle we have

$$[1, x_1, \ldots, x_j][2, y_1, \ldots, y_k],$$

so that we split the cycle with 1 and 2 into two cycles, with the split occurring just before the 2.

The alternative is that π is a permutation in S_n in which 1 and 2 appear in different cycles. Then these two cycles look like the following:

$$[1, x_1, \ldots, x_j][2, y_1, \ldots, y_k].$$

In this case, define $\phi(\pi)$ to be the same permutation as π except that instead of having 1 and 2 in different cycles we have

$$[1, x_1, \ldots, x_j, 2, y_1, \ldots, y_k],$$

so that we concatenate the cycle "beginning" with 2 to the end of the cycle "beginning" with 1.

Since ϕ maps permutations in S_n to permutations in S_n, ϕ is an involution. Also, ϕ changes the number of cycles in π by one, making ϕ sign-reversing. For $n \geq 2$, the elements 1 and 2 always appear in the permutation, and so ϕ is defined for all permutations in S_n. Since we have a sign-reversing involution on S_n that maps permutations with an even number of cycles to permutations with an odd number of cycles and vice versa, the alternating row sum of the Stirling numbers of the first kind must be zero for $n \geq 2$. $\qquad\square$

Our last three identities involving Stirling numbers feature both kinds. The first two are a pair of identities that express the idea that the two kinds of Stirling numbers are inverses, of sorts, of each other. The first identity is proved by converting from ordinary powers to factorial powers and back, and the second is proved using the reverse method.

Identity 232.

$$\sum_{k=0}^{n} \begin{Bmatrix} n \\ k \end{Bmatrix} \begin{bmatrix} k \\ m \end{bmatrix} (-1)^{k+m} = [n = m].$$

Proof. We have, by Identities 222 and 259,

$$x^n = \sum_{k=0}^{n} \begin{Bmatrix} n \\ k \end{Bmatrix} x^{\underline{k}} = \sum_{k=0}^{n} \begin{Bmatrix} n \\ k \end{Bmatrix} \sum_{m=0}^{k} \begin{bmatrix} k \\ m \end{bmatrix} x^m (-1)^{k-m}$$

$$= \sum_{m=0}^{n} x^m \sum_{k=m}^{n} \begin{Bmatrix} n \\ k \end{Bmatrix} \begin{bmatrix} k \\ m \end{bmatrix} (-1)^{k-m} = \sum_{m=0}^{n} x^m \sum_{k=0}^{n} \begin{Bmatrix} n \\ k \end{Bmatrix} \begin{bmatrix} k \\ m \end{bmatrix} (-1)^{k+m}.$$

This equation consists of two polynomials in x. Thus their coefficients must be equal. This means that

$$\sum_{k=0}^{n} \begin{Bmatrix} n \\ k \end{Bmatrix} \begin{bmatrix} k \\ m \end{bmatrix} (-1)^{k+m}$$

equals 1 precisely when $n = m$ and is 0 otherwise. $\square$

Identity 233.

$$\sum_{k=0}^{n} \begin{bmatrix} n \\ k \end{bmatrix} \begin{Bmatrix} k \\ m \end{Bmatrix} (-1)^{k+m} = [n = m].$$

Proof. See Exercise 23. The argument is similar to that for Identity 232. $\square$

Our final identity involving the Stirling numbers shows that they satisfy an inversion relationship like Theorem 1, the inversion formula for the binomial coefficients that we proved in Chapter 2:

$$f(n) = \sum_{k=0}^{n} \binom{n}{k} g(k)(-1)^k \iff g(n) = \sum_{k=0}^{n} \binom{n}{k} f(k)(-1)^k.$$

Theorem 26. *For functions f and g,*

$$f(n) = \sum_{k=0}^{n} \begin{Bmatrix} n \\ k \end{Bmatrix} g(k)(-1)^k \iff g(n) = \sum_{k=0}^{n} \begin{bmatrix} n \\ k \end{bmatrix} f(k)(-1)^k.$$

Proof. Suppose

$$f(n) = \sum_{k=0}^{n} \begin{Bmatrix} n \\ k \end{Bmatrix} g(k)(-1)^k.$$

Then we have

$$\sum_{k=0}^{n} \begin{bmatrix} n \\ k \end{bmatrix} f(k)(-1)^k = \sum_{k=0}^{n} \begin{bmatrix} n \\ k \end{bmatrix} \sum_{m=0}^{k} \begin{Bmatrix} k \\ m \end{Bmatrix} g(m)(-1)^{k+m}$$

$$= \sum_{m=0}^{n} g(m) \sum_{k=m}^{n} \begin{bmatrix} n \\ k \end{bmatrix} \begin{Bmatrix} k \\ m \end{Bmatrix} (-1)^{k+m}$$

$$= \sum_{m=0}^{n} g(m) \sum_{k=0}^{n} \begin{bmatrix} n \\ k \end{bmatrix} \begin{Bmatrix} k \\ m \end{Bmatrix} (-1)^{k+m} = g(n),$$

by Identity 233. The proof in the other direction uses Identity 232 and is similar. $\square$

8.3 Bell Numbers

Identity 230 says that the sum of the numbers in row n of the triangle of Stirling numbers of the first kind is $n!$. What about the row sums for the Stirling numbers of the second kind? These (more interestingly) do not have a simple formula. They are important enough that they have their own name, though: the *Bell numbers*. Then nth Bell number is denoted ϖ_n. (Another common notation is B_n, but we use that for the nth Bernoulli number.)

Identity 234.

$$\varpi_n = \sum_{k=0}^{n} \left\{ {n \atop k} \right\}.$$

Given the combinatorial interpretation of $\left\{ {n \atop k} \right\}$ as the number of ways to partition n elements into k nonempty subsets (Identity 24), we immediately obtain the following combinatorial interpretation of the nth Bell number.

Theorem 27. *The value of ϖ_n is the number of ways to partition a set of n elements.*

Since Identity 234 and Theorem 27 can be proved from the other, either can be taken as the definition of the Bell numbers. The first several Bell numbers are given in Table 8.3.

From the point of view of the binomial coefficients, the Bell numbers are interesting mostly because the binomial transform of $(\varpi_n)_{n=0}^{\infty}$ is the shifted sequence $(\varpi_{n+1})_{n=0}^{\infty}$.

Identity 235.

$$\sum_{k=0}^{n} \binom{n}{k} \varpi_k = \varpi_{n+1}.$$

Eric Temple Bell was born in Scotland in 1883 but lived most of his life in the United States. He held positions at the University of Washington and Cal-Tech. The numbers that bear his name were not first studied by him, but he helped popularize them in a pair of papers from the 1930s, in which he calls them "exponential numbers." Bell wrote several popular works on the history of mathematics, of which his lively *Men of Mathematics* is the most well-known. Bell also published science fiction novels under the pseudonym "John Taine" and served a term as president of the Mathematical Association of America.

Proof. Both sides count the number of partitions that can be formed from $n + 1$ elements. For the left side, condition on the number of elements not in

TABLE 8.3

First several Bell numbers

n	0	1	2	3	4	5	6	7	8	9	10
ϖ_n	1	1	2	5	15	52	203	877	4140	21147	115975

the set containing element $n + 1$. If there are k such elements, there are $\binom{n}{k}$ ways to choose them. Then there are ϖ_k ways to form a partition from those k elements. The remaining $n - k$ elements go in a set with element $n + 1$ in one way. Summing over all values of k gives the left side. □

Many identities involving the Bell numbers can be deduced from Identity 235 and the techniques in Chapter 7 on finite differences. For example, we have the following expression for the sum of the Bell numbers.

Identity 236.

$$\sum_{k=1}^{n} \binom{n}{k} \varpi_{k-1} = \sum_{k=1}^{n} \varpi_k.$$

Proof. Identity 186 says that if

$$b_n = \sum_{k=0}^{n} a_k$$

then

$$\sum_{k=0}^{n-1} b_k = \sum_{k=1}^{n} \binom{n}{k} a_{k-1}.$$

With $a_k = \varpi_k$, Identity 235 says that $b_n = \varpi_{n+1}$. The result follows from Identity 186 and reindexing. □

See the exercises for other examples, including the exponential generating function for the Bell numbers.

8.4 Lah Numbers

The two kinds of Stirling numbers are the coefficients used when converting from ordinary powers to factorial powers or vice versa. What about converting between the two kinds of factorial powers? The numbers needed to do this are called the *Lah numbers* and are denoted $L(n, k)$. They have properties similar to the two kinds of Stirling numbers, such as a combinatorial interpretation, a two-variable recurrence relation, and an inversion formula. There is one big (and nice) exception, though: Unlike both kinds of Stirling numbers, the Lah numbers have a very simple representation in terms of binomial coefficients.

Identity 237.

$$x^{\overline{n}} = \sum_{k=0}^{n} L(n, k) x^{\underline{k}}.$$

n	$L(n,0)$	$L(n,1)$	$L(n,2)$	$L(n,3)$	$L(n,4)$	$L(n,5)$	$L(n,6)$	$L(n,7)$
0	1							
1	0	1						
2	0	2	1					
3	0	6	6	1				
4	0	24	36	12	1			
5	0	120	240	120	20	1		
6	0	720	1800	1200	300	30	1	
7	0	5040	15120	12600	4200	630	42	1

FIGURE 8.4
Rows 0 through 7 of triangle of Lah numbers

Lah numbers can be used to convert from falling factorial powers to rising factorial powers as well; see Identity 273 in Exercise 36.

Figure 8.4 gives the first several rows of Lah numbers.

As with the two kinds of Stirling numbers, a combinatorial interpretation of the Lah numbers is key to proving many of their identities. A *tuple* is an ordered set or list.

Theorem 28. *The Lah numbers $L(n,k)$ count the number of ways that a set of size n can be partitioned into k nonempty tuples.*

Proof. Let $T(n,k)$ be the number of ways that a set of size n can be partitioned into k nonempty tuples.

Suppose we have x colors available, and we want to partition a set of size n into x lists of different colors (some of which may be empty). How many ways are there to do that?

One approach is to place the n elements, one-by-one, into the colored lists. The first element may be placed at the head of any of the colored lists, and so there are x choices for its placement. The second element may be placed at the head of any of the lists or after element 1, for a total of $x+1$ choices. The third element may be placed at the head of any of the lists or after elements 1 or 2, for a total of $x+2$ choices. This process continues until the $x+n-1$ choices for element n, yielding $x(x+1)\cdots(x+n-1) = x^{\overline{n}}$ as the number of ways to partition a set of size n into x differently-colored lists.

Another approach is to form the tuples first and then assign them to colors. Conditioning on the number of nonempty tuples, there are $T(n,k)$ ways to form k nonempty tuples from n elements without assigning colors. There is a natural lexicographic ordering to these $T(n,k)$ tuples based on the value of the first element in each tuple. Then there are x choices for the color for the first tuple, $x-1$ choices for the color for the second tuple, and so forth, down to $x-k+1$ choices for the color for the kth tuple. This gives $T(n,k)x^{\underline{k}}$ ways to form colored lists when k of the lists are nonempty. Summing over all possible values of k then yields $\sum_{k=0}^{n} T(n,k)x^{\underline{k}}$ as the number of ways to partition a set of size n into x differently-colored lists.

Therefore, $x^{\overline{n}} = \sum_{k=0}^{n} T(n,k)x^{\underline{k}}$. This equation is a polynomial in n, and we have proved it true for infinitely many values of x. Thus it must be true for all real x. However, Identity 237 says that $x^{\overline{n}} = \sum_{k=0}^{n} L(n,k)x^{\underline{k}}$. Therefore, $L(n,k) = T(n,k)$. $\square$

With a combinatorial interpretation in hand, we are now ready to prove the following explicit formula for the Lah numbers.

Identity 238. *For* $n \geq 1$,

$$L(n,k) = \binom{n-1}{k-1}\frac{n!}{k!}.$$

Proof. How many ways are there to partition a set of size n into a list of k nonempty tuples? One approach is to construct the tuples in $L(n,k)$ ways (by Theorem 28), and then order those tuples in a list in $k!$ ways. This gives a total of $L(n,k)k!$ ways.

Another approach is to construct a permutation of the n elements and then cut the permutation in $k-1$ places. For example, the permutation 314592687 on nine elements can be cut after the 3 and after the 9 to create a list of three tuples: 3, 1459, 2687. There are $n!$ ways to form the permutation and $\binom{n-1}{k-1}$ ways to select $k-1$ cut places from the $n-1$ spaces between numbers, for a total of $\binom{n-1}{k-1}n!$ ways.

Thus $L(n,k)k! = \binom{n-1}{k-1}n!$, and the identity follows. $\square$

Slovenian mathematician **Ivo Lah** (1896–1979) was born in what was then the Austro-Hungarian Empire. Lah worked as an actuary for much of his life, although he also published papers in mathematics, statistics, and demography. His most important contribution to mathematics is his proof that the numbers later named for him show how rising powers may be expressed in terms of falling powers.

Identity 238 means that every identity involving the Lah numbers is an identity involving the binomial coefficients.

We also mentioned that the Lah numbers satisfy a two-variable recurrence relation like those in Identities 225 and 229 for the Stirling numbers. In fact, the Lah recurrence is related to those for the Stirling numbers: The Stirling numbers of the first kind have $(n-1)\left[{n-1 \atop k}\right]$ as the term with $n-1$ and k, and those of the second kind have $k\left\{{n-1 \atop k}\right\}$, while the Lah recurrence has $(n+k-1)L(n-1,k)$. In other words, the coefficient of the $n-1,k$ term for the Lah numbers is the sum of the corresponding coefficients for the two kinds of Stirling numbers.

Identity 239.

$$L(n,k) = (n+k-1)L(n-1,k) + L(n-1,k-1) + [n=k=0], \text{ for } n,k \geq 0,$$

where $L(n,k)$ *is understood to be* 0 *for* $n < 0$ *or* $k < 0$.

Proof. There is one way to partition an empty set into zero nonempty tuples: Use the empty tuple. Thus, by Theorem 28, $L(0,0) = 1$. Otherwise, $L(0,k) = 0$.

To count the number of partitions of an n-element set into k nonempty tuples, we can condition on the location of element n. If n is in a tuple by itself, there are $L(n-1, k-1)$ ways to form the remaining tuples.

Now, suppose n is in a tuple with other elements. There are $L(n-1, k)$ ways to form k nonempty tuples from elements 1 through $n-1$. Then element n can be placed in the front of any of the k tuples or after any of the $n-1$ elements already placed, for a total of $n+k-1$ placements. Thus there are $(n+k-1)L(n-1,k)$ ways to form k nonempty tuples from n elements if n is in a tuple with other elements.

All together, then, there are $(n+k-1)L(n-1,k) + L(n-1,k-1)$ ways to form k nonempty tuples from n elements, which, by Theorem 28, is also $L(n,k)$. $\qquad\square$

The inversion formula for the Stirling numbers, Theorem 26, depends ultimately on the fact that the two kinds of Stirling numbers are used to convert between ordinary powers and factorial powers. Since the Lah numbers are used to convert between the two kinds of factorial powers, this implies that there should exist an inversion formula for the Lah numbers. Let's derive that formula now. First we need an identity for the Lah numbers like Identities 232 and 233 for the Stirling numbers.

Identity 240.

$$\sum_{k=0}^{n} L(n,k)L(k,m)(-1)^{k+m} = [n = m].$$

Proof. We have, by Identities 237 and 273,

$$
\begin{aligned}
x^{\overline{n}} &= \sum_{k=0}^{n} L(n,k)x^{\underline{k}} = \sum_{k=0}^{n} L(n,k) \sum_{m=0}^{k} L(k,m)x^{\overline{m}}(-1)^{k-m} \\
&= \sum_{m=0}^{n} x^{\overline{m}} \sum_{k=m}^{n} L(n,k)L(k,m)(-1)^{k-m} \\
&= \sum_{m=0}^{n} x^{\overline{m}} \sum_{k=0}^{n} L(n,k)L(k,m)(-1)^{k+m}.
\end{aligned}
$$

This equation consists of two polynomials in x; thus their coefficients must be equal. Therefore,

$$\sum_{k=0}^{n} L(n,k)L(k,m)(-1)^{k+m}$$

is 1 when $m = n$ and 0 otherwise. $\qquad\square$

Identity 240 leads directly to the inversion formula for the Lah numbers.

Theorem 29. *For functions f and g,*

$$f(n) = \sum_{k=0}^{n} L(n,k)g(k)(-1)^k \iff g(n) = \sum_{k=0}^{n} L(n,k)f(k)(-1)^k.$$

Proof. Suppose $f(n) = \sum_{k=0}^{n} L(n,k)g(k)(-1)^k$. Then

$$\sum_{k=0}^{n} L(n,k)f(k)(-1)^k = \sum_{k=0}^{n} L(n,k) \sum_{m=0}^{k} L(k,m)g(m)(-1)^{k+m}$$

$$= \sum_{m=0}^{n} g(m) \sum_{k=m}^{n} L(n,k)L(k,m)(-1)^{k+m}$$

$$= \sum_{m=0}^{n} g(m) \sum_{k=0}^{n} L(n,k)L(k,m)(-1)^{k+m} = g(n),$$

where we use Identity 240 in the last step. The proof in the other direction is nearly identical. $\square$

With the Lah numbers substituted for the Stirling numbers and Identity 240 for Identities 232 and 233, this proof is identical to that for the Stirling inversion formula. In fact, any identity of the form $\sum_{k=0}^{n} a(n,k)b(k,m) = [n = m]$ means that $a(n,k)$ and $b(n,k)$ are *inverse numbers* and satisfy an inversion formula. The binomial inversion formula can be proved using this approach, too; see Exercise 37. There are matrix consequences of these inversion formulas as well; see Chapter 9.

Our final identity with the Lah numbers relates them to the two kinds of Stirling numbers.

Identity 241.

$$L(n,m) = \sum_{k=m}^{n} \begin{bmatrix} n \\ k \end{bmatrix} \begin{Bmatrix} k \\ m \end{Bmatrix}.$$

Proof. Both expressions count the number of ways to partition n elements into m nonempty tuples. The left side follows from Theorem 28.

For the right side, partition the n elements into k nonempty cycles, in $\begin{bmatrix} n \\ k \end{bmatrix}$ ways. Then put the k cycles into m nonempty sets, in $\begin{Bmatrix} k \\ m \end{Bmatrix}$ ways. This approach yields a set of m nonempty tuples because each set of cycles uniquely defines a permutation (i.e., tuple) on the elements making up those cycles, as we argued in the proof of Theorem 25. Summing over all values of k must give the number of ways to partition n elements into m nonempty tuples. $\square$

See Exercise 38 for another proof of Identity 241.

Before we move on to the next section on Catalan numbers, let's group in one place the six formulas used when converting between rising, falling, and ordinary powers.

Identity 222.
$$x^n = \sum_{k=0}^{n} \left\{ {n \atop k} \right\} x^{\underline{k}}.$$

Identity 226.
$$x^n = \sum_{k=0}^{n} \left\{ {n \atop k} \right\} x^{\overline{k}} (-1)^{n-k}.$$

Identity 228.
$$x^{\overline{n}} = \sum_{k=0}^{n} \left[{n \atop k} \right] x^k.$$

Identity 259.
$$x^{\underline{n}} = \sum_{k=0}^{n} \left[{n \atop k} \right] x^k (-1)^{n-k}.$$

Identity 237.
$$x^{\overline{n}} = \sum_{k=0}^{n} L(n,k) x^{\underline{k}}.$$

Identity 273.
$$x^{\underline{n}} = \sum_{k=0}^{n} L(n,k) x^{\overline{k}} (-1)^{n-k}.$$

8.5   Catalan Numbers and More Lattice Path Counting

In Chapter 3 we discussed how the binomial coefficient $\binom{n+m}{n}$ gives the number of lattice paths from $(0,0)$ to (n,m). We also proved some binomial coefficient identities by conditioning on the location where such a lattice path crosses a particular line through the lattice.

In this section we investigate another lattice path problem: How many lattice paths are there from $(0,0)$ to (n,n) that do not go above the main diagonal $y = x$? See Figure 8.5 for such a path.

This problem is more difficult than the lattice path problems we considered in Chapter 3. We're going to tackle it in three different ways. The first is a brute-force approach that constructs a recurrence relation, produces a generating function from the recurrence, and then solves the recurrence. Methods two and three are more clever combinatorial arguments.

First, we define C_n to be answer to our question, so that C_n is the number of lattice paths from $(0,0)$ to (n,n) that do not go above the diagonal $y = x$. The number C_n is called the nth *Catalan number*. It is easy to calculate the first few Catalan numbers. For instance, $C_0 = 1$ (for the empty path), $C_1 = 1$, and $C_2 = 2$. The first several Catalan numbers are given in Table 8.4.

To construct the recurrence relation for the Catalan numbers, we first need the following.

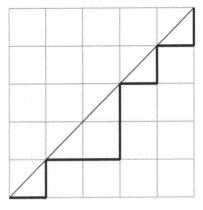

FIGURE 8.5
A lattice path below the main diagonal

TABLE 8.4
First several Catalan numbers

n	0	1	2	3	4	5	6	7	8	9	10	11
C_n	1	1	2	5	14	42	132	429	1430	4862	16796	58786

Theorem 30. *For $n \geq 1$, the number of lattice paths from $(0,0)$ to (n,n) that do not go above or touch the main diagonal except at the starting and ending points is C_{n-1}.*

Proof. A lattice path that does not go above or touch the main diagonal except at the starting and ending points must use the segment from $(0,0)$ to $(1,0)$ as its first step and the segment from $(n, n-1)$ to (n,n) as its last step. Other than those two steps, the restriction that the path does not even touch the main diagonal is equivalent to the restriction that the path cannot rise above the line $y = x - 1$, which is the dashed line in Figure 8.6.

This means that any lattice path from $(0,0)$ to (n,n) that stays below the main diagonal and only touches it at endpoints is equivalent to a path that runs from $(1,0)$ to $(n, n-1)$ that does not go above the line $y = x - 1$. But such a path, by re-orienting the coordinate axes, is equivalent to a lattice path from $(0,0)$ to $(n-1, n-1)$ that does not go above the line $y = x$. By definition of the Catalan numbers, then, the number of lattice paths that do not go above the main diagonal and only touch it at endpoints is C_{n-1}. $\square$

With Theorem 30 in hand, we can now prove the following recurrence relation for the Catalan numbers.

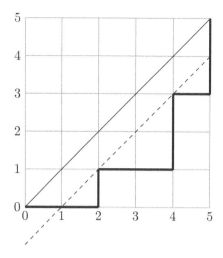

FIGURE 8.6
A lattice path that only touches the main diagonal at endpoints

Identity 242.

$$C_{n+1} = \sum_{k=0}^{n} C_k C_{n-k}.$$

Proof. Both sides count the number of lattice paths from $(0,0)$ to $(n+1, n+1)$ that do not go above the main diagonal. For the left side, this follows from the definition of the Catalan numbers. For the right side, condition on the last point (k, k) where a path touches the main diagonal before reaching $(n+1, n+1)$. There are C_k paths from $(0,0)$ to (k, k) that do not go above the main diagonal, and, by Theorem 30, there are $C_{n+1-k-1} = C_{n-k}$ paths from (k, k) to $(n+1, n+1)$ that do not go above the main diagonal and only touch it at (k, k) and $(n+1, n+1)$. For example, the path in Figure 8.5 has $k = 4$. Multiplying together and summing over all values of k yields the right side. □

As mentioned before, we're going to give three different derivations of the explicit formula for the Catalan numbers. The first uses their generating function. Deriving that allows us to review several concepts from Chapter 6, too.

Identity 243.

$$\sum_{n=0}^{\infty} C_n x^n = \frac{1 - \sqrt{1 - 4x}}{2x}.$$

Proof. Identity 242 tells us that the nth convolution of the Catalan numbers with themselves is the shifted Catalan number C_{n+1}. Multiplying both sides of Identity 242 by x^n and summing up yields

$$\sum_{n=0}^{\infty} C_{n+1} x^n = \sum_{n=0}^{\infty} \left(\sum_{k=0}^{n} C_k C_{n-k} \right) x^n.$$

Now, let $C(x) = \sum_{n=0}^{\infty} C_n x^n$, the generating function for the Catalan numbers. We know by the convolution property for ordinary generating functions (Theorem 13) that the generating function for the convolution of two sequences is the product of their generating functions. Thus we have

$$\frac{1}{x} \sum_{n=0}^{\infty} C_{n+1} x^{n+1} = C^2(x)$$

$$\implies \frac{1}{x} \sum_{n=1}^{\infty} C_n x^n = C^2(x)$$

$$\implies \frac{1}{x} \left(\sum_{n=0}^{\infty} C_n x^n - C_0 \right) = C^2(x)$$

$$\implies \frac{1}{x} C(x) - \frac{1}{x} = C^2(x)$$

$$\implies x C^2(x) - C(x) + 1 = 0$$

$$\implies C(x) = \frac{1 \pm \sqrt{1 - 4x}}{2x},$$

by the quadratic formula. Now, do we want the positive or the negative square root? To answer this, we need to consider the initial condition. Plugging $x = 0$ into the definition of $C(x)$, we have $C(0) = C_0$, the zeroth Catalan number. We know that $C_0 = 1$, so subbing zero into $C(x)$ must give 1. However, when $x = 0$, the expression

$$\frac{1 \pm \sqrt{1 - 4x}}{2x}$$

entails division by zero. To make this meaningful, then, we must consider the limit as $x \to 0^+$. This limit is clearly infinity for the positive square root. For the negative square root, we obtain the indeterminate form $0/0$. Let's take a closer look at the negative root, then. Applying l'Hôpital's rule, we have

$$\lim_{x \to 0^+} \frac{1 - \sqrt{1 - 4x}}{2x} = \lim_{x \to 0^+} \frac{-(1/2)(1 - 4x)^{-1/2}(-4)}{2} = 1.$$

Thus we want the negative square root, and we have $C(x) = \dfrac{1 - \sqrt{1 - 4x}}{2x}$. $\quad\square$

With the generating function for the Catalan numbers in hand, we are ready to derive an explicit formula for them. This formula turns out to be quite simple.

Identity 244.

$$C_n = \frac{1}{n+1}\binom{2n}{n}.$$

Generating function proof. Our main tool in this proof is the binomial series, Identity 18. We have

$$\frac{1 - \sqrt{1-4x}}{2x} = \frac{1}{2x}\left(1 - (1-4x)^{1/2}\right) = \frac{1}{2x}\left(1 - \sum_{n=0}^{\infty}\binom{1/2}{n}(-4x)^n\right)$$

$$= \frac{1}{2x}\left(1 - 1 - \sum_{n=1}^{\infty}\binom{1/2}{n}(-4x)^n\right) = \frac{-1}{2x}\sum_{n=1}^{\infty}\binom{1/2}{n}(-4x)^n.$$

Identity 30 says that

$$\binom{-1/2}{n} = \left(-\frac{1}{4}\right)^n\binom{2n}{n},$$

which is close to the binomial coefficient in the infinite series we are currently working with. Applying the absorption identity (Identity 6), then, we get

$$\frac{-1}{2x}\sum_{n=1}^{\infty}\binom{1/2}{n}(-4x)^n = \frac{-1}{2x}\sum_{n=1}^{\infty}\frac{1}{2n}\binom{-1/2}{n-1}(-4x)^n$$

$$= \frac{-1}{2x}\sum_{n=1}^{\infty}\frac{1}{2n}\binom{2(n-1)}{n-1}\left(-\frac{1}{4}\right)^{n-1}(-4x)^n$$

$$= \sum_{n=1}^{\infty}\frac{1}{n}\binom{2(n-1)}{n-1}x^{n-1}$$

$$= \sum_{n=0}^{\infty}\frac{1}{n+1}\binom{2n}{n}x^n.$$

By Identity 243, then, $C_n = \binom{2n}{n}/(n+1)$.   □

(For an alternate derivation that begins with Identity 150, the generating function for $\binom{2n}{n}$, see Exercise 11 in Chapter 6.)

We put in a lot of work to derive Identity 244. Since it's such a simple expression, though, it seems there should be a nice combinatorial argument for it. We give two such arguments in the next two proofs, although both require more sophisticated lattice path manipulations than the ones we saw in Chapter 3.

Identity 244.

$$C_n = \frac{1}{n+1}\binom{2n}{n}.$$

Reflection principle proof. We know that there are $\binom{2n}{n}$ paths from $(0,0)$ to (n,n) (Theorem 3), and C_n is the number of paths from $(0,0)$ to (n,n) that do not go above the diagonal $y = x$. We can find the value of C_n by counting the *bad paths*; i.e., those that actually do cross the diagonal, and subtracting that number from $\binom{2n}{n}$.

Since bad paths can go bad in so many places, we're going to perform a transformation on the bad paths that makes them easier to count. Given a bad path, find the first step that that path makes above the diagonal. Then, reflect across the line $y = x + 1$ the part of the bad path after that first step. See, for example, Figure 8.7.

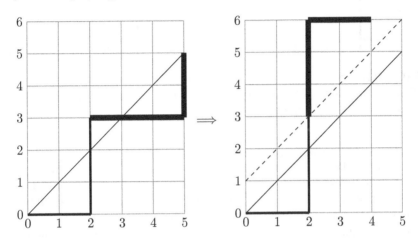

FIGURE 8.7
Transforming a bad path. The thick portion is reflected across $y = x + 1$.

Since the part of the path being reflected starts on the line $y = x + 1$, and the original path ends at (n, n), the transformed path must end at $(n-1, n+1)$. This is true regardless of which bad path we start with. Not only that, the transformation is reversible: Given a path from $(0,0)$ to $(n-1, n+1)$, it must cross the line $y = x + 1$ somewhere for the first time. Starting with the first place the path touches $y = x + 1$, reflecting the path across the line $y = x + 1$ results in a path from $(0,0)$ to (n, n) that goes above the line $y = x$ at least once. (See Figure 8.8.)

Since the process is reversible, the number of bad paths from $(0,0)$ to (n, n) is the same as the total number of paths from $(0,0)$ to $(n-1, n+1)$, which, by Theorem 3, is $\binom{2n}{n-1}$. Thus we have

$$C_n = \binom{2n}{n} - \binom{2n}{n-1} = \frac{(2n)!}{n!n!} - \frac{(2n)!}{(n-1)!(n+1)!} = \frac{(2n)!}{n!(n-1)!}\left(\frac{1}{n} - \frac{1}{n+1}\right)$$

$$= \frac{(2n)!}{n!(n-1)!}\frac{n+1-n}{n(n+1)} = \frac{1}{n+1}\binom{2n}{n}.$$

□

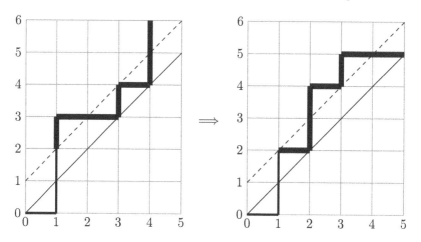

FIGURE 8.8
Transforming a path from $(0,0)$ to $(n-1, n+1)$ into a bad path. The thick portion is reflected across $y = x + 1$.

The reflection principle proof, while mostly combinatorial, still requires some manipulation with the factorials in the binomial coefficient formula to produce Identity 244. In particular, it does not explain the factor of $1/(n+1)$ in the identity combinatorially. Our third proof gives a direct combinatorial argument that the lattice paths from $(0,0)$ to (n,n) can be divided into $n+1$ groups of equal size, one of which is the set of paths counted by the Catalan number C_n. This implies Identity 244.

Identity 244.

$$C_n = \frac{1}{n+1}\binom{2n}{n}.$$

Counting flaws proof. Define a *flaw* of a lattice path from $(0,0)$ to (n,n) to be a vertical step that occurs above the line $y = x$. For example, the path in Figure 8.9 has four flaws (the thick segments).

Let $C_{n,k}$ be the number of lattice paths from $(0,0)$ to (n,n) with exactly k flaws. Of course, $C_{n,0} = C_n$. We will show a one-to-one mapping between $C_{n,k}$ and $C_{n,k+1}$ for $0 \le k \le n-1$.

Given a path P counted by $C_{n,k}$, with $0 \le k \le n-1$, let u denote the first up step that ends on the diagonal. Swap the portion of P that comes before u and the portion of P that comes after u, as in Figure 8.10, to create a new path P'. The steps in P' occurring before u are exactly as far from the diagonal as they were before. The step u is now above the diagonal and so is a flaw in P', whereas it is not a flaw in P. Finally, the steps after u in P' start on the line $y = x + 1$ and end on the line $y = x$. From the way u is defined, those same steps in P start on the line $y = x$ and end on the line $y = x - 1$, with no up steps from $y = x - 1$ to $y = x$. This means there are no up steps

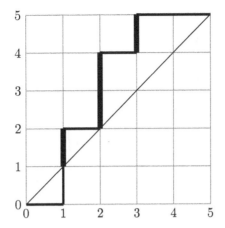

FIGURE 8.9
A path from $(0,0)$ to (n,n) with four flaws (the thick segments)

after u from $y = x$ to $y = x + 1$ in P'. Thus there are as many flawed steps after u in P' as there are before u in P'. All of this together means that P' has one more flawed step, u, than does P. Thus P' is counted by $C_{n,k+1}$. This process is reversible, too: Given a path P' counted by $C_{n,k+1}$, let u denote the last up step that begins on the diagonal. Swap the portion of P' that comes before u and the portion of P' that comes after u, as in Figure 8.11. The set of flawed steps before u in P' is the same as those after u in P, u is flawed in P' but not in P, and by the definition of u, the set of flawed steps after u in P' is the same as those before u in P. Thus P' is counted by $C_{n,k}$. Therefore, there is a one-to-one mapping between $C_{n,k}$ and $C_{n,k+1}$, for $0 \leq k \leq n-1$.

This means that $C_{n,n} = C_{n,n-1} = \cdots = C_{n,0}$. Since there are $n+1$ of these $C_{n,k}$ sets, and the sum of them must be $\binom{2n}{n}$, we have, for each k,

$$C_{n,k} = \frac{1}{n+1}\binom{2n}{n},$$

which, since $C_n = C_{n,0}$, implies Identity 244. □

This third proof of Identity 244 actually proves something stronger: The number of lattice paths from $(0,0)$ to (n,n) with k flaws is independent of the value of k. This result is known as the *Chung-Feller Theorem* [18].

According to Theorem 30, the Catalan number C_{n-1} counts the number of lattice paths from $(0,0)$ to (n,n) that stay below the diagonal and only touch it once after the starting point (namely, at the ending point). This means that $2C_{n-1}$ is the number of lattice paths from $(0,0)$ to (n,n) that touch the diagonal exactly once after $(0,0)$ (as there are as many of these that stay above the diagonal as stay below). How many lattice paths from $(0,0)$ to

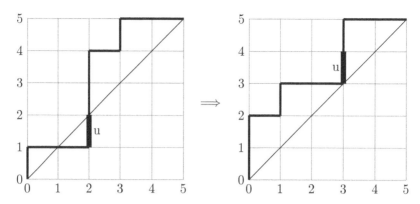

FIGURE 8.10
Transforming a path with four flaws into one with five flaws

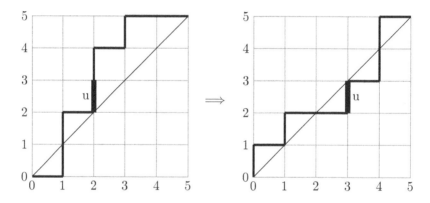

FIGURE 8.11
Transforming a path with four flaws into one with three flaws

(n, n) are there, though, that touch the diagonal exactly k times after $(0, 0)$, for $1 \leq k \leq n$? We're going to answer that question in three parts.

For the first part, we're going to prove what is known as the *ballot theorem*. In terms of lattice paths, the ballot theorem says the following.

Theorem 31 (Ballot Theorem). *The number of lattice paths from $(0,0)$ to (n, m), $n > m$, that touch the main diagonal only at $(0,0)$ is given by*

$$\frac{n - m}{n + m} \binom{n + m}{n}.$$

Proof. The proof uses an idea similar to the reflection principle proof of Identity 244. We define a *bad path* as one from $(0, 0)$ to (n, m) that touches the diagonal after $(0, 0)$. By Theorem 3, the total number of lattice paths from

$(0,0)$ to (n,m) is $\binom{n+m}{n}$. Thus the number we're looking for is $\binom{n+m}{n}$ minus the number of bad paths.

The bad paths fall into two categories: Those that start with a right step and those that start with an up step. Call these *right-bad paths* and *up-bad paths*, respectively. Every right-bad path must eventually return to the diagonal $y = x$ in order to be a bad path. Every up-bad path must also eventually return to the diagonal $y = x$, as each must cross this line in order to reach the ending point at (n,m). By reflecting the portion of the path before the first return to $y = x$ we can turn every right-bad path into an up-bad path and vice versa. (See Figure 8.12, where $n = 5$ and $m = 3$.) Since all up-bad paths must

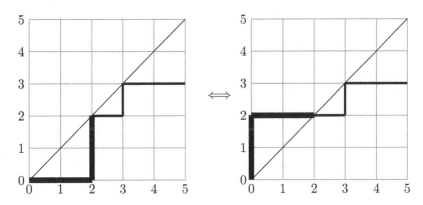

FIGURE 8.12
Transforming right-bad paths to up-bad paths and vice-versa by reflecting an initial portion of the path across $y = x$

include the segment from $(0,0)$ to $(0,1)$, and all paths from $(0,1)$ to (n,m) with the $(0,0)$ to $(0,1)$ segment attached at front are up-bad paths, the number of up-bad paths is the total number of paths from $(0,1)$ to (n,m). By Theorem 3, this is $\binom{n+m-1}{n}$. Thus the total number of bad paths is $2\binom{n+m-1}{n}$, and therefore the number of paths we are after is

$$\binom{n+m}{n} - 2\binom{n+m-1}{n} = \binom{n+m}{n} - 2\frac{(n+m-1)!}{n!\,(m-1)!}$$

$$= \binom{n+m}{n} - 2\frac{m(n+m)!}{(n+m)n!\,m!} = \binom{n+m}{n} - 2\frac{m}{n+m}\binom{n+m}{n}$$

$$= \binom{n+m}{n}\frac{n+m-2m}{n+m} = \frac{n-m}{n+m}\binom{n+m}{n}.$$

$\square$

The reason for the name "ballot theorem" is that the lattice path problem it answers was originally [10] something like the following: Suppose candidate A receives n votes in an election, and candidate B receives m votes, with $n >$

m. If the ballots in the election are counted one-by-one, how many orderings have candidate A always in the lead?

The second part of our derivation of the number of lattice paths from $(0,0)$ to (n,n) that touch the diagonal exactly k times after $(0,0)$ uses the ballot theorem to help prove the following.

Theorem 32. *The number of lattice paths from $(0,0)$ to (n,n) that touch the main diagonal exactly k times after $(0,0)$ without going above it is $\frac{k}{2n-k}\binom{2n-k}{n}$.*

When $k = 1$ the number of such paths is, by Theorem 30,

$$C_{n-1} = \frac{1}{n}\binom{2n-2}{n-1} = \frac{(2n-2)!}{(n)(n-1)!(n-1)!} = \frac{(2n-1)!}{(2n-1)n!(n-1)!}$$
$$= \frac{1}{2n-1}\binom{2n-1}{n},$$

which agrees with Theorem 32.

Proof. For any path that touches the diagonal k times after $(0,0)$ without going above it, there must be exactly k up steps that land on the diagonal. Remove these k steps from the path. The result is a path from $(0,0)$ to $(n, n-k)$ that does not touch the diagonal except at $(0,0)$. (See, for example, Figure 8.13.)

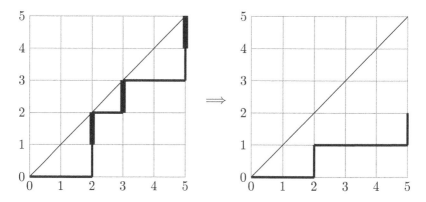

FIGURE 8.13
A path from $(0,0)$ to $(5,5)$ that stays below the diagonal and touches it three times after $(0,0)$ maps to one from $(0,0)$ to $(5,2)$ that never touches the diagonal after $(0,0)$.

To show that the mapping is reversible, let's start with a path from $(0,0)$ to $(n, n-k)$ that never touches the diagonal after $(0,0)$. Find the last place the path touches $y = x - 1$. There must be such a point because the path touches $(1,0)$. Insert an up step at this place in the path. This creates a new path that touches the line $y = x$ for a first time after $(0,0)$. This new path touches the line $y = x - 1$ directly to the right of where it touches $y = x$ and so must have a last place that it touches $y = x - 1$. Insert an up step at this place in the new path. Continue this process until the path touches (n, n). Since we always choose the last place a path touches a lattice point of $y = x - 1$ as the place we want to insert the up step, this is a reversal of the original mapping. (See Figure 8.14.) This means that the number of lattice paths from $(0, 0)$ to (n, n)

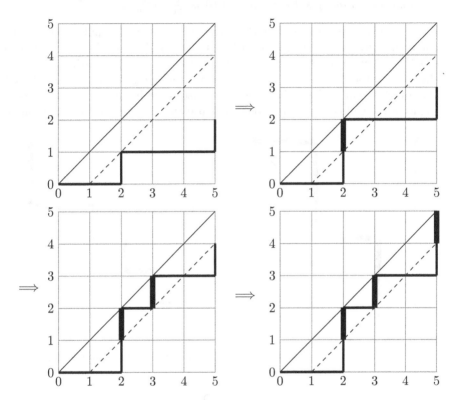

FIGURE 8.14
A path from $(0, 0)$ to $(5, 2)$ that never touches the diagonal after $(0, 0)$ maps to one from $(0, 0)$ to $(5, 5)$ that stays below the diagonal and touches it three times after $(0, 0)$.

that do not go above the diagonal and touch it exactly k times after $(0, 0)$ is the same as the number of lattice paths from $(0, 0)$ to $(n, n - k)$ that do not touch the diagonal after $(0, 0)$. The latter number is, by the ballot theorem (Theorem 31),

$$\frac{n - (n - k)}{2n - k} \binom{2n - k}{n} = \frac{k}{2n - k} \binom{2n - k}{n}.$$

$\square$

Theorem 32 doesn't tell us how many lattice paths from $(0,0)$ to (n,n) touch the diagonal exactly k times after $(0,0)$; it only tells us how many of those stay below the diagonal. We're almost there, though; the next theorem answers our original question.

Theorem 33. *The number of lattice paths from $(0,0)$ to (n,n) that touch the main diagonal exactly k times after $(0,0)$ is $\frac{k2^k}{2n-k}\binom{2n-k}{n}$.*

Proof. Call an *intersection point* a point where a lattice path touches the diagonal. Any lattice path from $(0,0)$ to (n,n) that touches the diagonal exactly k times after $(0,0)$ can be mapped to such a lattice path that always stays below the diagonal by reflecting about the line $y = x$ any segment that is above the diagonal and between two consecutive intersection points, as in Figure 8.15. Unlike previous proofs that use the reflection principle, though,

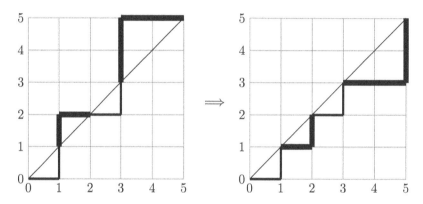

FIGURE 8.15
Reflecting across $y = x$ path segments that are between consecutive intersection points and above the line $y = x$

this mapping is not one-to-one. There are several paths, for example, that could be mapped to the path P on the right side of Figure 8.15. More precisely, there are exactly eight paths (counting P itself) that could be mapped to P, as each of the four segments between consecutive intersection points could be below the diagonal or reflected above it. Applying this idea in general, for any path from $(0,0)$ to (n,n) that does not go above the diagonal and touches it at exactly k points after $(0,0)$, there are k path segments between consecutive intersection points. For each of these segments, we can choose to reflect it across $y = x$ or not, producing 2^k distinct paths. Since there are, by Theorem 32, $\frac{k}{2n-k}\binom{2n-k}{n}$ paths from $(0,0)$ to (n,n) that touch the diagonal exactly k times after $(0,0)$ and never go above it, there are $2^k\frac{k}{2n-k}\binom{2n-k}{n}$ lattice paths from $(0,0)$ to (n,n) that intersect the diagonal exactly k times after $(0,0)$. □

Theorem 33 immediately implies the following rather complicated-looking binomial coefficient identity.

Identity 245.

$$\sum_{k=1}^{n} \binom{2n-k}{n} \frac{k2^k}{2n-k} = \binom{2n}{n}.$$

Proof. We know there are $\binom{2n}{n}$ lattice paths from $(0,0)$ to (n,n), thanks to Theorem 3. We could also count these by conditioning on the number k, $1 \leq k \leq n$, of times such a lattice path intersects the diagonal after $(0,0)$. By Theorem 33, there are $\frac{k2^k}{2n-k}\binom{2n-k}{n}$ paths from $(0,0)$ to (n,n) that intersect the diagonal exactly k times. Summing up over all possible values of k, we see that the left side of Identity 245 also counts the total number of lattice paths from $(0,0)$ to (n,n). □

In this section we have focused on lattice paths, but there are actually many, many other counting problems whose solutions involve the Catalan numbers. In fact, Stanley's *Catalan Numbers* [76] lists 214 of these! Before we leave this discussion of Catalan numbers let's look at a couple of them.

To explain our first one we need to define a few terms. An n-*gon* is a polygon with n vertices. A polygon is *convex* if no line segment between points in the polygon goes outside the polygon. Thus a convex 2-gon is a single line segment, a convex 3-gon is a triangle, a convex 4-gon is a quadrilateral, a convex 5-gon is a pentagon, and so forth. For $n \geq 3$, you *triangulate* a convex n-gon by drawing $n - 3$ straight, non-crossing line segments between vertices of the n-gon, creating $n - 2$ triangles in the process. (For the special case of the line segment, we consider it to have the single triangulation consisting of itself.)

Belgian mathematician **Eugène Catalan** (1814–1894) spent much of his life in France and for a time held French citizenship. In addition to the Catalan numbers that we discuss here, he is also known for the conjecture that 8 (as 2^3) and 9 (as 3^2) are the only consecutive positive integers that are powers of natural numbers. This conjecture was not proved until 2002. A fierce opponent of the monarchy, Catalan was turned down for positions in mathematics on several occasions because of his radical political commitments.

The question we wish to ask is this: How many ways are there to triangulate a convex n-gon? Figure 8.16 shows that the answers in the cases of the triangle, square (which holds for quadrilaterals in general), and pentagon are 1, 2, and 5, respectively. These are also the Catalan numbers C_1, C_2, and C_3.

Perhaps not surprisingly, we have the following.

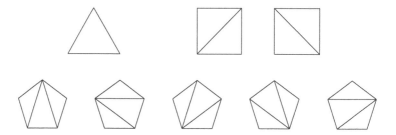

FIGURE 8.16
Triangulations of the triangle, square, and pentagon

Theorem 34. *For $n \geq 2$, the number of ways to triangulate a convex n-gon is C_{n-2}.*

Theorem 34 is due to Euler and was actually the first solved counting problem involving the Catalan numbers.

Proof. First, $C_{2-2} = C_0 = 1$, and there is one triangulation of a 2-gon.

Now, suppose $n \geq 3$, suppose that each n-gon has vertices labeled from 1 through n, and let T_n denote the number of triangulations of a convex n-gon. Given a triangulated convex n-gon P_n, remove the edge $(1, n)$ from P_n. This splits P_n into two triangulated convex polygons, one with k vertices (labeled 1 through k in P_n) and one with $n - k + 1$ vertices (labeled k through n in P_n).

This process is reversible. Given a value of k, $2 \leq k \leq n - 1$, you can construct a triangulation of a convex n-gon in the following fashion.

1. Take a convex k-gon with vertices numbered 1 through k and triangulate it.

2. Take another convex polygon with $n - k + 1$ vertices and triangulate it.

3. For the second polygon, add $k - 1$ to the labels on all of its vertices. This gives you a triangulated polygon with vertices labeled k through n.

4. Associate vertex k of the two polygons with each other.

5. Add an edge from vertex 1 in the first polygon to vertex n in the second polygon.

By conditioning on the value of k, all of this means that, for $n \geq 3$, T_n satisfies the recurrence

$$T_n = \sum_{k=2}^{n-1} T_k T_{n-k+1}, \tag{8.1}$$

with initial condition $T_2 = 1$. By Identity 242, C_{n-2} satisfies

$$C_{n-2} = \sum_{k=0}^{n-3} C_k C_{n-3-k} = \sum_{j=2}^{n-1} C_{j-2} C_{n-1-j} = \sum_{j=2}^{n-1} C_{j-2} C_{n-2-j+1}.$$

Thus C_{n-2} satisfies Equation (8.1), and $C_{n-2} = T_n$. Therefore, C_{n-2} is the number of ways to triangulate a convex n-gon. $\qquad\square$

Another problem whose solution features the Catalan numbers involves binary trees. A *binary tree* is a graph structure with a *root*, represented by a vertex at the top of the tree, and in which every vertex has at most two children, a *left child* and a *right child*. Our question is this: How many binary trees are there on n vertices? For 0 vertices, there is only one binary tree: the empty tree. Figure 8.17 shows that there are one, two, and five binary trees on, respectively, 1, 2, and 3 vertices.

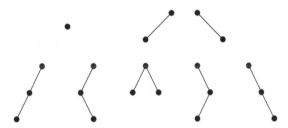

FIGURE 8.17
Binary trees on 1, 2, and 3 vertices

Theorem 35. *For $n \geq 0$, the number of binary trees on n vertices is C_n.*

Proof. Let A_n denote the number of binary trees on n vertices. The *left subtree* of a binary tree consists of the part of the tree below and to the left of the root. The *right subtree* is the part of the tree below and to the right of the root.

Every binary tree can be thought of as being composed of three parts: (1) the root, (2) the left subtree (which may be empty), and (3) the right subtree (which may also be empty). If there are k vertices in the left subtree of a binary tree on n vertices, then there are $n - k - 1$ vertices in the right subtree. Each subtree is itself a binary tree, and so there are A_k ways to construct the left subtree and A_{n-k-1} ways to construct the right subtree.

This process is reversible: To construct a binary tree on n vertices, we can simply take a tree on k vertices, make it the left subtree, take another tree on $n - k - 1$ vertices, make it the right subtree, and add a root.

All of this together means that $A_n = \sum_{k=0}^{n-1} A_k A_{n-k-1}$, which, with an index shift, could also be written as

$$A_{n+1} = \sum_{k=0}^{n} A_k A_{n-k}.$$

This is, of course, the Catalan recurrence in Identity 242. Since $A_0 = 1 = C_0$, we have that $C_n = A_n$. The theorem follows. $\square$

In the proofs of Theorems 34 and 35 we showed that the quantity we were trying to count satisfies the Catalan recurrence in Identity 242. This is a common method of proving that the Catalan numbers are the answer to a particular combinatorial question. In fact, it is likely that the recurrence is itself the reason that there are so many combinatorial interpretations of the Catalan numbers: It seems like a natural kind of recurrence for a collection of objects to obey. (The same thing could also be said about Identity 170 for why there are so many applications of the Fibonacci numbers.) Exercises 42 and 43 give examples of two other combinatorial interpretations of the Catalan numbers. And, of course, Stanley [76] gives a total of 214 objects counted by the Catalan numbers, including all the ones we give here.

8.6 Bernoulli Numbers

The Bell numbers have the interesting property that if you take their binomial transform you get the shifted sequence of Bell numbers. This is Identity 235:

$$\sum_{k=0}^{n} \binom{n}{k} \varpi_k = \varpi_{n+1}.$$

Is there a sequence whose binomial transform is itself? Well, not an interesting one; a few calculations show that such a sequence would have to be the sequence of all 0's. (See Exercise 52.) If we make the slight modification of subtracting 1 from the binomial transform only when $n = 1$, though, we get a very interesting sequence: the *Bernoulli numbers*. It also turns out that this slight modification means that the Bernoulli numbers are the sequence whose binomial transform is the alternating version of itself. (See Exercise 56.)

Identity 246. *The Bernoulli numbers B_n satisfy the following recurrence:*

$$B_n = \sum_{k=0}^{n} \binom{n}{k} B_k - [n = 1].$$

(As with other numbers we've seen, this is a definition rather than an identity, but we'll call it an identity.)

The first several Bernoulli numbers are given in Table 8.5.

Unlike any of the other numbers we have seen in this chapter, the Bernoulli numbers are not all integers. In addition, except for B_1, it appears all of the odd Bernoulli numbers are zero. (This is, in fact, the case; see Exercise 53.)

Despite their strange-looking appearance, the Bernoulli numbers have

TABLE 8.5

First several Bernoulli numbers

n	0	1	2	3	4	5	6	7	8	9	10
B_n	1	$-1/2$	$1/6$	0	$-1/30$	0	$1/42$	0	$-1/30$	0	$5/66$

many uses. We will mainly focus on their role in a formula for the power sum $\sum_{k=0}^{n-1} k^m$ (which was their first use, historically-speaking [9]). At the end we give a couple of formulas involving the Bernoulli numbers that follow from the fact that they are almost the binomial transform of themselves.

Our proof of the formula for the power sum involving the Bernoulli numbers begins with their exponential generating function.

Identity 247.

$$\frac{x}{e^x - 1} = \sum_{k=0}^{\infty} B_k \frac{x^k}{k!}.$$

Proof. Let $B(x)$ denote the exponential generating function for the Bernoulli numbers. The exponential generating function for the sequence given by $a_n = [n = 1]$ is just x. By Identity 246 and Theorem 16, then, we have

$$B(x) = e^x B(x) - x \implies B(x) = \frac{x}{e^x - 1}.$$

$\square$

(This argument is reversible, so that Identity 247 could also be taken as the definition of the Bernoulli numbers.)

With Identity 247, we can now prove the following. (Note that the sum only goes to $n - 1$. Some conventions define $B_1 = 1/2$, so that the formula on the right of Identity 248 gives the power sum up to n rather than $n - 1$.)

Identity 248.

$$\sum_{k=0}^{n-1} k^m = \sum_{k=0}^{m} \binom{m}{k} B_k \frac{n^{m+1-k}}{m+1-k} = \frac{1}{m+1} \sum_{k=0}^{m} \binom{m+1}{k} B_k n^{m+1-k}.$$

Proof. The exponential generating function for the power sum $\sum_{k=0}^{n-1} k^m$ as a function of n is

$$\sum_{m=0}^{\infty} \left(\sum_{k=0}^{n-1} k^m \right) \frac{x^m}{m!} = \sum_{k=0}^{n-1} \sum_{m=0}^{\infty} \frac{(kx)^m}{m!} = \sum_{k=0}^{n-1} e^{kx} = \sum_{k=0}^{n-1} (e^x)^k = \frac{1 - e^{nx}}{1 - e^x},$$

by the geometric sum formula. Letting $B(x)$ denote the exponential generating function for the Bernoulli numbers, we have, by Identity 247,

$$\frac{1 - e^{nx}}{1 - e^x} = \frac{e^{nx} - 1}{e^x - 1} = B(x) \frac{e^{nx} - 1}{x}.$$

We also have

$$\frac{e^{nx} - 1}{x} = \frac{1}{x} \sum_{k=1}^{\infty} \frac{(nx)^k}{k!} = \sum_{k=1}^{\infty} n^k \frac{x^{k-1}}{k!} = \sum_{k=0}^{\infty} \frac{n^{k+1}}{k+1} \frac{x^k}{k!}.$$

Thus the power sum $\sum_{k=0}^{n-1} k^m$ is, by Theorem 14, the binomial convolution of the Bernoulli numbers and the sequence $(n^{k+1}/(k+1))_{k=0}^{\infty}$. In other words,

$$\sum_{k=0}^{n-1} k^m = \sum_{k=0}^{m} \binom{m}{k} B_k \frac{n^{m+1-k}}{m+1-k} = \sum_{k=0}^{m} \frac{m!}{k!(m-k)!} B_k \frac{n^{m+1-k}}{m+1-k}$$

$$= \frac{1}{m+1} \sum_{k=0}^{m} \frac{(m+1)!}{k!(m+1-k)!} B_k n^{m+1-k}$$

$$= \frac{1}{m+1} \sum_{k=0}^{m} \binom{m+1}{k} B_k n^{m+1-k}.$$

$\square$

To end this section, let's take a look at a couple of formulas that are consequences of the fact that the Bernoulli numbers are almost their own binomial transform. First, we have the alternating binomial transform of the Bernoulli numbers.

Identity 249.

$$\sum_{k=0}^{n} \binom{n}{k} B_k (-1)^k = n + (-1)^n B_n.$$

Proof. Applying the variation of binomial inversion given in Theorem 2 to the Bernoulli recurrence in Identity 246 yields

$$B_n = \sum_{k=0}^{n} \binom{n}{k} (B_k + [k = 1])(-1)^{n-k}$$

$$= (-1)^n \sum_{k=0}^{n} \binom{n}{k} B_k (-1)^k + n(-1)^{n-1},$$

which implies

$$(-1)^n B_n = \sum_{k=0}^{n} \binom{n}{k} B_k (-1)^k - n.$$

$\square$

Finally, here is an expression for the sum of the first n Bernoulli numbers. The proof illustrates the use of one of the shift identities in Chapter 7.

Identity 250.

$$\sum_{k=0}^{n} B_k = \sum_{k=1}^{n+1} \binom{n+1}{k} B_{k-1} - 1.$$

Proof. Applying Identity 186 to the Bernoulli recurrence $B_n + [n = 1] = \sum_{k=0}^{n} \binom{n}{k} B_k$ (Identity 246), we have

$$\sum_{k=0}^{n} (B_k + [k = 1]) = \sum_{k=1}^{n+1} \binom{n+1}{k} B_{k-1}$$

$$\implies \sum_{k=0}^{n} B_k = \sum_{k=1}^{n+1} \binom{n+1}{k} B_{k-1} - 1.$$

$\square$

8.7 The On-Line Encyclopedia of Integer Sequences

In this final section, we briefly discuss a wonderful reference for sequences of numbers like the ones we've seen in this chapter. This reference is the *On-Line Encyclopedia of Integer Sequences*, or OEIS, available at www.oeis.org.

In 1964 mathematician Neil J. A. Sloane began collecting sequences that were interesting to him. He published 2372 of these in *A Handbook of Integer Sequences* [60] in 1973. Two decades later, in 1995, he and Simon Plouffe published an updated version, *The Encyclopedia of Integer Sequences* [61], containing 5487 sequences. The on-line version of the *Encyclopedia* began in 1996 and as of this writing has grown to encompass over 300,000 sequences.

A typical OEIS entry contains the first several numbers in a sequence, instances in which the numbers arise, formulas, references and online links, and code to generate the sequence. If you run across a sequence of integers and wonder what else might be known about it, just type it into the OEIS search field; it is quite likely that the sequence will be in there. For example, all of the numbers in this chapter appear in the OEIS; see Table 8.6. Even the Bernoulli numbers appear, despite the fact that they are frequently not integers!

TABLE 8.6

OEIS entries for numbers appearing in this chapter's text

Numbers	Main OEIS Entry
Binomial Coefficients	A007318
Fibonacci Numbers	A000045
Stirling Numbers of the First Kind	A008275
Stirling Numbers of the Second Kind	A008277
Bell Numbers	A000110
Lah Numbers	A008297
Catalan Numbers	A000108
Numerators of Bernoulli Numbers	A027641
Denominators of Bernoulli Numbers	A027642

8.8   Exercises

1. Prove Identity 221 by induction.

   Identity 221.

   $$\sum_{k=0}^{n} F_k = F_{n+2} - 1.$$

2. Use Identity 187, $\sum_{k=0}^{n} F_{2k} = F_{2n+1} - 1$, to give a different proof of the formula for the sum of the first n Fibonacci numbers:

   Identity 221.

   $$\sum_{k=0}^{n} F_k = F_{n+2} - 1.$$

3. Male honeybees come from eggs laid by a female that have not been fertilized by another male, while female honeybees come from eggs that have been fertilized by a male. This means that male honeybees have a mother but no father while female honeybees have both a mother and a father. If we consider a male honeybee's mother to be one generation back, his grandparents to be two generations back, and so forth, show that F_{n+1} is the number of ancestors n generations back for a male honeybee. (This is assuming that all of the bees in all the generations back are different bees, which is unrealistic. Thus this problem is really asking to show that F_{n+1} is the *maximum* number of ancestors the male bee could have n generations back.)

4. Prove that F_{n+2} is the number of subsets of $\{1, 2, \ldots, n\}$ that do not contain consecutive elements. For example, when $n = 3$ the allowable subsets are $\emptyset$, $\{1\}$, $\{2\}$, $\{3\}$, and $\{1, 3\}$.

5. Give a combinatorial proof of the following generalization of the binomial transform for the Fibonacci numbers:

Identity 182.

$$\sum_{k=0}^{n} \binom{n}{k} F_{m+k} = F_{2n+m}.$$

6. Give a combinatorial proof of the formula for the sum of the first n even Fibonacci numbers.

Identity 187.

$$\sum_{k=0}^{n} F_{2k} = F_{2n+1} - 1.$$

7. For each of the following special values of $\left\{{n \atop k}\right\}$, give a combinatorial proof.

   (a) $\left\{{n \atop 2}\right\} = 2^{n-1} - 1$, for $n \geq 2$.
   (b) $\left\{{n \atop n-1}\right\} = \binom{n}{2}$, for $n \geq 1$.
   (c) $\left\{{n \atop n-2}\right\} = 3\binom{n}{4} + \binom{n}{3}$, for $n \geq 2$.

8. Use Identity 201 to give a quick, non-combinatorial proof of Identity 230:

Identity 230.

$$\sum_{k=0}^{n} \left[{n \atop k}\right] = n!.$$

9. Use Identity 140,

$$\sum_{k=1}^{n} \binom{n}{k} p^k (1-p)^{n-k} k^{\underline{m}} = n^{\underline{m}} p^m$$

(valid for $0 \leq p \leq 1$), to prove the following generalization of Identity 223:

Identity 251. *For $0 \leq p \leq 1$,*

$$\sum_{k=0}^{n} \binom{n}{k} p^k (1-p)^{n-k} k^m = \sum_{j=0}^{m} \left\{{m \atop j}\right\} n^{\underline{j}} p^j.$$

This is the formula for the mth moment of the binomial distribution.

10. Use Identity 142,

$$\sum_{k=0}^{n} \binom{m}{k} \binom{n}{r-k} k^{\underline{p}} = \binom{m+n}{r} \frac{r^{\underline{p}} m^{\underline{p}}}{(m+n)^{\underline{p}}},$$

to prove the following formula involving the pth moment of the hypergeometric distribution.

Identity 252.

$$\sum_{k=0}^{n} \binom{m}{k} \binom{n}{r-k} k^p = \binom{m+n}{r} \sum_{j=0}^{p} \left\{ {p \atop j} \right\} \frac{r^{\underline{j}} m^{\underline{j}}}{(m+n)^{\underline{j}}}.$$

11. Use Identity 143,

$$\sum_{k=r}^{\infty} \binom{k-1}{r-1} p^r (1-p)^{k-r} k^{\overline{m}} = \frac{r^{\overline{m}}}{p^m}$$

(valid for $0 < p \leq 1$ and positive integers r), to prove the following formula for the mth moment of the negative binomial distribution.

Identity 253. *For $0 < p \leq 1$ and r a positive integer,*

$$\sum_{k=r}^{\infty} \binom{k-1}{r-1} p^r (1-p)^{k-r} k^m = \sum_{j=0}^{m} \left\{ {m \atop j} \right\} \frac{r^{\overline{j}}}{p^j}.$$

12. Use the recurrence relation for the Stirling numbers of the second kind (Identity 225),

$$\left\{ {n \atop k} \right\} = k \left\{ {n-1 \atop k} \right\} + \left\{ {n-1 \atop k-1} \right\} + [n = k = 0], \text{ for } n, k \geq 0,$$

to prove

Identity 222.

$$x^n = \sum_{k=0}^{n} \left\{ {n \atop k} \right\} x^{\underline{k}}.$$

13. Show that Identity 222 can be expressed as

$$n^m = \sum_{k=0}^{n} \binom{n}{k} \left\{ {m \atop k} \right\} k!,$$

and then use binomial inversion (Theorem 1) to prove the explicit formula for the Stirling numbers of the second kind:

Identity 224.

$$\left\{ {n \atop m} \right\} = \frac{1}{m!} \sum_{k=0}^{m} \binom{m}{k} (m-k)^n (-1)^k.$$

14. Give a combinatorial proof of the formula for the binomial power sum.

Identity 223.

$$\sum_{k=0}^{n}\binom{n}{k}k^m = \sum_{j=0}^{m}\left\{{m \atop j}\right\}\binom{n}{j}j!\,2^{n-j}.$$

15. Use Identity 38,

$$\sum_{k=0}^{n}\binom{n}{k}k^{\underline{m}}(-1)^k = (-1)^m m![n=m],$$

to prove the formula for the alternating binomial power sum,

Identity 254.

$$\sum_{k=0}^{n}\binom{n}{k}k^m(-1)^k = (-1)^n n!\left\{{m \atop n}\right\}.$$

16. Give a combinatorial proof of the formula for the alternating binomial power sum.

Identity 254.

$$\sum_{k=0}^{n}\binom{n}{k}k^m(-1)^k = (-1)^n n!\left\{{m \atop n}\right\}.$$

17. Give a combinatorial proof of the following.

Identity 255.

$$\sum_{k=0}^{n}\binom{n}{k}\left\{{k \atop m}\right\} = \left\{{n+1 \atop m+1}\right\}.$$

18. Give both a combinatorial and a non-combinatorial proof of the following.

Identity 256.

$$\sum_{k=0}^{n}\binom{n}{k}\left\{{k+1 \atop m+1}\right\}(-1)^{n-k} = \left\{{n \atop m}\right\}.$$

19. Give a combinatorial proof of the following.

Identity 257.

$$\sum_{i=0}^{n}\sum_{j=0}^{n}\left[{n \atop i+j}\right]\binom{i+j}{i} = (n+1)!.$$

20. Give a combinatorial proof of the following.

Identity 258.

$$\sum_{k=0}^{n}\binom{n}{k}\begin{bmatrix}k\\m\end{bmatrix}(n-k)! = \begin{bmatrix}n+1\\m+1\end{bmatrix}.$$

21. Prove that the Stirling numbers of the first kind can be used to convert from falling factorial powers to ordinary powers via the following formula.

Identity 259.

$$x^{\underline{n}} = \sum_{k=0}^{n}\begin{bmatrix}n\\k\end{bmatrix}x^k(-1)^{n-k}.$$

22. Generalize both Identity 228 and Identity 259 by proving the following, a binomial-like theorem for Stirling numbers of the first kind.

Identity 260.

$$\sum_{k=0}^{n}\begin{bmatrix}n\\k\end{bmatrix}x^k y^{n-k} = \prod_{i=0}^{n-1}(x+yi).$$

23. Prove the following.

Identity 233.

$$\sum_{k=0}^{n}\begin{bmatrix}n\\k\end{bmatrix}\begin{Bmatrix}k\\m\end{Bmatrix}(-1)^{k+m} = [n=m].$$

24. Prove the following.

Identity 261.

$$\sum_{k=0}^{n}\sum_{j=0}^{m}\begin{bmatrix}n\\k\end{bmatrix}\binom{m}{j}j^k(-1)^{n-k} = m^{\underline{n}}2^{m-n}.$$

25. Both kinds of Stirling numbers satisfy simple double generating functions. Prove them.

Identity 262.

$$(1-x)^y = \sum_{n=0}^{\infty}\sum_{k=0}^{n}\begin{bmatrix}n\\k\end{bmatrix}y^k\frac{x^n}{n!}.$$

Identity 263.

$$e^{xy} = \sum_{n=0}^{\infty}\sum_{k=0}^{n}\begin{Bmatrix}n\\k\end{Bmatrix}y^k\frac{x^n}{n!}.$$

26. Use Identity 255 in Exercise 17 to give a non-combinatorial proof of Identity 235:

    Identity 235.

$$\sum_{k=0}^{n} \binom{n}{k} \varpi_k = \varpi_{n+1}.$$

27. Prove the following, which generalizes both Identity 234 and Identity 235.

    Identity 264.

$$\sum_{k=0}^{n} \sum_{j=0}^{m} \binom{n}{k} \left\{ {m \atop j} \right\} j^{n-k} \varpi_k = \varpi_{n+m}.$$

28. Prove the following, which generalizes Identity 230 and is to the Stirling numbers of the first kind what Identity 264 is to the Stirling numbers of the second kind.

    Identity 265.

$$\sum_{k=0}^{n} \sum_{j=0}^{m} \binom{n}{k} \left[{m \atop j} \right] m^{\overline{n-k}} k! = (n+m)!.$$

29. Prove the following. (Both combinatorial and non-combinatorial proofs are possible.)

    Identity 266.

$$\sum_{k=0}^{n} \binom{n}{k} \varpi_{k+1} = \varpi_{n+2} - \varpi_{n+1} = \sum_{k=0}^{n+1} \left\{ {n+1 \atop k} \right\} k.$$

30. Prove the following.

    Identity 267.

$$\sum_{k=0}^{n} \binom{n}{k} \varpi_{k+1} (-1)^{n-k} = \varpi_n.$$

31. Prove

    Identity 268.

$$\sum_{k=0}^{n} \binom{n}{k} \varpi_{k+m} = \sum_{k=0}^{m} \varpi_{n+1+k} (-1)^{m-k}.$$

32. Prove the following expression relating the alternating binomial transform of the Bell numbers with their alternating sum.

   Identity 269.

$$\sum_{k=0}^{n}\binom{n}{k}\varpi_k(-1)^k = 1 + \sum_{k=0}^{n-1}\varpi_k(-1)^{k+1}.$$

33. Prove the following, the exponential generating function for the Bell numbers.

   Identity 270.

$$e^{e^x-1} = \sum_{n=0}^{\infty}\varpi_n\frac{x^n}{n!}.$$

34. In Exercises 9, 10, and 11 we used moments of the binomial, hypergeometric, and negative binomial distributions to prove some binomial coefficient identities. Another common distribution is the Poisson distribution, which is often used to count the number of events occurring for some random process in a particular time interval. If X has a Poisson(λ) distribution, then, for $k \geq 0$, $P(X = k) = \lambda^k e^{-\lambda}/k!$. Prove that if X is Poisson(1) then $E[X^n] = \varpi_n$. In other words, prove

   Identity 271 (Dobinski's Formula).

$$\frac{1}{e}\sum_{k=0}^{\infty}\frac{k^n}{k!} = \varpi_n.$$

35. The *ordered Bell numbers* O_n count the number of possible rankings of n people in a contest in which ties are allowed. For example, $O_3 = 13$ because there are 13 ways that three people $\{a, b, c\}$ could be ranked in a contest:

$a < b < c,$	$a < c < b,$	$a < b = c,$	$a = b < c,$
$a = c < b,$	$a = b = c,$	$b < a < c,$	$b < c < a,$
$b = c < a,$	$b < a = c,$	$c < a < b,$	$c < b < a,$
$c < a = b.$			

   Prove

   Identity 272.

$$O_n = \sum_{k=0}^{n}\left\{{n \atop k}\right\}k!.$$

36. Prove that the Lah numbers are also used to convert from falling factorial powers to rising factorial powers via the following identity.

Identity 273.

$$x^{\overline{n}} = \sum_{k=0}^{n} L(n,k) x^{\overline{k}} (-1)^{n-k}.$$

37. Use Identity 26,

$$\sum_{k=0}^{n} \binom{n}{k}\binom{k}{m}(-1)^{k+m} = [n=m],$$

to prove the binomial inversion formula,

Theorem 1.

$$f(n) = \sum_{k=0}^{n} \binom{n}{k} g(k)(-1)^k \iff g(n) = \sum_{k=0}^{n} \binom{n}{k} f(k)(-1)^k.$$

38. Use the interpretations of the Stirling and Lah numbers as the coefficients in the conversions between rising, falling, and ordinary powers of x to give a non-combinatorial proof of Identity 241.

Identity 241.

$$L(n,m) = \sum_{k=m}^{n} \begin{bmatrix} n \\ k \end{bmatrix} \begin{Bmatrix} k \\ m \end{Bmatrix}.$$

39. Prove that if we swap the Stirling numbers on the right side of Identity 241 and sum up we get the ordered Bell numbers O_n:

Identity 274.

$$O_n = \sum_{m=0}^{n} \sum_{k=0}^{n} \begin{Bmatrix} n \\ k \end{Bmatrix} \begin{bmatrix} k \\ m \end{bmatrix}.$$

40. Give a combinatorial proof of Identity 274. (See Exercise 39.)

41. Prove the following identity involving the Lah numbers, giving the nth derivative of $e^{1/x}$.

Identity 275.

$$\frac{d^n}{dx^n}(e^{1/x}) = (-1)^n e^{1/x} \sum_{k=0}^{n} L(n,k) x^{-n-k}.$$

42. Let P_n be the number of ways to arrange n pairs of parentheses correctly, so that every close parenthesis matches with an open parenthesis that precedes it. For example, when $n = 3$ there are five correct ways to arrange n pairs of parentheses: $((()))$, $(())()$, $()(())$, $(()())$, $()()()$. Prove that $P_n = C_n$, the nth Catalan number.

43. Let L_n be the number of sequences of n "+" symbols and n "−" symbols such that every partial sum is nonnegative. (Think of "+" as 1 and "−" as −1 for the purposes of addition.) Prove that $L_n = C_n$, the nth Catalan number. For example, when $n = 3$ there are five valid sequences.

$$+ + + - - - \qquad + + - + - - \qquad + + - - + -$$
$$+ - + + - - \qquad + - + - + -$$

44. Suppose $2n$ people are seated around a circular table. Let X_n denote the number of ways each of these people can shake hands with someone else without two people's arms crossing. Assuming that $X_0 = 1$, prove that $X_n = C_n$, the nth Catalan number. For example, when $n = 3$ there are five valid sets of three pairs of handshakes (where we have numbered the people in order around the table from 1 to 6):

$$\{(1,2),(3,4),(5,6)\}, \qquad \{(1,6),(2,3),(4,5)\},$$
$$\{(1,2),(3,6),(4,5)\}, \qquad \{(1,6),(2,5),(3,4)\},$$
$$\{(1,4),(2,3),(5,6)\}.$$

45. Prove

Identity 276.

$$\lim_{n \to \infty} \frac{C_{n+1}}{C_n} = 4.$$

46. Prove the following variation on the ballot theorem (Theorem 31).

Theorem 36 (Weak Ballot Theorem). *The number of lattice paths from $(0,0)$ to (n,m), $n \geq m$, that never go above the main diagonal is given by*

$$\frac{n+1-m}{n+1} \binom{n+m}{n}.$$

47. Prove

Identity 277.

$$\sum_{k=0}^{n} \binom{2k}{k} C_{n-k} = \binom{2n+1}{n}.$$

48. Use a lattice path argument to prove the following.

Identity 278. *For $r \geq 1$,*

$$\sum_{k=0}^{n} \binom{2k+r}{k} \binom{2n-2k}{n-k} \frac{r}{2k+r} = \binom{2n+r}{n}.$$

49. Use a lattice path argument to prove the following.

    Identity 279. *For $n \geq m$, $n \geq 1$,*

    $$\sum_{k=0}^{m} \binom{n+k}{k} \frac{n-k}{n+k} = \frac{n-m+1}{n+m+1} \binom{n+m+1}{m}.$$

50. Use a lattice path argument to prove the following.

    Identity 280. *For $n \geq m$,*

    $$\sum_{k=0}^{m} \binom{n+k}{k} \frac{n+1-k}{n+1} = \frac{n-m+2}{n+2} \binom{n+m+1}{m}.$$

51. Suppose we allow for "up-right" diagonal steps in our lattice as well as up steps and right steps. How many paths are there from $(0,0)$ to (n, m) when one can take a right step, an up step, or an up-right (diagonal) step at each stage? Let $D(n, m)$ denote the number of such paths; the $D(n, m)$ numbers are called the *Delannoy numbers* (sequence A008288 in the OEIS). Prove that the Delannoy numbers satisfy the following recurrence.

    Identity 281. *For $n, m \geq 1$,*

    $$D(n, m) = D(n, m-1) + D(n-1, m) + D(n-1, m-1).$$

52. Prove that if (a_n) is a sequence satisfying

    $$a_n = \sum_{k=0}^{n} \binom{n}{k} a_k,$$

    then (a_n) is the zero sequence.

53. Prove that B_1 is the only odd Bernoulli number that is nonzero.

54. Prove the following exponential generating function involving the even Bernoulli numbers.

    Identity 282.

    $$\sum_{k=0}^{\infty} B_{2k} \frac{x^{2k}}{(2k)!} = \frac{x}{2} \coth \frac{x}{2}.$$

    Here, $\coth x$ is the *hyperbolic cotangent* of x, defined as $\coth x = \cosh x / \sinh x$. (See Exercise 23 in Chapter 6 for definitions of $\cosh x$ and $\sinh x$.)

55. Prove the following, showing that the Bernoulli numbers and the sequence of reciprocals of the positive integers are inverse sequences under binomial convolution.

Identity 283.

$$\sum_{k=0}^{n} \binom{n}{k} \frac{B_{n-k}}{k+1} = 1.$$

56. Prove the following, an alternate expression for the binomial transform of the Bernoulli numbers.

Identity 284.

$$\sum_{k=0}^{n} \binom{n}{k} B_k = (-1)^n B_n.$$

8.9 Notes

Two good sources on the Fibonacci numbers (both identities and properties) are texts by Grimaldi [33] and Koshy [40]. Four of the nine chapters in Benjamin and Quinn's *Proofs that Really Count* [7] discuss combinatorial proofs of identities involving Fibonacci numbers or their generalizations. In particular, Exercises 1 through 9 of Chapter 1 of *Proofs that Really Count* feature combinatorial interpretations of Fibonacci numbers.

Chapter 8 of Charalambides's *Enumerative Combinatorics* [15] contains a wealth of identities involving the Stirling numbers, including (on p. 291) an explicit formula for the Stirling numbers of the first kind involving a double sum. There are several Stirling identities given on pages 264–265 of *Concrete Mathematics* [32] as well. See also Chapter 7 of *Proofs that Really Count* for combinatorial proofs of many Stirling number identities. Identity 264 appears in Spivey [64] and Identity 265 in Mező [47].

Spivey [63] applies the techniques in Section 7.2 using finite differences to find identities for the Stirling numbers of the first kind.

The reflection principle used in the proof of the ballot theorem (Theorem 31) has been generally attributed to André [2]. However, Renault [56] points out that André actually used a different but related method. The flaws (not "flawed") proof of Identity 244 is adapted from Chen [16].

Spivey [66] contains a generalization of Identity 245 and Theorems 32 and 33 using the line $y = rx$, where $r \in \mathbb{Z}^+$.

There are many lattice path problems that have been studied besides the ones presented here. For example, the number of paths from $(0,0)$ to (n,n) that allow for right steps, up steps, and up-right steps but that do not go above the main diagonal are counted by the *Schröder numbers* (sequence A006318 in the OEIS). Thus the Schröder numbers are to the Delannoy numbers (see

Exercise 51) what the Catalan numbers are to the binomial coefficients in terms of lattice path counting. The number of Catalan paths that have a fixed number of valleys (where a *valley* is a right step followed by an up step) is counted by the *Narayana numbers* (sequence A001263 in the OEIS). Thus the row sums of the Narayana numbers are the Catalan numbers. In addition, the number of paths in a lattice from $(0,0)$ to $(n,0)$ that allow for up-right, down-right, or right steps but that do not go below the horizontal axis are counted by the *Motzkin numbers* (sequence A001006 in the OEIS).

General references for lattice path problems include Feller [26], Mohanty [49], and Humphreys [37]. (These were also mentioned in the notes to Chapter 3.)

It is worth repeating that Stanley's text [76] contains 214 combinatorial interpretations of the Catalan numbers. Other Catalan number references include Grimaldi [33] and Koshy [41].

9

Miscellaneous Techniques

In this chapter we show how ideas from two mostly unrelated areas of mathematics—complex numbers and linear algebra—can be used to prove binomial identities. Complex numbers turn out to be a useful tool for proving a variety of identities involving aerated, alternating binomial sums. (For example, $\sum_{k \geq 0} \binom{n}{2k}(-1)^k$ is such a sum because it alternates and because it features every other binomial coefficient from row n of Pascal's triangle.) Proving an identity involving such a sum is often accomplished by substituting i or, more generally, a complex root of unity into the binomial theorem. We review basic properties of complex numbers before diving into the details of this process.

Placing Pascal's triangle into a matrix is our starting point for the use of linear algebra in proving binomial identities. Once this is done, several binomial identities have very simple matrix proofs. In addition, the binomial inversion technique we saw in Chapter 2 turns out to be matrix inversion in disguise. Matrices containing Stirling and Lah numbers will give us more insight into the relationships between these numbers as well. We end the chapter with a matrix proof of Cassini's identity for Fibonacci numbers.

For the section on linear algebra the reader will need to have seen some concepts normally encountered in a first course in that subject. Understanding how to perform basic matrix operations like transposition, multiplication, and inversion should be sufficient to follow most of the arguments. Some determinant properties are used as well. Concepts associated with linear independence are used in one proof (Theorem 37) and in one of the exercises (Exercise 17).

9.1 Complex Numbers

Before we take a look at how to prove binomial identities using complex numbers, let's review some basic facts about them. Any complex number can be represented in the form $a + bi$, where a and b are real numbers. In the usual representation of the complex plane the horizontal axis is the real axis, while the vertical axis is the imaginary axis. Thus the number $a + bi$ appears in the complex plane as the point (a, b). This allows for yet another representation of $a + bi$: via its *modulus*, which is the distance $\sqrt{a^2 + b^2}$ from the

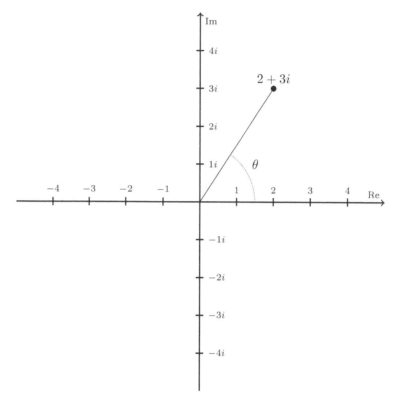

FIGURE 9.1
The complex number $2 + 3i$ in the complex plane

origin to the point (a, b) in the complex plane, and its *argument*, the angle θ that the point makes with the positive real axis. Thanks to Euler's formula $e^{ix} = \cos x + i \sin x$, the modulus-argument (or exponential) representation of $a + bi$ is given by $a + bi = \sqrt{a^2 + b^2}\,e^{i\theta}$. (See Exercise 2 for an explanation.) Many of our identities will involve $b = 1$ and $a > 0$, and in this case θ satisfies $0 < \theta < \pi/2$. For this reason we will often denote θ by $\arctan(b/a)$.

For example, in Figure 9.1 we have the complex number $2 + 3i$ plotted at $(2, 3)$ in the complex plane. The length of the line segment from $2 + 3i$ to the origin is $\sqrt{4 + 9} = \sqrt{13}$, and the angle θ that the point $(2, 3)$ makes with the positive real axis is $\theta = \arctan(3/2)$. Thus the modulus-argument representation of $2 + 3i$ is given by $2 + 3i = \sqrt{13}\,e^{i\arctan(3/2)}$.

Let's now look at how to use complex numbers to prove some binomial identities. In the first chapter of this book we proved the binomial theorem (Identity 3): $(x+y)^n = \sum_{k=0}^{n} \binom{n}{k} x^k y^{n-k}$. There is nothing about the binomial theorem or its proof that requires x or y to be real; rather, the binomial theorem tells us something about the number of times certain powers of x

and y occur together when $(x + y)^n$ is expanded. (In fact, we even used the shift operator E in place of x when we proved Identity 174 in Chapter 7.) If we replace x with i in the binomial theorem then we get the following.

Identity 285. *For real $y > 0$,*

$$\sum_{k=0}^{n} \binom{n}{k} i^k y^{n-k} = (y^2 + 1)^{n/2} e^{i\,n\,\arctan(1/y)}.$$

Proof. Substituting i for x in the binomial theorem yields

$$\sum_{k=0}^{n} \binom{n}{k} i^k y^{n-k} = (i + y)^n.$$

Since the real part of $y + i$ is y and the imaginary part is 1, the modulus of $y + i$ is $\sqrt{y^2 + 1}$ and the argument is $\arctan(1/y)$. Rewriting $y + i$ in exponential form, then, we have

$$(y + i)^n = \left(\sqrt{y^2 + 1}\, e^{i\arctan(1/y)} \right)^n = (y^2 + 1)^{n/2} e^{i\,n\,\arctan(1/y)}.$$

$\square$

At first glance Identity 285 may not look particularly appealing, but it can actually be used to derive some interesting binomial identities—including ones that involve only real numbers. In particular, taking values of y for which $\arctan(1/y)$ evaluates to a common angle works well. For example, the following identities result from letting $y = 1$ in Identity 285.

Identity 286.

$$\sum_{k \geq 0} \binom{n}{2k} (-1)^k = 2^{n/2} \cos \frac{n\pi}{4}.$$

Identity 287.

$$\sum_{k \geq 0} \binom{n}{2k + 1} (-1)^k = 2^{n/2} \sin \frac{n\pi}{4}.$$

Since $\cos(n\pi/4)$ and $\sin(n\pi/4)$ each break into multiple cases depending on the value of $n \bmod 8$, it is simplest to leave the identities in terms of trigonometric functions. Even when n is odd, though, the expressions on the right side of Identities 286 and 287 evaluate to rational numbers because the resulting values of cosine and sine are either $1/\sqrt{2}$ or $-1/\sqrt{2}$. (Also, compare Identity 286 with Identity 196 and Identity 287 with Identity 215.)

Proof. With $y = 1$ in Identity 285, as well as Euler's formula $e^{ix} = \cos x + i \sin x$, we have

$$\sum_{k=0}^{n} \binom{n}{k} i^k = 2^{n/2} e^{i\,n\,\arctan(1)} = 2^{n/2} e^{i\,n\pi/4} = 2^{n/2} \left(\cos \frac{n\pi}{4} + i \sin \frac{n\pi}{4} \right).$$

Since the powers of i cycle through $1, i, -1$, and $-i$, the even powers of i result in the pattern $1, -1, 1, -1, \ldots$, and the odd powers of i result in the pattern $i, -i, i, -i, \ldots$. Thus the real and imaginary parts of $\sum_{k=0}^{n} \binom{n}{k} i^k$ consist of those terms containing, respectively, the even and odd powers of i. Since these powers each alternate in sign, separating $\sum_{k=0}^{n} \binom{n}{k} i^k$ into real and imaginary parts yields the identities we want:

$$\sum_{k \geq 0} \binom{n}{2k} (-1)^k = 2^{n/2} \cos \frac{n\pi}{4}, \text{ and}$$

$$\sum_{k \geq 0} \binom{n}{2k+1} (-1)^k = 2^{n/2} \sin \frac{n\pi}{4}.$$

$\square$

Identities 286 and 287 give a good sense of the kinds of binomial identities for which complex numbers can be useful in proving. Substituting a number like i into the binomial theorem and separating into real and imaginary parts yields a binomial sum that is aerated (it includes every second binomial coefficient in a particular row) and that alternates. While we have seen both aerated sums and alternating sums in this text, Identities 286 and 287 feature our first binomial sums that are both aerated and alternating. In fact, without using complex numbers it is difficult to see how else one might prove such identities. (Although, again, see Identity 196 and its proof, as well as Identity 215.)

We can generalize Identities 286 and 287 as well.

Swiss mathematician **Leonhard Euler** (1707–1783) was the most prolific mathematician of the 1700s. He made important discoveries in virtually every area he touched, from calculus and infinite series to graph theory, number theory, and mechanics. For the last seventeen years of his life he was almost completely blind, yet his incredible memory (for example, he could recite the complete text of Virgil's *Aeneid*) was such that his lack of eyesight did not slow his mathematical output. French mathematician Pierre-Simon Laplace is said to have referred to Euler as "the master of us all."

Identity 288. *For real $x > 0$,*

$$\sum_{k \geq 0} \binom{n}{2k} x^k (-1)^k = (x+1)^{n/2} \cos(n \arctan \sqrt{x}).$$

Identity 289. *For real $x > 0$,*

$$\sum_{k \geq 0} \binom{n}{2k+1} x^k (-1)^k = \frac{(x+1)^{n/2}}{\sqrt{x}} \sin(n \arctan \sqrt{x}).$$

Proof. Let $y = 1/\sqrt{x}$ in Identity 285. This yields

$$\sum_{k=0}^{n} \binom{n}{k} i^k x^{(k-n)/2} = (1 + 1/x)^{n/2} e^{in \arctan \sqrt{x}}$$

$$= (1 + 1/x)^{n/2} \left(\cos(n \arctan \sqrt{x}) + i \sin(n \arctan \sqrt{x}) \right).$$

Multiplying both sides by $x^{n/2}$ produces

$$\sum_{k=0}^{n} \binom{n}{k} i^k x^{k/2} = x^{n/2}(1 + 1/x)^{n/2} \left(\cos(n \arctan \sqrt{x}) + i \sin(n \arctan \sqrt{x}) \right)$$

$$= (x + 1)^{n/2} \left(\cos(n \arctan \sqrt{x}) + i \sin(n \arctan \sqrt{x}) \right).$$

Taking real and imaginary parts of this equation, as in the proofs of Identities 286 and 287, we have

$$\sum_{k \geq 0} \binom{n}{2k} x^{2k/2}(-1)^k = (x + 1)^{n/2} \cos(n \arctan \sqrt{x}),$$

$$\sum_{k \geq 0} \binom{n}{2k+1} x^{(2k+1)/2}(-1)^k = (x + 1)^{n/2} \sin(n \arctan \sqrt{x}),$$

from whence we get our identities. $\square$

Identity 285 uses the binomial theorem, but in truth many of the identities in Chapter 4 can be used with complex numbers to produce aerated binomial coefficient identities, too. Let's apply this idea to Identity 89,

$$\sum_{k=0}^{n} \binom{n}{k} k x^{k-1} y^{n-k} = n(x + y)^{n-1}.$$

First, we obtain the following.

Identity 290. *For real $y > 0$,*

$$\sum_{k=0}^{n} \binom{n}{k} k i^k y^{n-k} = in(1 + y^2)^{(n-1)/2} e^{i(n-1) \arctan(1/y)}.$$

Proof. Substituting i for x in Identity 89 and then multiplying by i yields

$$\sum_{k=0}^{n} \binom{n}{k} k i^k y^{n-k} = in(y + i)^{n-1}.$$

Rewriting $(y + i)^{n-1}$ as a complex exponential as in the proof of Identity 285, the right side becomes

$$in(y + i)^{n-1} = in(y^2 + 1)^{(n-1)/2} e^{i(n-1) \arctan(1/y)}.$$

$\square$

If we separate the real and imaginary parts of Identity 290 we can get the following two identities.

Identity 291. *For real $x > 0$,*

$$\sum_{k \geq 0} \binom{n}{2k} kx^k(-1)^k = -\frac{nx^{1/2}(x+1)^{(n-1)/2}}{2} \sin((n-1)\arctan\sqrt{x}).$$

Identity 292. *For real $x > 0$,*

$$\sum_{k \geq 0} \binom{n}{2k+1} kx^k(-1)^k$$

$$= \frac{(x+1)^{(n-1)/2}}{2} \left(n\cos((n-1)\arctan\sqrt{x}) - \sqrt{1+1/x} \sin(n\arctan\sqrt{x}) \right).$$

Proof. Let $y = 1/\sqrt{x}$ in Identity 290. This produces

$$\sum_{k=0}^{n} \binom{n}{k} ki^k x^{(k-n)/2} = in(1+1/x)^{(n-1)/2} e^{i\,(n-1)\arctan(\sqrt{x})}$$

$$= in(1+1/x)^{(n-1)/2} \left(\cos((n-1)\arctan\sqrt{x}) + i\sin((n-1)\arctan\sqrt{x}) \right)$$

$$= n(1+1/x)^{(n-1)/2} \left(i\cos((n-1)\arctan\sqrt{x}) - \sin((n-1)\arctan\sqrt{x}) \right).$$

Multiplying both sides by $x^{n/2}$ gives us

$$\sum_{k=0}^{n} \binom{n}{k} ki^k x^{k/2}$$

$$= nx^{1/2}(x+1)^{(n-1)/2} \left(i\cos((n-1)\arctan\sqrt{x}) - \sin((n-1)\arctan\sqrt{x}) \right).$$

The real part of this equation is

$$\sum_{k \geq 0} \binom{n}{2k} 2kx^k(-1)^k = -nx^{1/2}(x+1)^{(n-1)/2} \sin((n-1)\arctan\sqrt{x}),$$

and dividing by 2 yields Identity 291.

The imaginary part is

$$\sum_{k \geq 0} \binom{n}{2k+1} (2k+1)x^{(2k+1)/2}(-1)^k$$

$$= nx^{1/2}(x+1)^{(n-1)/2} \cos((n-1)\arctan\sqrt{x}).$$

Dividing by $\sqrt{x}$, subtracting $\sum_{k \geq 0} \binom{n}{2k+1} x^k(-1)^k$ (via Identity 289), dividing by 2, and then factoring yields Identity 292. $\qquad\square$

Substituting different values for x in Identities 291 and 292 produces different identities; for example, the following two result from letting $x = 1$.

Identity 293.

$$\sum_{k\geq 0}\binom{n}{2k}k(-1)^k = -n2^{(n-3)/2}\sin\frac{(n-1)\pi}{4}.$$

Identity 294.

$$\sum_{k\geq 0}\binom{n}{2k+1}k(-1)^k = 2^{(n-3)/2}\left(n\cos\frac{(n-1)\pi}{4} - \sqrt{2}\sin\frac{n\pi}{4}\right).$$

See Exercises 8 and 9 for variations of Identities 293 and 294. Also, Exercise 10 generalizes Identity 290.

Another identity we proved in Chapter 4 is Identity 95, obtained by integrating the binomial theorem:

$$\sum_{k=0}^{n}\binom{n}{k}\frac{(x_2^{k+1} - x_1^{k+1})y^{n-k}}{k+1} = \frac{(x_2+y)^{n+1} - (x_1+y)^{n+1}}{n+1}.$$

We can obtain aerated, alternating binomial sum identities from Identity 95. As with previous examples in this chapter, we will start with an identity that contains powers of i.

Identity 295. *For real $y > 0$,*

$$\sum_{k=0}^{n}\binom{n}{k}\frac{(i^{k+1} - x^{k+1})y^{n-k}}{k+1} = \frac{(y^2+1)^{(n+1)/2}e^{i(n+1)\arctan(1/y)} - (x+y)^{n+1}}{n+1}.$$

Proof. In Identity 95, let $x_1 = x$ and $x_2 = i$. Then convert $(i+y)^{n+1}$ to a complex exponential as in the proofs of Identities 285 and 290. □

From Identity 295 we can derive the following two identities.

Identity 296. *For real $x > 0$,*

$$\sum_{k\geq 0}\binom{n}{2k}\frac{x^k(-1)^k}{2k+1} = \frac{(x+1)^{(n+1)/2}\sin\left((n+1)\arctan\sqrt{x}\right)}{\sqrt{x}(n+1)}.$$

Identity 297. *For real $x > 0$,*

$$\sum_{k\geq 0}\binom{n}{2k+1}\frac{x^k(-1)^k}{k+1}$$

$$= \frac{2}{x(n+1)}\left(1 - (x+1)^{(n+1)/2}\cos\left((n+1)\arctan\sqrt{x}\right)\right).$$

Proof. Let $x = 0$ in Identity 295; then let $y = 1/\sqrt{x}$. Multiply both sides by $x^{n/2}$ and divide by i. The real part of the result is Identity 296. Multiply both sides of the imaginary part by $2/\sqrt{x}$ to obtain Identity 297. □

Letting $x = 1$ in Identities 296 and 297 yields the following.

Identity 298.

$$\sum_{k \geq 0} \binom{n}{2k} \frac{(-1)^k}{2k+1} = \frac{2^{(n+1)/2}}{n+1} \sin \frac{(n+1)\pi}{4}.$$

Identity 299.

$$\sum_{k \geq 0} \binom{n}{2k+1} \frac{(-1)^k}{k+1} = \frac{2}{n+1} \left(1 - 2^{(n+1)/2} \cos \frac{(n+1)\pi}{4} \right).$$

Thus far we have seen complex numbers used only with the binomial theorem and some of its variations. Now we will take a look at a different application of complex numbers; namely, in proving the following multiple-angle formulas for sine and cosine. We'll still use the binomial theorem—just in a different way than we have seen thus far.

Identity 300.

$$\cos nx = \sum_{k \geq 0} \binom{n}{2k} \sin^{2k} x \cos^{n-2k} x \, (-1)^k.$$

Identity 301.

$$\sin nx = \sum_{k \geq 0} \binom{n}{2k+1} \sin^{2k+1} x \cos^{n-2k-1} x \, (-1)^k.$$

Proof. By Euler's formula, the complex exponential e^{inx} is equal to $\cos nx + i \sin nx$. However, we have $e^{inx} = (e^{ix})^n = (\cos x + i \sin x)^n$ as well. Applying the binomial theorem yields

$$\cos nx + i \sin nx = (\cos x + i \sin x)^n$$

$$= \sum_{k=0}^{n} \binom{n}{k} i^k \sin^k x \cos^{n-k} x.$$

Finally, separating this equation into its real and imaginary parts produces the identities we are after:

$$\cos nx = \sum_{k \geq 0} \binom{n}{2k} \sin^{2k} x \cos^{n-2k} x \, (-1)^k, \text{ and}$$

$$\sin nx = \sum_{k \geq 0} \binom{n}{2k+1} \sin^{2k+1} x \cos^{n-2k-1} x \, (-1)^k.$$

$\square$

Thus far in this chapter we have seen sum identities involving even-index and odd-index binomial coefficients. For the rest of this section we consider the problem of proving sum identities involving every third, fourth, or, in general, every rth index binomial coefficient. For identities like these we need what are called *roots of unity*.

An *rth root of unity* is a complex number ζ such that $\zeta^r = 1$. A *primitive rth root of unity* is an rth root of unity that is not also a kth root of unity for a positive integer value of k smaller than r. For example, $e^{2\pi i/3}$ is a primitive third root of unity because $\left(e^{2\pi i/3}\right)^3 = e^{2\pi i} = 1$, and no smaller positive integer power of r has $\left(e^{2\pi i/3}\right)^r = 1$. We can also write $e^{2\pi i/3} = \cos(2\pi/3) + i\sin(2\pi/3) = -\frac{1}{2} + \frac{i\sqrt{3}}{2}$.

Roots of unity are useful for proving binomial coefficient identities involving every rth index because of the following property.

Identity 302. *Let $\zeta = e^{2\pi i/r}$, where ζ is a primitive rth root of unity. Then*

$$\sum_{k=0}^{r-1} \zeta^{sk} = \begin{cases} r, & r \mid s; \\ 0, & otherwise. \end{cases}$$

Proof. If $r \mid s$, then $\zeta^s = 1$, and the sum becomes $1 + 1 + \cdots + 1 = r$. If r does not divide s, then $\zeta^s \neq 1$, and we have, by the geometric sum formula,

$$\sum_{k=0}^{r-1} \zeta^{sk} = \sum_{k=0}^{r-1} (\zeta^s)^k = \frac{1 - \zeta^{rs}}{1 - \zeta^s} = \frac{1 - (\zeta^r)^s}{1 - \zeta^s} = 0.$$

$\square$

With Identity 302 in mind, we can prove the following result about sums involving every rth index binomial coefficients.

Identity 303. *Let $\zeta = e^{2\pi i/r}$, where ζ is a primitive rth root of unity. Then*

$$\sum_{k \geq 0} \binom{n}{rk+m} x^{n-rk-m} = \frac{1}{r} \sum_{j=0}^{r-1} (\zeta^j + x)^n (\zeta^j)^{-m}.$$

Proof. Starting with the right side, use the binomial theorem and swap the order of summation. This yields

$$\frac{1}{r} \sum_{j=0}^{r-1} (\zeta^j + x)^n (\zeta^j)^{-m} = \frac{1}{r} \sum_{j=0}^{r-1} \sum_{l=0}^{n} \binom{n}{l} \zeta^{jl} x^{n-l} (\zeta^j)^{-m}$$

$$= \frac{1}{r} \sum_{l=0}^{n} \binom{n}{l} x^{n-l} \sum_{j=0}^{r-1} \zeta^{(l-m)j}.$$

By Identity 302, we have

$$\frac{1}{r}\sum_{l=0}^{n}\binom{n}{l}x^{n-l}\sum_{j=0}^{r-1}\zeta^{(l-m)j} = \sum_{l=0}^{n}\binom{n}{l}x^{n-l}\,[r|(l-m)],$$

where $[r|(l-m)]$ is the Iverson bracket denoting whether r divides $l-m$.

However, the requirement $r|(l-m)$ is equivalent to saying $l = rk + m$ for some k, and the identity follows. □

Identity 303 is the key to proving identities that involve row sums and evenly-spaced binomial coefficients. As an example, Identity 303 can be used to prove the following.

Identity 304.

$$\sum_{k\geq 0}\binom{n}{3k} = \frac{1}{3}\left(2^n + 2\cos\frac{n\pi}{3}\right).$$

Identity 305.

$$\sum_{k\geq 0}\binom{n}{3k+1} = \frac{1}{3}\left(2^n + 2\cos\frac{(n-2)\pi}{3}\right).$$

Identity 306.

$$\sum_{k\geq 0}\binom{n}{3k+2} = \frac{1}{3}\left(2^n + 2\cos\frac{(n+2)\pi}{3}\right).$$

Proof. We prove Identity 304, saving Identities 305 and 306 for Exercise 12. Letting $r = 3$, $m = 0$, and $x = 1$ in Identity 303, we have

$$\sum_{k\geq 0}\binom{n}{3k} = \frac{1}{3}\sum_{j=0}^{2}(\zeta^j + 1)^n,$$

where

$$\zeta = e^{2\pi i/3} = \cos(2\pi/3) + i\sin(2\pi/3) = -\frac{1}{2} + \frac{i\sqrt{3}}{2},$$

$$\zeta^2 = e^{4\pi i/3} = \cos(4\pi/3) + i\sin(4\pi/3) = -\frac{1}{2} - \frac{i\sqrt{3}}{2}.$$

Thus

$$\zeta + 1 = \frac{1}{2} + \frac{i\sqrt{3}}{2} = e^{i\pi/3},$$

$$\zeta^2 + 1 = \frac{1}{2} - \frac{i\sqrt{3}}{2} = e^{-i\pi/3}.$$

This means we have

$$\frac{1}{3}\sum_{j=0}^{2}(\zeta^{j}+1)^{n} = \frac{1}{3}\left(2^{n}+e^{n\pi i/3}+e^{-n\pi i/3}\right) = \frac{1}{3}\left(2^{n}+2\cos\frac{n\pi}{3}\right),$$

where in the last step we use the identity $\cos x = (e^{ix}+e^{-ix})/2$. (See Exercise 3.) $\qquad\square$

The exercises contain other examples of the use of Identity 303.

9.2 Linear Algebra

In this section we consider the use of linear algebra to prove binomial identities, including identities involving the two kinds of Stirling numbers, the Lah numbers, and the Fibonacci numbers.

Many of the identities we have considered in this book are of the form $b_n = \sum_{k=0}^{n}\binom{n}{k}a_k$ for some sequences (a_n) and (b_n). This relationship can actually be expressed in matrix form. To see this, take the left-justified version of Pascal's triangle up through row m and place it into a square matrix P_m. This matrix is known as the *Pascal matrix*. For example, P_3 looks like the following:

$$P_3 = \begin{bmatrix} 1 & 0 & 0 & 0 \\ 1 & 1 & 0 & 0 \\ 1 & 2 & 1 & 0 \\ 1 & 3 & 3 & 1 \end{bmatrix}.$$

More formally, the (i,j) entry in matrix P_m is $\binom{i}{j}$. (In this section we index matrices starting with row and column 0, as it makes the indexing consistent with Pascal's triangle. We index vectors starting with 0 as well.)

Let $\mathbf{a} = (a_0, a_1, a_2, a_3)$, and let $\mathbf{b} = (b_0, b_1, b_2, b_3)$. Then the relation $b_n = \sum_{k=0}^{n}\binom{n}{k}a_k$, for $n \in \{0,1,2,3\}$, is captured by the matrix equation

$$\mathbf{b} = P_3\mathbf{a}.$$

Of course, by increasing the value of m and including more of the (a_n) and (b_n) sequences we can express the binomial transform relationship between (a_n) and (b_n) up through any fixed value of n.

Let's now take a look at some binomial coefficient identities we can derive from the relationship $\mathbf{b} = P_m\mathbf{a}$ (for vectors $\mathbf{a}$ and $\mathbf{b}$ in $\mathbb{R}^{m+1}$). Since P_m is lower-triangular with all 1's on its diagonal, it is invertible. Thus we can calculate P_m^{-1}, which must also be lower triangular. We can take powers of P_m, too. We can also transpose P_m. Each of these matrices, when combined with the original matrix P_m, will give us a binomial identity.

Let's start with P_m^{-1}, which is given by the next theorem.

Theorem 37. P_m^{-1} *is the* $(m+1) \times (m+1)$ *matrix whose* (i,j) *entry is* $\binom{i}{j}(-1)^{i-j}$.

For example,

$$P_3^{-1} = \begin{bmatrix} 1 & 0 & 0 & 0 \\ -1 & 1 & 0 & 0 \\ 1 & -2 & 1 & 0 \\ -1 & 3 & -3 & 1 \end{bmatrix}.$$

Proof. Suppose we multiply P_m by the column vector $\mathbf{u} = (1, x, x^2, \ldots, x^m)$ to obtain some vector $\mathbf{v}$. Since row n of P_m consists of row n of Pascal's triangle, we have, by the binomial theorem, $v_n = \sum_{k=0}^{n} \binom{n}{k} x^k = (x+1)^n$. Thus $\mathbf{v} = (1, x+1, (x+1)^2, \ldots, (x+1)^m)$. For example, when $m = 3$ we have

$$\begin{bmatrix} 1 & 0 & 0 & 0 \\ 1 & 1 & 0 & 0 \\ 1 & 2 & 1 & 0 \\ 1 & 3 & 3 & 1 \end{bmatrix} \begin{bmatrix} 1 \\ x \\ x^2 \\ x^3 \end{bmatrix} = \begin{bmatrix} 1 \\ x+1 \\ (x+1)^2 \\ (x+1)^3 \end{bmatrix}.$$

Now, let's find a lower-triangular matrix that maps $\mathbf{v} = (1, x+1, (x+1)^2, \ldots, (x+1)^m)$ back to $\mathbf{u} = (1, x, x^2, \ldots, x^m)$. From the binomial theorem we know that

$$\sum_{k=0}^{n} \binom{n}{k}(x+1)^k(-1)^{n-k} = (x+1-1)^n = x^n = u_n.$$

Since this is true for each n, the lower-triangular matrix that has entry (i,j) as $\binom{i}{j}(-1)^{i-j}$ maps $\mathbf{v}$ to $\mathbf{u}$. Let's call this matrix A_m for now. For example,

$$A_3 = \begin{bmatrix} 1 & 0 & 0 & 0 \\ -1 & 1 & 0 & 0 \\ 1 & -2 & 1 & 0 \\ -1 & 3 & -3 & 1 \end{bmatrix}.$$

But is it the case that $A_m = P_m^{-1}$? To find the inverse of a matrix, it's not sufficient to find a matrix that reverses the original matrix's mapping on a single vector. It's not even sufficient to find a matrix that reverses the original matrix's mapping on an infinite number of vectors. If we can show that A_m reverses P_m's mapping on a *basis* of $\mathbb{R}^{m+1}$, though, that will be sufficient to prove that $A_m = P_m^{-1}$. It turns out that $\mathbf{u} = \{(1, x, x^2, x^3, \ldots, x^m)\}$ for $m+1$ distinct real values of x is actually a basis for $\mathbb{R}^{m+1}$. (See Exercises 17 and 18.) This means that the same set of vectors with $x+1$ in place of x is also a basis for $\mathbb{R}^{m+1}$. Thus we have already shown that A_m reverses P_m's mapping on a basis of $\mathbb{R}^{m+1}$. Therefore, $A_m = P_m^{-1}$. □

Theorem 37 immediately leads to a matrix proof of Identity 26.

Identity 26.

$$\sum_{k=0}^{n} \binom{n}{k}\binom{k}{m}(-1)^k = (-1)^m[n = m].$$

Proof. Let $r \geq m, n$. Clearly, $P_r P_r^{-1} = I_r$. This means that multiplying row n of P_r by column m of P_r^{-1} gives 1, if $n = m$, and 0, otherwise. In other words,

$$\sum_{k=0}^{n} \binom{n}{k}\binom{k}{m}(-1)^{k-m} = [n = m].$$

Multiplying both sides by $(-1)^m$ yields the identity. □

The same idea can be used to prove the binomial inversion formula. Moreover, this argument actually goes both ways, in the sense that P_m^{-1} can also be derived from the variation of the binomial inversion formula in Theorem 2. See Exercises 19 and 20.

Let's now consider powers of P_m. First, define $P_m(x)$ to be the matrix whose (i, j) entry is $\binom{i}{j}x^{i-j}$. For example,

$$P_m(x) = \begin{bmatrix} 1 & 0 & 0 & 0 \\ x & 1 & 0 & 0 \\ x^2 & 2x & 1 & 0 \\ x^3 & 3x^2 & 3x & 1 \end{bmatrix}.$$

We see that $P_m(1) = P_m$, and $P_m(-1) = P_m^{-1}$. In general, $P_m(x)$ gives the xth power of P_m, as we see in the following theorem.

Identity 307. *For $x \in \mathbb{Z}$, $P_m(x) = P_m^x$.*

Proof. When $x = 0$ the identity states that $I_{m+1} = I_{m+1}$. Now, suppose $x \in \mathbb{Z}^+$. If we multiply $P_m(x)$ by the vector $\mathbf{u} = (1, y, \ldots, y^m)$ we get the vector $\mathbf{v}$ with $v_n = \sum_{k=0}^{n} \binom{n}{k}x^{n-k}y^k = (x + y)^n$.

We also know that $P_m\mathbf{u} = (1, y + 1, \ldots, (y + 1)^m)$. Then $P_m^2\mathbf{u} = P_m(P_m\mathbf{u}) = (1, y + 2, \ldots (y + 2)^m)$. If we continue this process of successively left-multiplying by P_m, after x total multiplications we end with $P_m^x\mathbf{u} = (1, x + y, \ldots, (x + y)^m) = \mathbf{v}$. This holds regardless of the value of y. As in the proof of Theorem 37, then, we have $P_m(x)$ and P_m^x mapping a set of basis vectors for $\mathbb{R}^{m+1}$ to the same vectors. Thus $P_m(x) = P_m^x$ when $x \in \mathbb{Z}^+$.

Since $P^m(-1) = P_m^{-1}$ as well, a similar argument shows that $P_m(x) = P_m^x$ for all $x \in \mathbb{Z}$. □

Identity 307 immediately gives us a matrix proof of the following identity.

Identity 49.

$$\sum_{k=0}^{n} \binom{n}{k}\binom{k}{m} = 2^{n-m}\binom{n}{m}.$$

Proof. Let $r \geq m, n$. The left side of Identity 49 is the product of row n of P_r and column m of P_r and thus is entry (n, m) in P_r^2. Entry (n, m) in $P_r(2)$ is $2^{n-m}\binom{n}{m}$. By Identity 307, $P_r^2 = P_r(2)$, completing the proof.   □

We can use the same idea to generalize Identity 49, too.

Identity 50. *For $x, y \in \mathbb{Z}$,*

$$\sum_{k=0}^{n} \binom{n}{k}\binom{k}{m} x^{n-k} y^{k-m} = (x+y)^{n-m} \binom{n}{m}.$$

Proof. Let $r \geq m, n$. The left side of Identity 50 is the product of row n of $P_r(x)$ and column m of $P_r(y)$ and thus is entry (n, m) in $P_r(x)P_r(y)$. Entry (n, m) in $P_r(x+y)$ is $(x+y)^{n-m}\binom{n}{m}$. By Identity 307, $P_r(x)P_r(y) = P_r^x P_r^y = P_r^{x+y} = P_r(x+y)$, completing the proof.   □

Next, we consider P_m^T, which we denote R_m. Thus the (i, j) entry in R_m is $\binom{j}{i}$. For example,

$$R_3 = \begin{bmatrix} 1 & 1 & 1 & 1 \\ 0 & 1 & 2 & 3 \\ 0 & 0 & 1 & 3 \\ 0 & 0 & 0 & 1 \end{bmatrix}.$$

If we let $\mathbf{u} = (1, x, \ldots, x^m)$, multiplying $R_m \mathbf{u}$ gives us a vector $\mathbf{v}$ with $v_n = \sum_{k=0}^{m} \binom{k}{n} x^k$. Unlike $P_m \mathbf{u}$, this summation cannot be simplified by the binomial theorem. However, the corresponding infinite sum does have a nice expression; it's the variant of the negative binomial series given as Identity 129: For $0 \leq x < 1$ and integers $n \geq 0$,

$$\sum_{k=0}^{\infty} \binom{k}{n} x^k = \frac{x^n}{(1-x)^{n+1}}.$$

This formula can also be viewed as a generating function. Doing so means, as we discussed in Chapter 6, that we do not have to worry about convergence issues.

If we are going to use Identity 129 to help us understand the transpose of the Pascal matrix, we will need the infinite versions of P_m and R_m. Let P denote the infinite Pascal matrix and R its transpose, so that we have

$$P = \begin{bmatrix} 1 & 0 & 0 & 0 & \cdots \\ 1 & 1 & 0 & 0 & \cdots \\ 1 & 2 & 1 & 0 & \cdots \\ 1 & 3 & 3 & 1 & \cdots \\ \vdots & \vdots & \vdots & \vdots & \ddots \end{bmatrix}, \quad R = \begin{bmatrix} 1 & 1 & 1 & 1 & \cdots \\ 0 & 1 & 2 & 3 & \cdots \\ 0 & 0 & 1 & 3 & \cdots \\ 0 & 0 & 0 & 1 & \cdots \\ \vdots & \vdots & \vdots & \vdots & \ddots \end{bmatrix}.$$

Also, let $\mathbf{u} = (1, x, x^2, \ldots)$. Then we have, by Identity 129,

$$Ru = \begin{bmatrix} 1 & 1 & 1 & 1 & \cdots \\ 0 & 1 & 2 & 3 & \cdots \\ 0 & 0 & 1 & 3 & \cdots \\ 0 & 0 & 0 & 1 & \cdots \\ \vdots & \vdots & \vdots & \vdots & \ddots \end{bmatrix} \begin{bmatrix} 1 \\ x \\ x^2 \\ x^3 \\ \vdots \end{bmatrix} = \begin{bmatrix} 1/(1-x) \\ x/(1-x)^2 \\ x^2/(1-x)^3 \\ x^3/(1-x)^4 \\ \vdots \end{bmatrix} = \frac{1}{1-x} \begin{bmatrix} 1 \\ y \\ y^2 \\ y^3 \\ \vdots \end{bmatrix},$$

where $y = x/(1-x)$.

Let's see if we can find an expression for the matrix PR. We have

$$(PR)\mathbf{u} = P(R\mathbf{u}) = \frac{1}{1-x} \begin{bmatrix} 1 & 0 & 0 & 0 & \cdots \\ 1 & 1 & 0 & 0 & \cdots \\ 1 & 2 & 1 & 0 & \cdots \\ 1 & 3 & 3 & 1 & \cdots \\ \vdots & \vdots & \vdots & \vdots & \ddots \end{bmatrix} \begin{bmatrix} 1 \\ y \\ y^2 \\ y^3 \\ \vdots \end{bmatrix} = \frac{1}{1-x} \begin{bmatrix} 1 \\ y+1 \\ (y+1)^2 \\ (y+1)^3 \\ \vdots \end{bmatrix},$$

by the binomial theorem. However,

$$y + 1 = \frac{x}{1-x} + 1 = \frac{x+1-x}{1-x} = \frac{1}{1-x}.$$

This means that we can simplify $(PR)\mathbf{u}$ to

$$(PR)\mathbf{u} = \begin{bmatrix} 1/(1-x) \\ 1/(1-x)^2 \\ 1/(1-x)^3 \\ 1/(1-x)^4 \\ \vdots \end{bmatrix}.$$

In other words, the nth entry in $(PR)\mathbf{u}$ is $1/(1-x)^{n+1}$. Therefore, the nth row of PR consists of the sequence of numbers that generates $1/(1-x)^{n+1}$. Since a generating function determines its corresponding sequence, if we can find the sequence that generates $1/(1-x)^{n+1}$ we will have found the nth row of PR.

We do not have to look hard for this sequence; it is given by the negative binomial series formula, Identity 149:

$$\sum_{k=0}^{\infty} \binom{n+k}{n} x^k = \frac{1}{(1-x)^{n+1}}.$$

This means that PR is the matrix whose (i, j) entry is $\binom{i+j}{i}$. Let's denote this matrix Q. We have

$$Q = \begin{bmatrix} 1 & 1 & 1 & 1 & \cdots \\ 1 & 2 & 3 & 4 & \cdots \\ 1 & 3 & 6 & 10 & \cdots \\ 1 & 4 & 10 & 20 & \cdots \\ \vdots & \vdots & \vdots & \vdots & \ddots \end{bmatrix}.$$

So $PR = Q$, where Q itself is a kind of Pascal matrix. In fact, Q is known as the *symmetric Pascal matrix*.

Since we have done all of this work, we might as well call the conclusion a theorem:

Identity 308. *Let P, R, and Q, be the infinite matrices whose (i, j) entries are $\binom{i}{j}$, $\binom{j}{i}$, and $\binom{i+j}{j}$, respectively. Then $PR = Q$.*

Identity 308 gives us the following identity.

Identity 309.

$$\sum_{k=0}^{n} \binom{n}{k}\binom{m}{k} = \binom{n+m}{n}.$$

Proof. Multiply the nth row of P by the mth column of R, and you have the left side. The right side is entry (n, m) in Q. Since $PR = Q$, the proof is complete. $\qquad\square$

Identity 309 is a slight variation on Vandermonde's identity, Identity 57. See Exercise 22.

We can also ask about properties of matrices consisting of the other triangular sets of numbers we have studied: the two kinds of Stirling numbers and the Lah numbers. Define S_m, T_m, and L_m to be the square matrices consisting of rows 0 to m of the Stirling numbers of the first kind, the Stirling numbers of the second kind, and the Lah numbers, respectively. For example, we have

$$S_4 = \begin{bmatrix} 1 & 0 & 0 & 0 & 0 \\ 0 & 1 & 0 & 0 & 0 \\ 0 & 1 & 1 & 0 & 0 \\ 0 & 2 & 3 & 1 & 0 \\ 0 & 6 & 11 & 6 & 1 \end{bmatrix}, T_4 = \begin{bmatrix} 1 & 0 & 0 & 0 & 0 \\ 0 & 1 & 0 & 0 & 0 \\ 0 & 1 & 1 & 0 & 0 \\ 0 & 1 & 3 & 1 & 0 \\ 0 & 1 & 7 & 6 & 1 \end{bmatrix}, L_4 = \begin{bmatrix} 1 & 0 & 0 & 0 & 0 \\ 0 & 1 & 0 & 0 & 0 \\ 0 & 2 & 1 & 0 & 0 \\ 0 & 6 & 6 & 1 & 0 \\ 0 & 24 & 36 & 12 & 1 \end{bmatrix}.$$

What about inverses and powers of the S_m, T_m, and L_m matrices? The inversion formulas for the Stirling and Lah numbers answer the questions of S_m^{-1}, T_m^{-1}, and L_m^{-1}. First, though, we need to define the following matrices.

- $S_m(x)$ is the $(m+1) \times (m+1)$ matrix whose (i, j) entry is $\left[\begin{smallmatrix} i \\ j \end{smallmatrix}\right] x^{i-j}$.

- $T_m(x)$ is the $(m+1) \times (m+1)$ matrix whose (i, j) entry is $\left\{\begin{smallmatrix} i \\ j \end{smallmatrix}\right\} x^{i-j}$.

- $L_m(x)$ is the $(m+1) \times (m+1)$ matrix whose (i, j) entry is $L(i, j) x^{i-j}$.

Identity 310. $S_m^{-1} = T_m(-1)$.

For example,

$$S_4^{-1} = T_m(-1) = \begin{bmatrix} 1 & 0 & 0 & 0 & 0 \\ 0 & 1 & 0 & 0 & 0 \\ 0 & -1 & 1 & 0 & 0 \\ 0 & 1 & -3 & 1 & 0 \\ 0 & -1 & 7 & -6 & 1 \end{bmatrix}.$$

Proof. Identities 232 and 233 state that

$$\sum_{k=0}^{n} \left\{ {n \atop k} \right\} \left[{k \atop m} \right] (-1)^{k+m} = [n = m], \text{ and}$$

$$\sum_{k=0}^{n} \left[{n \atop k} \right] \left\{ {k \atop m} \right\} (-1)^{k+m} = [n = m].$$

Thus Identity 232 says that $(T_m(-1))S_m = I_{m+1}$ (as $(-1)^{k+m} = (-1)^{k-m}$), and Identity 233 says that $S_m(T_m(-1)) = I_{m+1}$. $\square$

Identity 311. $T_m^{-1} = S_m(-1)$.

For example,

$$T_4^{-1} = S_4(-1) = \begin{bmatrix} 1 & 0 & 0 & 0 & 0 \\ 0 & 1 & 0 & 0 & 0 \\ 0 & -1 & 1 & 0 & 0 \\ 0 & 2 & -3 & 1 & 0 \\ 0 & -6 & 11 & -6 & 1 \end{bmatrix}.$$

Proof. The proof is virtually identical to that for Identity 310. $\square$

Identity 312. $L_m^{-1} = L_m(-1)$.

For example,

$$L_4^{-1} = L_4(-1) = \begin{bmatrix} 1 & 0 & 0 & 0 & 0 \\ 0 & 1 & 0 & 0 & 0 \\ 0 & -2 & 1 & 0 & 0 \\ 0 & 6 & -6 & 1 & 0 \\ 0 & -24 & 36 & -12 & 1 \end{bmatrix}.$$

Proof. The proof is similar to that for Identity 310, but use Identity 240 instead of Identities 232 and 233. $\square$

The Lah matrices thus have the same inverse property that the Pascal matrices do, as $P_m^{-1} = P_m(-1)$.

Back in Chapter 8 we mentioned that any identity of the form $\sum_{k=0}^{n} a(n,k)b(k,m) = [n = m]$ means that $a(n,k)$ and $b(n,k)$ are in a sense inverse numbers and satisfy an inversion relation like Theorem 1. In light of Theorem 37 and Identities 310, 311, and 312, it appears that an inverse identity of the form $\sum_{k=0}^{n} a(n,k)b(k,m) = [n = m]$ means that lower-triangular matrices consisting of the $a(n,k)$ and $b(n,k)$ numbers, respectively, are inverse matrices. This is, in fact, the case. (See Exercise 24.)

What about powers of the Stirling and Lah matrices? Powers of the Stirling matrices are harder to compute, but powers of the Lah matrices satisfy the same property that powers of the Pascal matrices do. To prove this, we need a Lah version of Identity 50,

$$\sum_{k=0}^{n} \binom{n}{k}\binom{k}{m} x^{n-k} y^{k-m} = (x+y)^{n-m} \binom{n}{m};$$

namely

Identity 313.

$$\sum_{k=0}^{n} L(n,k)L(k,m) x^{n-k} y^{k-m} = (x+y)^{n-m} L(n,m).$$

Proof. Exercise 26. □

We can now prove an expression for powers of the Lah matrix.

Identity 314. *For $x \in \mathbb{Z}$,*

$$L_m^x = L_m(x).$$

Proof. We already know that Identity 314 is true when $x = 1$, $x = -1$, or $x = 0$. (The last holds when we remember our definition of $0^0 = 1$.) Entry (i,j) in L_m^2 is given by

$$\sum_{k=0}^{n} L(i,k)L(k,j) = 2^{i-j} L(i,j),$$

by Identity 313, proving the theorem true in the case $x = 2$.

Now, assume Identity 314 is true for a fixed positive value of x. Then $L_m^{x+1} = L_m^x L_m$. Entry (i,j) in $L_m^x L_m$ is given by

> **Giovanni Domenico Cassini** (1625–1712) was born in Italy but became a naturalized French citizen as an adult (and changed his name to Jean-Dominique Cassini). He was head of the Paris Observatory for many years, and his most noteworthy contributions to science are in astronomy. For example, he discovered the division in Saturn's rings as well as four of Saturn's moons. With Robert Hooke, he is also credited with discovering Jupiter's great red spot.

$$\sum_{k=0}^{n} L(i,k) x^{i-k} L(k,j) = (x+1)^{i-j} L(i,j),$$

by Identity 313. Thus Identity 314 is true for all positive integer values of x.

A similar argument for negative integer values of x, starting with $x = -2$, shows that Identity 314 is true for all $x \in \mathbb{Z}$. □

Thus far we have only placed doubly-indexed numbers like the binomial coefficients, the two kinds of Stirling numbers, and the Lah numbers into matrices. Can we do the same with singly-indexed numbers? Certainly! We'll end this chapter by looking at a couple of examples with the Fibonacci numbers.

Define the *Fibonacci matrix* $\mathcal{F}$ via

$$\mathcal{F} = \begin{bmatrix} F_2 & F_1 \\ F_1 & F_0 \end{bmatrix} = \begin{bmatrix} 1 & 1 \\ 1 & 0 \end{bmatrix}.$$

Then we have the following.

Identity 315. *For integers* n,

$$\mathcal{F}^n = \begin{bmatrix} F_{n+1} & F_n \\ F_n & F_{n-1} \end{bmatrix}.$$

Proof. We'll prove this by induction for nonnegative integers n. The claim is certainly true for $n = 1$. Since $F_{-1} = 1$, it is also true for $n = 0$.

Now, suppose that Identity 315 is true for some fixed value of n. Then we have

$$\mathcal{F}^{n+1} = \mathcal{F}^n \mathcal{F} = \begin{bmatrix} F_{n+1} & F_n \\ F_n & F_{n-1} \end{bmatrix} \begin{bmatrix} 1 & 1 \\ 1 & 0 \end{bmatrix} = \begin{bmatrix} F_{n+1} + F_n & F_{n+1} \\ F_n + F_{n-1} & F_n \end{bmatrix}$$

$$= \begin{bmatrix} F_{n+2} & F_{n+1} \\ F_{n+1} & F_n \end{bmatrix}.$$

The proof for negative values of n is similar; see Exercise 27. $\square$

With Identity 315 in hand, it becomes quite easy to prove the following, one of the more well-known Fibonacci identities.

Identity 316 (Cassini's Identity). *For integers* n,

$$F_{n+1}F_{n-1} - F_n^2 = (-1)^n.$$

Proof. Two basic properties of determinants are that $\det(AB) = (\det A)(\det B)$ and $\det(A^{-1}) = (\det A)^{-1}$. This means that, for any integer n,

$$F_{n+1}F_{n-1} - F_n^2 = \det(\mathcal{F}^n) = (\det \mathcal{F})^n$$
$$= (-1)^n.$$

$\square$

9.3 Exercises

1. With $f(x) = e^{-ix}(\cos x + i \sin x)$, show that $f'(x) = 0$. Then use this result to prove Euler's formula, $e^{ix} = \cos x + i \sin x$.

2. Given a complex number $a + bi$, use Euler's formula to show

$$a + bi = \sqrt{a^2 + b^2}\, e^{i\theta},$$

where θ is the angle that the point (a, b) makes with the positive horizontal axis in the complex plane.

3. Use Euler's formula to prove the following representations of sine and cosine in terms of complex exponentials:

$$\sin x = \frac{e^{ix} - e^{-ix}}{2i}, \qquad \cos x = \frac{e^{ix} + e^{-ix}}{2}.$$

4. Use Euler's formula to prove the sum-angle formulas for sine and cosine:

$$\cos(A + B) = \cos A \cos B - \sin A \sin B,$$
$$\sin(A + B) = \sin A \cos B + \cos A \sin B.$$

5. Prove the following identities.

Identity 317.

$$\sum_{k \geq 0} \binom{n}{2k} 3^k (-1)^k = 2^n \cos \frac{n\pi}{3}.$$

Identity 318.

$$\sum_{k \geq 0} \binom{n}{2k+1} 3^k (-1)^k = \frac{2^n}{\sqrt{3}} \sin \frac{n\pi}{3}.$$

(The right side of Identity 318 is an integer because $\sin(n\pi/3)$ is divisible by $\sqrt{3}$ for any value of n.)

6. From Identity 196,

$$\sum_{k \geq 0} \binom{n}{2k} x^k = \frac{1}{2}\left((1 + \sqrt{x})^n + (1 - \sqrt{x})^n\right),$$

derive Identity 288:

Identity 288. *For real $x > 0$,*

$$\sum_{k \geq 0} \binom{n}{2k} x^k (-1)^k = (x + 1)^{n/2} \cos(n \arctan \sqrt{x}).$$

7. From Identity 215,

$$\sum_{k \geq 0} \binom{n}{2k+1} x^k = \frac{1}{2\sqrt{x}}\left((1 + \sqrt{x})^n - (1 - \sqrt{x})^n\right).$$

derive Identity 289:

Identity 289. *For real $x > 0$,*

$$\sum_{k\geq 0} \binom{n}{2k+1} x^k (-1)^k = \frac{(x+1)^{n/2}}{\sqrt{x}} \sin(n \arctan \sqrt{x}).$$

8. Prove the following variant of Identity 293.

Identity 319.

$$\sum_{k\geq 0} \binom{n}{2k} k(-1)^k = n2^{n/2-2} \left(\cos \frac{n\pi}{4} - \sin \frac{n\pi}{4} \right).$$

9. Prove the following variant of Identity 294.

Identity 320.

$$\sum_{k\geq 0} \binom{n}{2k+1} k(-1)^k = 2^{n/2-2} \left(n \cos \frac{n\pi}{4} + (n-2) \sin \frac{n\pi}{4} \right).$$

10. Prove Identity 321, a generalization of Identity 290:

Identity 321. *For real $y > 0$,*

$$\sum_{k=0}^{n} \binom{n}{k} k^{\underline{m}} i^k y^{n-k} = i^m n^{\underline{m}} (1+y^2)^{(n-m)/2} e^{i(n-m)\arctan(i/y)}.$$

11. Prove the following two identities.

Identity 322.

$$\sum_{k\geq 0} \binom{n}{2k} \frac{3^k (-1)^k}{2k+1} = \frac{2^{n+1}}{\sqrt{3}(n+1)} \sin \frac{(n+1)\pi}{3}.$$

Identity 323.

$$\sum_{k\geq 0} \binom{n}{2k+1} \frac{3^k(-1)^k}{k+1} = \frac{2}{3(n+1)} \left(1 - 2^{n+1} \cos \frac{(n+1)\pi}{3} \right).$$

12. Prove the remaining two identities that give row sums for every third binomial coefficient.

Identity 305.

$$\sum_{k\geq 0} \binom{n}{3k+1} = \frac{1}{3} \left(2^n + 2\cos \frac{(n-2)\pi}{3} \right).$$

Identity 306.

$$\sum_{k\geq 0}\binom{n}{3k+2} = \frac{1}{3}\left(2^n + 2\cos\frac{(n+2)\pi}{3}\right).$$

13. Prove the following identities that give alternating row sums for every third binomial coefficient.

Identity 324.

$$\sum_{k\geq 0}\binom{n}{3k}(-1)^k = 2\cdot 3^{n/2-1}\cos\frac{n\pi}{6}.$$

Identity 325.

$$\sum_{k\geq 0}\binom{n}{3k+1}(-1)^k = -2\cdot 3^{n/2-1}\cos\frac{(n+4)\pi}{6}.$$

Identity 326.

$$\sum_{k\geq 0}\binom{n}{3k+2}(-1)^k = 2\cdot 3^{n/2-1}\cos\frac{(n-4)\pi}{6}.$$

14. Prove the following identity giving a formula for row sums involving every fourth binomial coefficient.

Identity 327. *For* $m \in \{0, 1, 2, 3\}$,

$$\sum_{k\geq 0}\binom{n}{4k+m} = 2^{n-2} + 2^{n/2-1}\cos\frac{(n-2m)\pi}{4} + \frac{(-1)^m}{4}[n=0].$$

15. In this exercise we derive the formula for the alternating sum of the reciprocals of the central binomial coefficients. This requires us to determine what it means to evaluate the arcsine of an imaginary number. There are different ways to do this; our approach uses the relationship between sine and hyperbolic sine. Here we assume that x is real.

   (a) With the definition of hyperbolic sine as

   $$\sinh x = \frac{e^x - e^{-x}}{2},$$

   prove that $\sin(ix) = i\sinh x$.

   (b) Use the result in part (a) to prove that $\arcsin(ix) = i\sinh^{-1}x$.

   (c) Using the definition of hyperbolic sine in part (a), derive the formula

   $$\sinh^{-1}x = \ln\left(x + \sqrt{x^2 + 1}\right).$$

(d) Finally, use the result in part (c), together with Identity 163, to prove

Identity 328.

$$\sum_{n=0}^{\infty} \frac{(-1)^n}{\binom{2n}{n}} = \frac{4}{5} - \frac{4\sqrt{5}}{25} \ln\left(\frac{1+\sqrt{5}}{2}\right).$$

(Interestingly enough, Identity 328 features yet another appearance of the golden ratio $(1+\sqrt{5})/2$.)

16. For a review of some basic matrix operations, find the following, where $A = \begin{bmatrix} 1 & 2 \\ 3 & 4 \end{bmatrix}$ and $B = \begin{bmatrix} 5 & 6 \\ 7 & 8 \end{bmatrix}$.

   (a) AB.
   (b) BA.
   (c) A^{-1}.
   (d) $\det A$.

17. Let V be the $(m+1) \times (m+1)$ *Vandermonde matrix*, given by

$$V = \begin{bmatrix} 1 & x_0 & \cdots & x_0^m \\ \vdots & & \ddots & \vdots \\ 1 & x_m & \cdots & x_m^m \end{bmatrix}.$$

Prove that if $x_0, x_1, \ldots, x_m$ are distinct then V is invertible by showing that if $V\mathbf{c} = \mathbf{0}$ then $\mathbf{c} = \mathbf{0}$. Next, argue that the invertibility of V implies that the set $\{(1, x_0, \ldots, x_0^m), \ldots, (1, x_m, \ldots, x_m^m)\}$ is linearly independent.

18. Let V be the $(m+1) \times (m+1)$ Vandermonde matrix (introduced in Exercise 17). Prove that the determinant of V is given by

$$\prod_{0 \le i < j \le m} (x_j - x_i).$$

If $x_i \ne x_j$ when $i \ne j$, then $\det V \ne 0$, which is another way of proving that V is invertible.

19. Use matrices to prove the following variation of the binomial inversion formula.

Theorem 2.

$$f(n) = \sum_{k=0}^{n} \binom{n}{k} g(k) \iff g(n) = \sum_{k=0}^{n} \binom{n}{k} f(k)(-1)^{n-k}.$$

20. Use Theorem 2 (see Exercise 19) to prove that P_m^{-1} is the $(m+1) \times (m+1)$ matrix with $\binom{i}{j}(-1)^{i-j}$ in entry (i, j) (Theorem 37).

21. Let P'_m and P''_m be the $(m+1) \times (m+1)$ matrices whose (i,j) entries are $\binom{i}{j}(-1)^j$ and $\binom{i}{j}(-1)^i$, respectively. Prove that $(P'_m)^{-1} = P'_m$ and $(P''_m)^{-1} = P''_m$.

22. Starting with Vandermonde's identity (Identity 57),

$$\sum_{k=0}^{n} \binom{n}{k}\binom{m}{r-k} = \binom{n+m}{r},$$

give a short proof of the following.

Identity 309.

$$\sum_{k=0}^{n} \binom{n}{k}\binom{m}{k} = \binom{n+m}{n}.$$

23. Prove that $\det Q_m = 1$, where Q_m is the $(m+1)\times(m+1)$ symmetric Pascal matrix. (This requires knowledge of some properties of determinants.)

24. Suppose $\{a(n,k)\}$ and $\{b(n,k)\}$ are sets of numbers that have the property that $a(n,k) = b(n,k) = 0$ when $k > n$ and that satisfy the inverse relation $\sum_{k=0}^{i} a(i,k)b(k,j) = [i=j]$. Let A_m and B_m be the $(m+1) \times (m+1)$ matrices with entry (i,j) equal to $a(i,j)$ and $b(i,j)$, respectively. Prove that $A_m B_m = I_{m+1}$.

25. Prove

Identity 329.

$$S_m(x)T_m(x) = L_m(x),$$

where $S_m(x)$, $T_m(x)$, and $L_m(x)$ are the $(m+1) \times (m+1)$ matrices whose (i,j) entries are given by $\begin{bmatrix} i \\ j \end{bmatrix}x^{i-j}$, $\begin{Bmatrix} i \\ j \end{Bmatrix}x^{i-j}$, and $L(i,j)x^{i-j}$, respectively.

26. Prove the Lah version of Identity 50:

Identity 313.

$$\sum_{k=0}^{n} L(n,k)L(k,m)x^{n-k}y^{k-m} = (x+y)^{n-m}L(n,m).$$

27. Prove Identity 315 in the case where n is negative. In other words, prove that if

$$\mathcal{F} = \begin{bmatrix} 1 & 1 \\ 1 & 0 \end{bmatrix}$$

then, for $n \geq 1$,

$$\mathcal{F}^{-n} = \begin{bmatrix} F_{-n+1} & F_{-n} \\ F_{-n} & F_{-n-1} \end{bmatrix}.$$

28. Here are two more properties of determinants.

- Adding a row or column to another row or column in a matrix does not change the matrix's determinant.
- Swapping two rows or columns in a matrix changes the sign of the matrix's determinant.

Let $A_0 = \begin{bmatrix} 1 & 0 \\ 0 & 1 \end{bmatrix}$, the identity matrix. Then, for $n \geq 1$, let A_n be constructed from A_{n-1} by adding the second row of A_{n-1} to the first and then interchanging the two rows.

With this process for constructing A_n, use induction and the two determinant properties to show that $A_n = \begin{bmatrix} F_{n-1} & F_n \\ F_n & F_{n+1} \end{bmatrix}$ and that $\det A_n = (-1)^n$. This gives a different determinant-based proof of Cassini's identity than the one described in the text. (The proof only holds only for $n \geq 0$, though; see Exercise 29.)

Identity 316. (Cassini's Identity) *For integers n,*

$$F_{n+1}F_{n-1} - F_n^2 = (-1)^n.$$

29. The argument in Exercise 28 only holds for nonnegative integers n, so it really only gives a partial proof of Cassini's identity. Show how to complete the proof by modifying the argument in Exercise 28 to apply to integers $n < 0$.

9.4   Notes

Benjamin and Scott [8] give combinatorial proofs of Identity 304 and the $m = 0$ case of Identity 327. Benjamin, Chen, and Kindred [6] describe a combinatorial proof of Identity 303 in the case $x = 1$. Loehr and Michael [44] give a combinatorial interpretation of Identity 303 when $x = 1$ as well.

Guichard [35] takes Identity 303 in the case $x = 1$ and develops it further, until the right side is a sum of cosines. He also proves that the sum $\sum_k \binom{n}{rk+m}$ for $m \in \{0, 1, \ldots, r-1\}$ is asymptotically $2^n/r$.

Deng and Pan [20] contains some additional expressions for evenly-spaced binomial sums, although their primary focus is on divisibility properties of the sums.

Many authors, such as Brawer and Pirovino [13], Call and Velleman [14], and Zhang [83] have investigated properties of the Pascal matrix. These include factorizations, inverses, and generalizations of various kinds. In addition, Edelman and Strang [25] explicitly consider the $Q = PR$ factorization. They give four proofs of it; the one in this chapter is their fourth proof. Other authors, such as Cheon and Kim [17] and Yang and Qiao [81], have studied properties of the Stirling matrices, which consist of the two kinds of Stirling numbers. Spivey and Zimmer [72] show that many of these results on Pascal matrices and Stirling matrices are special cases of results on matrices containing symmetric polynomials.

Exercise 23 asks for a proof that the determinant of the symmetric Pascal matrix is 1. Benjamin and Cameron [4] give a combinatorial interpretation of the determinant that, when combined with the lattice path interpretation of the binomial coefficient (Theorem 3), can be used to show that $\det Q = 1$.

The matrix-and-determinants proof of Cassini's identity (Identity 316) is classic. The argument in Exercise 28, as well as related proofs giving two generalizations of Cassini, appears in Spivey [62]. Also, Werman and Zeilberger [78] give a combinatorial proof of Cassini's identity.

Benjamin, Cameron, Quinn, and Yerger [5] have, from a combinatorial perspective, a nice discussion of the determinants of matrices consisting of Catalan numbers. A particularly interesting such results is that, regardless of the value of n,

$$
\det \begin{bmatrix} C_0 & C_1 & \cdots & C_n \\ C_1 & C_2 & \cdots & C_{n+1} \\ \vdots & \vdots & \ddots & \vdots \\ C_n & C_{n+1} & \cdots & C_{2n} \end{bmatrix} = \det \begin{bmatrix} C_1 & C_2 & \cdots & C_{n+1} \\ C_2 & C_3 & \cdots & C_{n+2} \\ \vdots & \vdots & \ddots & \vdots \\ C_{n+1} & C_{n+2} & \cdots & C_{2n+1} \end{bmatrix} = 1.
$$

10

Mechanical Summation

In our final chapter we take a look at one last method for proving binomial identities: mechanical summation. Unlike the rest of the rather general mathematical ideas discussed in this text, mechanical summation exists solely to prove identities (both binomial and otherwise).

The *mechanical* aspect of mechanical summation is that nearly all of the work involved in using the methods can be automated. Thus mechanical summation should perhaps be considered more properly a part of the *science* (rather than the *art*) of proving binomial identities. Nevertheless, mechanical summation is an extremely powerful—and general—technique for proving binomial identities and thus deserves its own chapter in this text. The approach we take is to present the two major algorithms and work through some simpler examples. The algebra even in these "simpler" examples can be quite involved, however.

The two algorithms apply to sums featuring what are known as *hypergeometric terms*. The first section discusses what hypergeometric terms are. In the second section we give our first algorithm, which finds antidifferences for hypergeometric terms and is due to Gosper, and work through some examples. The last section contains Zeilberger's extension to Gosper's algorithm, which allows one to evaluate certain definite sums involving hypergeometric terms—even when Gosper's original algorithm fails to find an antidifference.

10.1 Hypergeometric Series

The mechanical summation techniques we examine in this chapter apply to sums involving hypergeometric terms, which are often encountered in the context of hypergeometric series. In this section we define hypergeometric series and hypergeometric terms and discuss how to recognize them.

Formally, a *hypergeometric series* is an infinite series of the form

$$
{}_pF_q\!\left[\begin{matrix} a_1, & \dots, & a_p \\ b_1, & \dots, & b_q \end{matrix}; x\right] = \sum_{k=0}^{\infty} \frac{(a_1)^{\overline{k}} \cdots (a_p)^{\overline{k}}}{(b_1)^{\overline{k}} \cdots (b_q)^{\overline{k}}} \frac{x^k}{k!}. \tag{10.1}
$$

If the reader finds this notation frightening at first glance, the author understands. (He found it frightening at the first several glances.) It is, however, standard hypergeometric series notation, and, once you come to understand what it's saying, it actually isn't difficult to work with.

Let's take a look more closely at Equation (10.1). With its $x^k/k!$ factor, the right side looks like the exponential generating function for the sequence (s_k), where

$$s_k = \frac{(a_1)^{\overline{k}} \cdots (a_p)^{\overline{k}}}{(b_1)^{\overline{k}} \cdots (b_q)^{\overline{k}}}.$$

This sequence is just p rising factorial powers, $(a_1)^{\overline{k}}, (a_2)^{\overline{k}}, \ldots, (a_p)^{\overline{k}}$, divided by q rising factorial powers, $(b_1)^{\overline{k}}, (b_2)^{\overline{k}}, \ldots, (b_q)^{\overline{k}}$. The rather formidable-looking notation on the left side of Equation (10.1) collects this information, along with the value of x: The $_pF_q$ tells us that we have p rising factorial powers in the numerator and q rising factorial powers in the denominator, the a_i's and the b_j's tell us which rising powers those are, and x tells us which ordinary powers appear in the series.

One more piece of terminology: The expression

$$\frac{(a_1)^{\overline{k}} \cdots (a_p)^{\overline{k}}}{(b_1)^{\overline{k}} \cdots (b_q)^{\overline{k}}} \frac{x^k}{k!}$$

is called a *hypergeometric term*. If t_k is a hypergeometric term, then $\sum_{k=0}^{\infty} t_k$ is a hypergeometric series, and vice versa.

Let's now take a look at some examples of hypergeometric series. The simplest one that retains the variable x and has nonempty lists for the a_i's and b_j's has $a_1 = b_1 = 1$. Thus $p = q = 1$, and we get

$$_1F_1 \begin{bmatrix} 1 \\ 1 \end{bmatrix}; x \end{bmatrix} = \sum_{k=0}^{\infty} \frac{1^{\overline{k}}}{1^{\overline{k}}} \frac{x^k}{k!} = \sum_{k=0}^{\infty} \frac{x^k}{k!},$$

which is just the Maclaurin series for e^x. Thus we could also say

$$_1F_1 \begin{bmatrix} 1 \\ 1 \end{bmatrix}; x \end{bmatrix} = e^x.$$

Since $1^{\overline{k}} = 1 \cdot 2 \cdot 3 \cdots k = k!$, we could cancel the $k!$ in the hypergeometric series (and thus get another simple example) by making sure there is an extra factor of $1^{\overline{k}}$ in the numerator. Thus, for example, we have

$$_2F_1 \begin{bmatrix} 1, \, 1 \\ 1 \end{bmatrix}; x \end{bmatrix} = \sum_{k=0}^{\infty} \frac{1^{\overline{k}} 1^{\overline{k}}}{1^{\overline{k}}} \frac{x^k}{k!} = \sum_{k=0}^{\infty} \frac{k! \, x^k}{k!} = \sum_{k=0}^{\infty} x^k = \frac{1}{1-x}.$$

This is a *geometric series*. Thus we can see that the general geometric series $\sum_{k=0}^{\infty} ax^k$ has the hypergeometric series representation $a \, _2F_1 \begin{bmatrix} 1, \, 1 \\ 1 \end{bmatrix}; x \end{bmatrix}$.

Many binomial sums can also be represented as hypergeometric series—even identities that involve finite sums! The binomial theorem, Identity 3, presents us with a nice example of this. We have, by Identity 32,

$$y^n \, {}_2F_1\left[\begin{matrix} 1, \, -n \\ 1 \end{matrix}; \frac{-x}{y}\right] = y^n \sum_{k=0}^{\infty} \frac{1^{\overline{k}}(-n)^{\overline{k}}}{1^{\overline{k}}} \frac{(-x/y)^k}{k!}$$

$$= y^n \sum_{k=0}^{\infty} n^{\underline{k}}(-1)^k(-1)^k \frac{(x/y)^k}{k!} = \sum_{k=0}^{\infty} \frac{n^{\underline{k}}}{k!} x^k y^{n-k}$$

$$= \sum_{k=0}^{n} \binom{n}{k} x^k y^{n-k} = (x+y)^n.$$

With this derivation we can see that $y^n \, {}_2F_1\left[\begin{matrix} 1, \, -n \\ 1 \end{matrix}; \frac{-x}{y}\right]$ does, in fact, represent the binomial sum $\sum_{k=0}^{n} \binom{n}{k} x^k y^{n-k}$.

In general, though, how can you tell whether a series is hypergeometric? (The derivation we just did starts with a hypergeometric representation, but we're not usually going to be given that.) It turns out there is an easy test to determine whether a given series is hypergeometric. Theorem 38 gives the characterization we need; afterwards we'll discuss how to apply that characterization in practice.

Theorem 38. *The series $\sum_{k=0}^{\infty} t_k$ is hypergeometric if and only if $t_0 = 1$ and there exist nonzero polynomials $P(k)$ and $Q(k)$ such that, for all k, $Q(k)t_{k+1} = P(k)t_k$.*

Proof. ($\Longrightarrow$) Suppose t_k is a hypergeometric term with

$$t_k = \frac{(a_1)^{\overline{k}} \cdots (a_p)^{\overline{k}}}{(b_1)^{\overline{k}} \cdots (b_q)^{\overline{k}}} \frac{x^k}{k!}.$$

Then $t_0 = 1$. In addition, since

$$a^{\overline{k+1}} = a(a+1)\cdots(a+k-1)(a+k) = (k+a)a^{\overline{k}},$$

we have

$$t_{k+1} = \frac{(k+a_1)\cdots(k+a_p)x}{(k+b_1)\cdots(k+b_q)(k+1)}t_k.$$

Thus

$$(k+b_1)\cdots(k+b_q)(k+1)t_{k+1} = (k+a_1)\cdots(k+a_p)xt_k.$$

With $P(k) = (k+a_1)\cdots(k+a_p)x$ and $Q(k) = (k+b_1)\cdots(k+b_q)(k+1)$, the claim follows.

($\Longleftarrow$) Suppose we have a sequence (t_k) and nonzero polynomials $P'(k)$ and $Q'(k)$ such that $t_0 = 1$ and $Q'(k)t_{k+1} = P'(k)t_k$. If -1 is not a root of $Q'(k)$, then let $Q(k) = Q'(k)(k+1)$ and $P(k) = P'(k)(k+1)$. Otherwise, let $P(k) = P'(k)$ and $Q(k) = Q'(k)$. Either way, we have $Q(k)t_{k+1} = P(k)t_k$.

Now, factor the polynomials $P(k)$ and $Q(k)$ over the complex numbers so that we have $P(k) = c_1(k + a_1) \cdots (k + a_p)$ and $Q(k) = c_2(k + b_1) \cdots (k + b_q)(k + 1)$. With $x = c_1/c_2$, we have

$$(k + b_1) \cdots (k + b_q)(k + 1)t_{k+1} = (k + a_1) \cdots (k + a_p)xt_k.$$

Then
$$t_{k+1} = \frac{(k + a_1) \cdots (k + a_p)x}{(k + b_1) \cdots (k + b_q)(k + 1)} t_k.$$

With $t_0 = 1$, the solution to this recurrence is

$$t_k = \frac{(a_1)^{\overline{k}} \cdots (a_p)^{\overline{k}}}{(b_1)^{\overline{k}} \cdots (b_q)^{\overline{k}}} \frac{x^k}{k!}.$$

Therefore, $\sum_{k=0}^{\infty} t_k$ is hypergeometric.   □

Let's summarize what Theorem 38 and its proof mean about how to determine whether a given series $\sum_{k=0}^{\infty} t_k$ is hypergeometric, and, if so, how to represent that series in the $_pF_q$ notation.

- First, form the ratio t_{k+1}/t_k. If the result can be expressed as the rational function $P(k)/Q(k)$, where P and Q are polynomials in k, then t_k is a hypergeometric term, and the series $\sum_{k=0}^{\infty} t_k$ is hypergeometric.

- Factor the polynomials $P(k)$ and $Q(k)$ over the complex numbers, into $P(k) = (k + a_1) \cdots (k + a_p)x$ and $Q(k) = (k + b_1) \cdots (k + b_q)(k + 1)$, so that the leading coefficient of Q is 1 and the leading coefficient of P is x. If $(k + 1)$ is not a factor of $Q(k)$, then include it as a factor of $Q(k)$ and include $(k + 1)$ as a factor of $P(k)$ as well (thus retaining the value of the ratio $P(k)/Q(k)$).

- We then have the representation

$$\sum_{k=0}^{\infty} t_k = {_pF_q}\left[\begin{matrix} a_1, & \ldots, & a_p \\ b_1, & \ldots, & b_q \end{matrix}; x\right] = \sum_{k=0}^{\infty} \frac{(a_1)^{\overline{k}} \cdots (a_p)^{\overline{k}}}{(b_1)^{\overline{k}} \cdots (b_q)^{\overline{k}}} \frac{x^k}{k!}.$$

- If b_i is a negative integer for some i, then $_pF_q\left[\begin{smallmatrix} a_1, & \ldots, & a_p \\ b_1, & \ldots, & b_q \end{smallmatrix}; x\right]$ is not well-defined. We can see this because $k + b_i = 0$ results in division by zero in the definition of t_k for all $k \geq -b_i$.

- Otherwise, if a_i is a negative integer for some i, then $_pF_q\left[\begin{smallmatrix} a_1, & \ldots, & a_p \\ b_1, & \ldots, & b_q \end{smallmatrix}; x\right]$ terminates as a finite sum. This is because $k + a_i = 0$ means that $t_k = 0$ for all $k \geq -a_i$. We see this in the binomial theorem example, as we have $a_2 = -n$ (and n is a nonnegative integer).

Let's look at some examples using this procedure. The row sum for the binomial coefficients, $\sum_{k=0}^{n} \binom{n}{k}$, is hypergeometric because it is a special case of the binomial theorem we did before. Let's try the aerated row sum $\sum_{k\geq 0} \binom{n}{2k}$. First, $t_0 = \binom{n}{0} = 1$. Also, we have

$$\frac{t_{k+1}}{t_k} = \frac{\binom{n}{2k+2}}{\binom{n}{2k}} = \frac{n^{2k+2}(2k)!}{n^{2k}(2k+2)!} = \frac{(n-2k)(n-2k-1)}{(2k+1)(2k+2)}.$$

The right side is a rational function of k, but we need to do some more work to identify the a_i's and b_j's for the $_pF_q$ notation. Factor out the coefficients of each of the k terms to obtain

$$\frac{(n-2k)(n-2k-1)}{(2k+1)(2k+2)} = \frac{(-2)(-2)(k-n/2)(k-n/2+1/2)}{(2)(2)(k+1/2)(k+1)}$$
$$= \frac{(k-n/2)(k-n/2+1/2)}{(k+1/2)(k+1)}.$$

We have $a_1 = -n/2$, $a_2 = -n/2+1/2$, $b_1 = 1/2$. Therefore,

$$\sum_{k\geq 0} \binom{n}{2k} = {}_2F_1\left[\begin{array}{c} -n/2, \ -n/2+1/2 \\ 1/2 \end{array} ; 1\right] = \sum_{k=0}^{\infty} \frac{(-n/2)^{\overline{k}}(-n/2+1/2)^{\overline{k}}}{(1/2)^{\overline{k}} k!}.$$

If n is a nonnegative integer, then either $-n/2$ or $-n/2+1/2$ must be a nonnegative integer. Thus when n is a nonnegative integer $\sum_{k\geq 0} \binom{n}{2k}$ terminates (although, of course, we already knew this).

What about series involving some of the other numbers that we have seen in this book? Well, we have no way of expressing sums involving Stirling numbers as hypergeometric series because we have no simple explicit formulas for $\left\{{n \atop k}\right\}$ and $\left[{n \atop k}\right]$ that would allow us to simplify the ratios $\left\{{n \atop k+1}\right\}/\left\{{n \atop k}\right\}$ and $\left[{n \atop k+1}\right]/\left[{n \atop k}\right]$. (Identity 224, for example, is far too complicated to be of help here.) The same thing holds for the Bell numbers and the Bernoulli numbers.

A Fibonacci sum looks like it might be promising, though, thanks to Binet's formula $F_n = (\phi^n - \psi^n)/\sqrt{5}$, where $\phi = (1+\sqrt{5})/2$ and $\psi = (1-\sqrt{5})/2$ (Identity 173). We have

$$\frac{F_{k+1}}{F_k} = \frac{(\phi^{k+1} - \psi^{k+1})/\sqrt{5}}{(\phi^k - \psi^k)/\sqrt{5}} = \frac{\phi^{k+1} - \psi^{k+1}}{\phi^k - \psi^k}.$$

Using Exercise 12 in Chapter 4, we can simplify this as

$$\frac{\phi^{k+1} - \psi^{k+1}}{\phi^k - \psi^k} = \frac{(\phi-\psi)(\phi^k + \phi^{k-1}\psi + \cdots + \psi^k)}{(\phi-\psi)(\phi^{k-1} + \phi^{k-2}\psi + \cdots + \psi^{k-1})}$$
$$= \frac{\phi^k + \phi^{k-1}\psi + \cdots + \psi^k}{\phi^{k-1} + \phi^{k-2}\psi + \cdots + \psi^{k-1}}.$$

This is not a rational function of k, and so F_k is not a hypergeometric term. Thus sums involving Fibonacci numbers cannot be hypergeometric.

Moving on, let's try the Lah numbers. They also look promising because of the explicit formula $L(n, k) = \binom{n-1}{k-1}\frac{n!}{k!}$, valid for $n \geq 1$ (Identity 238). We have

$$\frac{L(n, k+1)}{L(n, k)} = \frac{\binom{n-1}{k}n!k!}{(k+1)!\binom{n-1}{k-1}n!} = \frac{(n-1)^{\underline{k}}(k-1)!}{(k+1)k!(n-1)^{\underline{k-1}}} = \frac{n-1-k+1}{k(k+1)}$$

$$= \frac{(-1)(k-n)}{k(k+1)}.$$

Thus $L(n, k)$ is a hypergeometric term.

We need to do a little more work to find the $_pF_q$ representation of, say, $\sum_{k=0}^{n} L(n, k)$, though, since $t_0 = L(n, 0) = 0 \neq 1$ when $n \geq 1$. When this kind of situation occurs the solution is to reindex the series so that the $k = 0$ term corresponds to the first nonzero term. In this case that's $L(n, 1)$. Thus we instead consider the Lah row sum in the form $\sum_{k=0}^{n-1} L(n, k+1)$. The ratio is

$$\frac{L(n, k+2)}{L(n, k+1)} = \frac{(-1)(k-n+1)}{(k+1)(k+2)}.$$

Unfortunately, we're still not done. The first term t_0 in the sum $\sum_{k=0}^{n-1} L(n, k+1)$ is $t_0 = L(n, 1) = n!$, which is 1 only when $n = 1$. Thus in order to represent the row sum of the Lah numbers as a hypergeometric series we must factor out an $n!$. This doesn't affect the calculation of the ratio $L(n, k+2)/L(n, k+1)$, though, and so we're finally left with

$$\sum_{k=0}^{n-1} L(n, k+1) = n! \; _1F_1\left[\begin{matrix} -n+1 \\ 2 \end{matrix}; -1\right].$$

Next, the Catalan numbers. These are also promising because of the formula $C_n = \binom{2n}{n}/(n+1)$ (Identity 244). We have

$$\frac{C_{k+1}}{C_k} = \frac{\binom{2k+2}{k+1}(k+1)}{\binom{2k}{k}(k+2)} = \frac{(2k+2)^{\underline{k+1}}k!(k+1)}{(k+1)!(2k)^{\underline{k}}(k+2)} = \frac{(2k+2)(2k+1)}{(2k-k+1)(k+2)}$$

$$= \frac{4(k+1/2)(k+1)}{(k+1)(k+2)}.$$

Thus C_k is a hypergeometric term.

We can use this calculation to find the hypergeometric representation of the Catalan generating function $\sum_{k=0}^{\infty} C_k x^k$. Since $t_0 = C_0 x^0 = 1$ and

$$\frac{C_{k+1}x^{k+1}}{C_k x^k} = \frac{4x(k+1/2)(k+1)}{(k+1)(k+2)},$$

we have

$$\sum_{k=0}^{\infty} C_k x^k = \; _2F_1\left[\begin{matrix} 1/2, \; 1 \\ 2 \end{matrix}; 4x\right].$$

10.2 Antidifferences of Hypergeometric Terms

Now that we're familiar with hypergeometric series, let's turn to our two algorithms that can be used to evaluate sums with hypergeometric terms. The first is *Gosper's algorithm.*

At a very high level, Gosper's algorithm looks for an antidifference of a given hypergeometric term t_k. As we mentioned in Chapter 7, an *antidifference* for a sequence (t_k) is a sequence (T_k) satisfying $\Delta T_k = T_{k+1} - T_k = t_k$. Such an antidifference would be wonderful for evaluating any sum involving t_k: Thanks to telescoping, we would have (for example)

$$\sum_{k=a}^{b} t_k = \sum_{k=a}^{b} (T_{k+1} - T_k) = T_{b+1} - T_a.$$

(The situation here is entirely analogous to having an antiderivative F for a function f so that we could evaluate an integral via $\int_a^b f(x)\,dx = F(b) - F(a)$.)

More specifically, given a hypergeometric term t_k, Gosper's algorithm either (1) finds a closed-form antidifference of t_k as another hypergeometric term or (2) shows that t_k has no closed-form antidifference as a hypergeometric term. The steps for Gosper's algorithm are given in Figure 10.1.

Let's look at a few examples with Gosper's algorithm. These will also help us flesh out some of the details in the algorithm.

A natural first question is this: Is there an antidifference for the binomial coefficient? There are actually two questions here, though, as we can view the binomial coefficient as a function of k through its lower index, via $t_k = \binom{n}{k}$, or through its upper index, via $t_k = \binom{k}{m}$. Let's take the lower index question first.

Step 1 of Gosper's algorithm considers the ratio t_{k+1}/t_k. With $t_k = \binom{n}{k}$ (which we already know is hypergeometric), we have

American computer scientist and mathematician **Bill Gosper** (1943–) is one of the pioneers in the field of symbolic computation. In addition to the algorithm for finding antidifferences of hypergeometric sums that we discuss here, he also helped create an early dialect of the functional programming language Lisp. For about three months in the mid-1980s he held the world record for computing the most number of digits of π.

$$\frac{t_{k+1}}{t_k} = \frac{\binom{n}{k+1}}{\binom{n}{k}} = \frac{n^{\underline{k+1}}\,k!}{(k+1)!\,n^{\underline{k}}} = \frac{n-k}{k+1}.$$

Let's try to satisfy the rest of Step 1 by setting $q(k) = n - k$, $r(k) = k$, and $p(k) = 1$, and see if this fits the form we need in Step 1. (In general, setting

---

Gosper's Algorithm

1. Given a hypergeometric term t_k, express the ratio t_{k+1}/t_k as

$$\frac{t_{k+1}}{t_k} = \frac{p(k+1)q(k)}{p(k)r(k+1)},$$

   where p, q, and r are polynomials satisfying the following condition: If $k + a$ is a factor of $q(k)$ and $k + b$ is a factor of $r(k)$, then $a - b$ is not a positive integer.

2. Look for a polynomial $s(k)$ satisfying

$$p(k) = q(k)s(k+1) - r(k)s(k).$$

3. If such a polynomial $s(k)$ does not exist, then t_k does not have a closed-form antidifference that is also a hypergeometric term.

4. If such a polynomial $s(k)$ does exist, then let T_k be the hypergeometric term

$$T_k = \frac{r(k)s(k)t_k}{p(k)}.$$

   Then T_k is an antidifference for t_k, and $\sum_{k=a}^{b} t_k = T_{b+1} - T_a$.

FIGURE 10.1
Gosper's algorithm for finding an antidifference of a hypergeometric term

$p(k) = 1$, $q(k)$ equal to the numerator, and $r(k)$ equal to the appropriate offset of the denominator is a good place to start.)

Does this satisfy our condition? Well, the only factor $k + a$ of $n - k$ has $a = -n$, and the only factor $k + b$ of k has $b = 0$. Thus as long as n is not a negative integer (we'll come back to that case later), $a - b$ will not be a positive integer.

Step 2 of Gosper's algorithm has us looking for a polynomial $s(k)$ satisfying $p(k) = q(k)s(k+1) - r(k)s(k)$. Thus we want to find $s(k)$ such that

$$1 = (n - k)s(k+1) - ks(k).$$

Supposing $s(k)$ exists, let's denote its highest term by $\alpha_d k^d$. Then the highest term in $(n-k)s(k+1)$ is $-\alpha_d k^{d+1}$, and the highest term in $-ks(k)$ is $-\alpha_d k^{d+1}$. Thus the highest term on the right side of this equation is $-2\alpha_d k^{d+1}$. The highest term on the left side is k^0. Thus in order for the two sides to be equal we must have $d + 1 = 0$ and $-2\alpha_d = 1$. The second equation is fine, but the first equation yields $d = -1$. Thus the polynomial $s(k)$ has degree -1. This is impossible (even if we take the convention that the zero polynomial has degree -1), and so the polynomial $s(k)$ does not exist.

Step 3 of Gosper's algorithm thus says that, as long as n is not a negative integer, $t_k = \binom{n}{k}$ has no closed-form antidifference that is also a hypergeometric term. This seems entirely plausible given our discussion thus far in the text: There is no identity we have seen that allows us to sum $\sum_{k=a}^{b} \binom{n}{k}$ for general a and b. (Of course, we have repeatedly seen a formula for $\sum_{k=0}^{n} \binom{n}{k}$, but that formula is only for the specific case $a = 0$, $b = n$.)

Let's now consider the situation where n is a negative integer. With $t_k = \binom{n}{k}$, we still need to find $p(k)$, $q(k)$, and $r(k)$ such that

$$\frac{t_{k+1}}{t_k} = \frac{n-k}{k+1}.$$

We've already discussed the fact that setting $q(k) = n - k$, $r(k) = k$, and $p(k) = 1$ doesn't satisfy the condition on $q(k)$ and $r(k)$ in Step 1 of Gosper's algorithm. Again, the reason is that $q(k) = n-k$ has $k-n$ as a factor, $r(k) = k$ has $k + 0$ as a factor, and $-n - 0$ is a positive integer when n is a negative integer.

However, it's easy to modify these values of $p(k)$, $q(k)$, and $r(k)$ so that they do work. In general, if we have polynomials p, q, and r satisfying the equation in Step 1 of Gosper's algorithm, $k + a$ is a factor of $q(k)$, $k + b$ is a factor of $r(k)$, and $a - b$ is a positive integer, do the following.

1. Multiply $p(k)$ by $(k + a - 1)^{\underline{a-b-1}}$.

2. Divide $q(k)$ by $k + a$.

3. Divide $r(k)$ by $k + b$.

If there remain any other values for a and b that cause the condition in Step 1 of Gosper's algorithm to be violated, repeat this process until the condition is satisfied.

With $t_k = \binom{n}{k}$ and n a negative integer, these three steps give us $p(k) = (k - n - 1)^{\underline{-n-1}}$, $q(k) = -1$, and $r(k) = 1$. Let's check to make sure this assignment of values still causes the right side of the equation in Step 1 of Gosper's algorithm to equal t_{k+1}/t_k:

$$\frac{(k-n)^{\underline{-n-1}}(-1)}{(k-n-1)^{\underline{-n-1}}} = \frac{-(k-n)(k-n-1)\cdots(k-n-(-n-1)+1)}{(k-n-1)(k-n-2)\cdots(k-n-1-(-n-1)+1)}$$

$$= \frac{-(k-n)(k-n-1)\cdots(k+2)}{(k-n-1)(k-n-2)\cdots(k+1)} = \frac{n-k}{k+1}.$$

Sure enough, it does.

Moving on to Step 2 of Gosper's algorithm, we need to find a polynomial $s(k)$ satisfying

$$(k-n-1)^{\underline{-n-1}} = -s(k+1) - s(k).$$

If one of the minus signs in front of $s(k)$ or $s(k + 1)$ was a plus sign we could use Identity 177 to find $s(k)$ for general n. Since it's not, the process of finding

$s(k)$ will be much harder. Let's work through this process for a specific value of n: $n = -3$. We end up deriving the following identity.

Identity 330.

$$\sum_{k=0}^{n} \binom{-3}{k} = (-1)^n \left(\frac{n^2}{4} + n + \frac{7}{8} \right) + \frac{1}{8}.$$

Proof. Follow the argument we just made for finding an antidifference of $t_k = \binom{n}{k}$ when n is a negative integer until the place where we're looking for a polynomial $s(k)$ satisfying $(k - n - 1)\underline{-n-1} = -s(k+1) - s(k)$. Since $n = -3$ this equation becomes

$$(k + 2)\underline{2} = (k + 2)(k + 1) = k^2 + 3k + 2 = -s(k+1) - s(k). \qquad (10.2)$$

If we posit that the leading term of $s(k)$ is $\alpha_d k^d$, then the leading terms of both $-s(k+1)$ and $-s(k)$ are $-\alpha_d k^d$. Thus the leading term on the right side of Equation (10.2) is $-2\alpha_d k^d$. Since this must equal k^2 (the leading term on the left side of Equation (10.2)), we have $d = 2$ and $\alpha_2 = -1/2$. Thus $s(k) = -k^2/2 + \alpha_1 k + \alpha_0$, and

$$s(k+1) = \frac{-(k+1)^2}{2} + \alpha_1(k+1) + \alpha_0 = \frac{-k^2}{2} - k - \frac{1}{2} + \alpha_1 k + \alpha_1 + \alpha_0$$

$$= \frac{-k^2}{2} + (\alpha_1 - 1)k + \alpha_1 + \alpha_0 - \frac{1}{2}.$$

We now want $s(k)$ satisfying

$$k^2 + 3k + 2 = -s(k+1) - s(k)$$

$$= \frac{k^2}{2} + (-\alpha_1 + 1)k - \alpha_1 - \alpha_0 + \frac{1}{2} + \frac{k^2}{2} - \alpha_1 k - \alpha_0$$

$$= k^2 + (-2\alpha_1 + 1)k - \alpha_1 - 2\alpha_0 + \frac{1}{2}.$$

Equating coefficients of k, we have $3 = -2\alpha_1 + 1$, and so $\alpha_1 = -1$. For the constant term we have $2 = 1 - 2\alpha_0 + 1/2$, and so $\alpha_0 = -1/4$. We're left with $s(k) = -k^2/2 - k - 1/4$.

We've found an $s(k)$ that works, so we can move to Step 4 of Gosper's algorithm. The antidifference T_k of $\binom{-3}{k}$ is thus

$$T_k = \frac{-k^2/2 - k - 1/4}{(k+2)(k+1)} \binom{-3}{k} = \frac{-k^2/2 - k - 1/4}{(k+2)(k+1)} \binom{2+k}{k}(-1)^k$$

$$= \frac{-k^2/2 - k - 1/4}{(k+2)(k+1)} \cdot \frac{(k+2)(k+1)}{2}(-1)^k = (-1)^{k-1} \left(\frac{1}{4}k^2 + \frac{k}{2} + \frac{1}{8} \right),$$

where we have used Identity 19 to rewrite $\binom{-3}{k}$ as $\binom{2+k}{k}(-1)^k$.

Since T_k is an antidifference of $\binom{-3}{k}$, we have $\sum_{k=0}^{n} \binom{-3}{k} = T_{n+1} - T_0$, which simplifies to Identity 330. $\qquad \square$

Let's now look at the case $t_k = \binom{k}{m}$. According to Identity 17, this is defined only when m is an integer and nonzero only when m is nonnegative; thus let's place these two restrictions on m. Gosper's algorithm should give us something here, thanks to Identity 58, so let's just place our work inside a proof of this identity.

Identity 58. *For integers $m \geq 0$,*

$$\sum_{k=0}^{n} \binom{k}{m} = \binom{n+1}{m+1}.$$

Proof. With $t_k = \binom{k}{m}$, we have

$$\frac{t_{k+1}}{t_k} = \frac{\binom{k+1}{m}}{\binom{k}{m}} = \frac{(k+1)^{\underline{m}} \, m!}{m! \, k^{\underline{m}}} = \frac{k+1}{k-m+1}.$$

For fixed m, this is a rational function of k, and so $\binom{k}{m}$ is a hypergeometric term.

By Step 1 of Gosper's algorithm, we want to find $p(k)$, $q(k)$, and $r(k)$ such that

$$\frac{p(k+1)q(k)}{p(k)r(k+1)} = \frac{k+1}{k-m+1}.$$

As before, let's try $q(k) = k+1$ (taking care of the numerator), $r(k) = k - m$ (taking care of the denominator), and $p(k) = 1$.

Once again, does this satisfy the condition in Step 1 of Gosper's algorithm? The only factor $k+a$ of $k+1$ has $a = 1$, and the only factor $k+b$ of $k-m$ has $b = -m$. We're assuming m is a nonnegative integer, so $a - b = 1 - (-m) = 1 + m$, which is a positive integer. This is exactly the condition we must avoid in Step 1 of Gosper's algorithm, so our initial attempt to find values for $p(k)$, $q(k)$, and $r(k)$ doesn't work.

Next, we'll try our three-step process for modifying the values of $p(k)$, $q(k)$, and $r(k)$. This gives us $p(k) = k^{\underline{m}}$, $q(k) = 1$, and $r(k) = 1$.

According to Step 2 of Gosper's algorithm, we must now find a polynomial $s(k)$ satisfying

$$k^{\underline{m}} = s(k+1) - s(k) = \Delta s(k).$$

Identity 177 is helpful here. It says that

$$\Delta x^{\underline{n}} = n x^{\underline{n-1}}.$$

Letting $n = m + 1$ and $x = k$ in Identity 177 and applying the linearity of the finite difference operator (see Exercise 9 of Chapter 7) to move the $m+1$ term inside the Δ gives us

$$\Delta \left(\frac{k^{\underline{m+1}}}{m+1} \right) = k^{\underline{m}},$$

so that $s(k) = k^{m+1}/(m+1)$.

Moving on to Step 4 of Gosper's algorithm, we now have an antidifference of $t_k = \binom{k}{m}$:

$$
\begin{aligned}
T_k &= \frac{k^{m+1}}{(m+1)k^{\underline{m}}}\binom{k}{m} = \frac{k-(m+1)+1}{(m+1)}\binom{k}{m} = \frac{(k-m)\,k!}{(m+1)\,m!\,(k-m)!} \\
&= \frac{k!}{(m+1)!\,(k-m-1)!} = \binom{k}{m+1}.
\end{aligned}
$$

Therefore,

$$
\sum_{k=0}^{n}\binom{k}{m} = T_{n+1} - T_0 = \binom{n+1}{m+1} - \binom{0}{m+1} = \binom{n+1}{m+1}.
$$

$\square$

10.3   Definite Summation of Hypergeometric Terms

If a hypergeometric term t_k has an antidifference T_k that is also hypergeometric, Gosper's algorithm will produce T_k for you. Much like antiderivatives and integration, you can then use T_k to evaluate any sum of the form $\sum_{k=a}^{b} t_k$, as we saw in the previous section.

But what if Gosper's algorithm tells you that no such antidifference exists (at least within the class of hypergeometric terms)? What do you do then? In this section we'll discuss an extension of Gosper's algorithm developed by Zeilberger that allows us to evaluate many more sums involving hypergeometric terms—even if Gosper's original algorithm tells us that there is no hypergeometric antidifference.

Before we get to the details of Zeilberger's extension, let's give a brief overview about how it works. There are several points to be made here, so we'll list them.

1. Zeilberger's extension applies to sums of the form $S(n) = \sum_k t_{n,k}$, where the sum is taken over all integer values of k.

2. The term $t_{n,k}$ needs to be hypergeometric in both n and k; i.e., $t_{n+1,k}/t_{n,k}$ and $t_{n,k+1}/t_{n,k}$ must both be rational functions of n and k.

3. Instead of finding an antidifference for $t_{n,k}$, Zeilberger's extension finds an antidifference for a *recurrence* involving $t_{n,k}$. When successful, it produces something of the form

$$
\beta_0(n)t_{n,k} + \beta_1(n)t_{n+1,k} + \cdots + \beta_l(n)t_{n+l,k} = \Delta_k T_{n,k} = T_{n,k+1} - T_{n,k}.
$$
$$\tag{10.3}$$

4. If $\lim_{k \to \infty} T_{n,k} = A(n)$, and $\lim_{k \to -\infty} T_{n,k} = B(n)$, summing both sides of Equation (10.3) over all integer values of k produces the recurrence

$$\beta_0(n)S(n,k) + \beta_1(n)S(n+1,k) + \cdots + \beta_l(n)S(n+l,k) = A(n) - B(n),$$

which you can then solve to find $S(n)$. (Frequently there are only finitely many values k for which $T_{n,k}$ is nonzero, in which case $A(n) = B(n) = 0$.)

The full Gosper-Zeilberger algorithm is given in Figures 10.2 and 10.3. Gosper's algorithm is basically the iteration with $l = 0$; Zeilberger's extension is the rest of it. (See Step 3 in Figure 10.2, though, for a difference between Gosper's algorithm and the $l = 0$ iteration of Gosper-Zeilberger.) Also, because of the recurrence produced by Gosper-Zeilberger, in this section we'll assume that n is a nonnegative integer.

Let's try the Gosper-Zeilberger algorithm on an example: $\sum_k \binom{n}{k}$. Gosper's algorithm has already shown us that there is no closed-form anti-difference for $\binom{n}{k}$ among hypergeometric terms, yet we already know an incredibly simple formula for this sum (at least when n is a nonnegative integer). Thus this sum seems like a perfect first example with Gosper-Zeilberger. As with some examples in the previous section, we'll place our work inside a proof.

> **Doron Zeilberger** (1950–) is an Israeli mathematician who has spent much of his career in the United States. He is a strong advocate for the use of computers in doing mathematical research, as evidenced by his work extending Gosper's algorithm and the extensive computations featured in his proof of the alternating sign matrix conjecture.

Identity 4.

$$\sum_{k=0}^{n} \binom{n}{k} = 2^n.$$

Proof. We know from the previous section that the iteration $l = 0$ fails (as it's basically just Gosper's algorithm), so let's start with the iteration $l = 1$. With Step 1, we set $\hat{t}_{n,k} = \beta_0(n)t_{n,k} + \beta_1(n)t_{n+1,k}$.

Step 2 says to look at the ratio

$$\frac{t_{n+1,k}}{t_{n,k}} = \frac{\binom{n+1}{k}}{\binom{n}{k}} = \frac{(n+1)!\, k!\, (n-k)!}{k!\, (n+1-k)!\, n!} = \frac{n+1}{n+1-k}.$$

Then we use this ratio to set

$$p(n,k) = (n+1-k)\beta_0(n) + (n+1)\beta_1(n),$$

and

$$\bar{t}_{n,k} = \frac{t_{n,k}}{n+1-k},$$

Gosper-Zeilberger Algorithm, Part 1

Suppose we want to evaluate the sum $S(n) = \sum_k t_{n,k}$, where $t_{n,k}$ is a hyper-geometric term in both n and k.

0. Set $l = 0$.

1. Set $\hat{t}_{n,k} = \beta_0(n)t_{n,k} + \beta_1(n)t_{n+1,k} + \cdots + \beta_l(n)t_{n+l,k}$.

2. Consider the ratio $t_{n+i,k}/t_{n,k}$ for each i, $1 \le i \le l$. Use these to express $\hat{t}_{n,k}$ as
$$\hat{t}_{n,k} = p(n,k)\bar{t}_{n,k}.$$

   Here, $p(n,k)$ is a linear combination of $\beta_0(n), \ldots, \beta_l(n)$ with coefficients that are polynomials in n and k, and $\bar{t}_{n,k}$ is a hypergeometric term in k.

3. Similar to Step 1 of Gosper's algorithm, express the ratio $\bar{t}_{n,k+1}/\bar{t}_{n,k}$ as
$$\frac{\bar{t}_{n,k+1}}{\bar{t}_{n,k}} = \frac{\bar{p}(n,k+1)q(n,k)}{\bar{p}(n,k)r(n,k+1)},$$

   where $\bar{p}$, q, and r are polynomials in n and k satisfying the following condition: If $k + a$ is a factor of $q(n,k)$ and $k + b$ is a factor of $r(n,k)$, then $a - b$ is not a positive integer. (One crucial difference between this step and Step 1 of Gosper's algorithm: Since we're treating n as a variable, we only need to be concerned about situations in which the quantity $a - b$ is a constant independent of n.)

4. Similar to Step 2 of Gosper's algorithm, look for a nonzero polynomial $s(n,k)$ in k and values of $\beta_0(n), \ldots, \beta_l(n)$ satisfying
$$\bar{p}(n,k)p(n,k) = q(n,k)s(n,k+1) - r(n,k)s(n,k),$$

   where the coefficients of $s(n,k)$ are functions of n.

FIGURE 10.2

The Gosper-Zeilberger algorithm for evaluating certain hypergeometric sums

so that we have $\hat{t}_{n,k} = p(n,k)\bar{t}_{n,k}$, as required in the remainder of Step 2.

For Step 3, we calculate the ratio $\bar{t}_{n,k+1}/\bar{t}_{n,k}$:

$$\frac{\bar{t}_{n,k+1}}{\bar{t}_{n,k}} = \frac{\binom{n}{k+1}(n+1-k)}{\binom{n}{k}(n-k)} = \frac{n!\,k!\,(n-k)!\,(n+1-k)}{(k+1)!\,(n-1-k)!\,n!\,(n-k)} = \frac{n+1-k}{k+1}.$$

Then we look for polynomials $\bar{p}(n,k)$, $q(n,k)$ and $r(n,k)$ satisfying

$$\frac{n+1-k}{k+1} = \frac{\bar{p}(n,k+1)q(n,k)}{\bar{p}(n,k)r(n,k+1)}.$$

A simple solution is $\bar{p}(n,k) = 1$, $q(n,k) = n+1-k$, and $r(n,k) = k$. With

---

Gosper-Zeilberger Algorithm, Part 2

5. If a nonzero polynomial $s(n, k)$ does not exist, increment l. Return to Step 1.

6. If a nonzero polynomial $s(n, k)$ does exist, then let $T_{n,k}$ be the hypergeometric term

$$T_{n,k} = \frac{r(n, k)s(n, k)\bar{t}_{n,k}}{\bar{p}(n, k)}.$$

Then $T_{n,k}$ is an antidifference in k for $\hat{t}_{n,k}$.

7. Sum the equation $T_{n,k+1} - T_{n,k} = \hat{t}_{n,k}$ over all integer values of k. If $\lim_{k \to \infty} T_{n,k} = A(n)$ and $\lim_{k \to -\infty} T_{n,k} = B(n)$ (very often $T_{n,k}$ is nonzero for only finitely many values of k, which means $A(n) = B(n) = 0$), then we'll have, thanks to the equation in Step 1, the recurrence

$$A(n) - B(n) = \beta_0(n)S(n) + \beta_1(n)S(n + 1) + \cdots + \beta_l(n)S(n + l). \quad (10.4)$$

8. Solve the recurrence in Equation (10.4) to find $S(n) = \sum_k t_{n,k}$.

FIGURE 10.3
The Gosper-Zeilberger algorithm for evaluating certain hypergeometric sums

$k - n - 1$ the only factor of $q(n, k)$ and k the only factor of $r(n, k)$, the condition in Step 3 has us checking $a - b = -n - 1 - 0 = -n - 1$. Since this is not a positive integer independent of n, we've satisfied the condition.

Moving on to Step 4, we need to find a polynomial $s(n, k)$ satisfying

$$(n + 1 - k)\beta_0(n) + (n + 1)\beta_1(n) = (n + 1 - k)s(n, k + 1) - ks(n, k). \quad (10.5)$$

Let's guess a solution of the form $s(n, k) = \alpha_d(n)k^d + O(k^{d-1})$, where $O(k^{d-1})$ just means that all the remaining powers of k are of order $d - 1$ or smaller. (This *Big Oh* notation is very useful for handling terms that exist but don't affect the part of the calculation that you're focusing on.) The leading term on the right side of Equation (10.5) is $-2\alpha_d(n)k^{d+1}$. Since the leading term on the left side of Equation (10.5) is $-k\beta_0(n)$, we have $d = 0$. Thus $s(n, k) = \alpha_0(n)$, and Equation (10.5) becomes

$$(n + 1 - k)\beta_0(n) + (n + 1)\beta_1(n) = (n + 1 - k)\alpha_0(n) - k\alpha_0(n).$$

Equating coefficients of k in this equation yields the following system:

$$-\beta_0(n) = -2\alpha_0(n),$$
$$(n + 1)\beta_0(n) + (n + 1)\beta_1(n) = (n + 1)\alpha_0(n).$$

This is a homogeneous system (i.e., no constant terms appear by themselves in the equation) with two equations and three unknowns. The zero solution—in which all three unknowns are identically zero—works, but because there are more unknowns than equations linear algebra theory tells us that we are guaranteed an infinite family of nonzero solutions as well. To find one, let's set $\alpha_0(n) = 1$ and solve for the two β values. This yields $\beta_0(n) = 2$ and $\beta_1(n) = -1$.

Moving on to Step 6, we have

$$T_{n,k} = \frac{k\binom{n}{k}}{n+1-k} = \frac{k\,n!}{k!\,(n-k)!\,(n+1-k)} = \frac{n!}{(k-1)!\,(n+1-k)!}$$
$$= \binom{n}{k-1}.$$

Since $T_{n,k} = \binom{n}{k-1}$ is nonzero for only finitely many values of k, we have $A(n) = B(n) = 0$. Thus, in Step 7, we get

$$0 = \sum_k (T_{n,k+1} - T_{n,k}) = \sum_k \hat{t}_{n,k} = \sum_k (2t_{n,k}) - \sum_k t_{n+1,k}.$$

With $t_{n,k} = \binom{n}{k}$ and $S(n) = \sum_k \binom{n}{k} = \sum_{k=0}^n \binom{n}{k}$, we obtain the recurrence

$$0 = 2S(n) - S(n+1) \implies S(n+1) = 2S(n).$$

Step 8 is quite simple: Since we know $S(0) = 1$, the solution to this recurrence is $S(n) = 2^n$. $\qquad\square$

This proof of Identity 4 is fairly complicated, especially compared to, well, any of the other proofs of Identity 4 that we've seen in this text. But the value of the Gosper-Zeilberger algorithm is that it can be *automated*: Plug the sum $\sum_{k=0}^n \binom{n}{k}$ into a program running Gosper-Zeilberger, and it will output the answer 2^n. Our proof here just works through the details of what a computer would normally do.

Let's try another example with a more difficult sum—one that we proved in Chapter 6 using generating functions.

Identity 154.
$$\sum_{k=0}^n \binom{2k}{k}\binom{2n-2k}{n-k} = 4^n.$$

Proof. Start by letting $l = 0$ and $\hat{t}_{n,k} = \beta_0(n)\binom{2k}{k}\binom{2n-2k}{n-k}$. Step 2 of Gosper-Zeilberger is easy, then; just set $p(n,k) = \beta_0(n)$ and $\bar{t}_{n,k} = \binom{2k}{k}\binom{2n-2k}{n-k}$.

Moving on to Step 3, we have

$$\frac{\bar{t}_{n,k+1}}{\bar{t}_{n,k}} = \frac{\binom{2k+2}{k+1}\binom{2n-2k-2}{n-k-1}}{\binom{2k}{k}\binom{2n-2k}{n-k}}$$

$$= \frac{(2k+2)!\,(2n-2k-2)!\,k!\,k!\,(n-k)!\,(n-k)!}{(k+1)!\,(k+1)!\,(n-k-1)!\,(n-k-1)!\,(2k)!\,(2n-2k)!}$$

$$= \frac{(2k+2)(2k+1)(n-k)(n-k)}{(k+1)(k+1)(2n-2k)(2n-2k-1)}$$

$$= \frac{(2k+1)(n-k)}{(k+1)(2n-2k-1)}.$$

We can set $\bar{p}(n,k) = 1$, $q(n,k) = (2k+1)(n-k)$, and $r(n,k) = k(2n-2k+1)$. This satisfies the condition in Step 3: The only choices for a are $1/2$ and $-n$, and the only choices for b are 0 and $-n-1/2$. No combination of these a and b values gives $a - b$ to be a positive integer constant independent of n.

According to Step 4, we now need to find a nonzero polynomial $s(n,k)$ satisfying

$$\beta_0(n) = (2k+1)(n-k)s(n,k+1) - k(2n-2k+1)s(n,k). \qquad (10.6)$$

With $s(n,k) = \alpha_d(n)k^d + \alpha_{d-1}(n)k^{d-1} + O(k^{d-2})$, we have $s(n,k+1) = \alpha_d(n)k^d + (d\alpha_d(n) + \alpha_{d-1}(n))k^{d-1} + O(k^{d-2})$. The leading term of $(2k+1)(n-k)s(n,k+1)$ is $-2\alpha_d(n)k^{d+2}$, and the leading term of $-k(2n-2k+1)s(n,k)$ is $2\alpha_d(n)k^{d+2}$, which means these terms cancel. It takes a bit more algebra to show this, but the k^{d+1} term on the right side of Equation (10.6) is given by

$$(2n-1)\alpha_d(n)k^{d+1} - 2d\alpha_d(n)k^{d+1} - 2\alpha_{d-1}(n)k^{d+1}$$

$$- (2n+1)\alpha_d(n)k^{d+1} + 2\alpha_{d-1}(n)k^{d+1}$$

$$= -2(d+1)\alpha_d(n)k^{d+1}.$$

Since this must equal $\beta_0(n)$, we have that $d = -1$ or $s(n,k) = \beta_0(n) = 0$. The former isn't possible, and the latter is ruled out by the fact that $s(n,k)$ must be nonzero. Thus the $l = 0$ iteration of Gosper-Zeilberger fails.

Now, on to the $l = 1$ iteration (with hope that we'll be more successful this time!). Step 1 of the Gosper-Zeilberger algorithm says to let $\hat{t}_{n,k} = \beta_0(n)t_{n,k} + \beta_1(n)t_{n+1,k}$, where $\beta_0(n)$ and $\beta_1(n)$ will be determined later.

In Step 2, we look at the ratio

$$\frac{t_{n+1,k}}{t_{n,k}} = \frac{\binom{2k}{k}\binom{2n+2-2k}{n+1-k}}{\binom{2k}{k}\binom{2n-2k}{n-k}} = \frac{(2n+2-2k)!\,(n-k)!\,(n-k)!}{(n+1-k)!\,(n+1-k)!\,(2n-2k)!}$$

$$= \frac{(2n+2-2k)(2n+1-2k)}{(n+1-k)(n+1-k)} = \frac{2(2n+1-2k)}{n+1-k}.$$

Continuing with Step 2, we let

$$p(n,k) = (n+1-k)\beta_0(n) + 2(2n+1-2k)\beta_1(n)$$

and

$$\bar{t}_{n,k} = \frac{t_{n,k}}{n+1-k},$$

so that $\hat{t}_{n,k} = p(n,k)\bar{t}_{n,k}$.

Moving on to Step 3, we need to calculate

$$\frac{\bar{t}_{n,k+1}}{\bar{t}_{n,k}} = \frac{\binom{2k+2}{k+1}\binom{2n-2k-2}{n-k-1}(n+1-k)}{\binom{2k}{k}\binom{2n-2k}{n-k}(n-k)}.$$

Most of the work of simplifying this expression was done in the previous iteration, though, so we can jump to

$$\frac{\bar{t}_{n,k+1}}{\bar{t}_{n,k}} = \frac{(2k+1)(n-k)(n+1-k)}{(k+1)(2n-2k-1)(n-k)} = \frac{(2k+1)(n+1-k)}{(k+1)(2n-2k-1)}.$$

Let's try $\bar{p}(n,k) = 1$, $q(n,k) = (2k+1)(n+1-k)$, and $r(n,k) = k(2n-2k+1)$. This satisfies the condition in Step 3: The only choices for a are $1/2$ and $-n-1$, and the only choices for b are 0 and $-n-1/2$. No combination of a and b gives $a-b$ to be a positive integer constant independent of n.

Proceeding to Step 4, we want a nonzero polynomial $s(n,k)$ such that

$$(n+1-k)\beta_0(n) + 2(2n+1-2k)\beta_1(n)$$
$$= (2k+1)(n+1-k)s(n,k+1) - k(2n-2k+1)s(n,k).$$

Multiplying out the factors, this becomes

$$(n+1-k)\beta_0(n) + (4n+2-4k)\beta_1(n)$$
$$= (-2k^2 + (2n+1)k + n + 1)s(n,k+1) + (2k^2 - (2n+1)k)s(n,k). \quad (10.7)$$

Interestingly enough, the coefficients of $s(n,k+1)$ and $s(n,k)$ are the same— except for the sign difference and the extra $n+1$ term in the coefficient of $s(n,k+1)$. What this means is that if $s(n,k)$ is constant with respect to k, most of the right side of Equation (10.7) will cancel. Since all we need is *one* polynomial $s(n,k)$ that works, let's try $s(n,k) = \alpha_0(n)$. If it doesn't work, we can try something more general for $s(n,k)$.

With $s(n,k) = \alpha_0(n)$, Equation (10.7) becomes

$$(n+1-k)\beta_0(n) + (4n+2-4k)\beta_1(n) = (n+1)\alpha_0(n).$$

Setting powers of k equal, we have the following system:

$$-\beta_0(n) - 4\beta_1(n) = 0$$
$$(n+1)\beta_0(n) + (4n+2)\beta_1(n) = (n+1)\alpha_0(n).$$

We have a homogeneous system with two equations and three unknowns; thus we are guaranteed a nonzero solution.

The first equation tells us that $\beta_0(n) = -4\beta_1(n)$, and the fact that we have one more variable than equations means we have a free parameter. So let's set $\alpha_0(n) = 1$ as well. The second equation then becomes

$$(n+1)(-4\beta_1(n)) + (4n+2)\beta_1(n) = (n+1)$$

$$\Longrightarrow \beta_1(n) = \frac{n+1}{-2}.$$

Thus $\beta_0(n) = 2(n+1)$. This gives us our polynomial $s(n,k) = 1$ and our values for $\beta_0(n)$ and $\beta_1(n)$.

In Step 6, we set

$$T_{n,k} = \frac{k(2n-2k+1)\binom{2k}{k}\binom{2n-2k}{n-k}}{n+1-k} = \binom{2k}{k}\binom{2n+1-2k}{n+1-k}k,$$

by Identity 6.

Since $T_{n,k}$ is nonzero for only finitely many values of k, by Step 7 we get $A(n) = B(n) = 0$ and the recurrence

$$0 = 2(n+1)S(n) - (n+1)S(n+1)/2 \Longrightarrow S(n+1) = 4S(n),$$

where $S(n) = \sum_{k=0}^{n} \binom{2k}{k}\binom{2n-2k}{n-k}$. (Since we're taking n to be a positive integer, the possibility that $n+1 = 0$ is ruled out.)

Finally, because $S(0) = \sum_{k=0}^{0} \binom{0}{0}\binom{0}{0} = 1$, the recurrence $S(n+1) = 4S(n)$ gives us $S(n) = 4^n$. $\qquad\square$

The two identities we've picked thus far to illustrate Gosper-Zeilberger both have right sides that satisfy very simple recurrences. For our third example with Gosper-Zeilberger, let's see what it can do with an identity that doesn't have such a simple recurrence—an identity, in fact, that features an Iverson bracket.

Identity 13.

$$\sum_{k\geq 0} \binom{n}{2k} = 2^{n-1} + \frac{1}{2}[n=0].$$

Proof. First, $t_{n,k} = \binom{n}{2k}$. It turns out that the $l = 0$ iteration fails in a manner similar to the way it fails for the proof of Identity 154; the details are in Exercise 16.

For the $l = 1$ iteration, set $\hat{t}_{n,k} = \beta_0(n)t_{n,k} + \beta_1(n)t_{n+1,k}$. Then, in Step 2, we consider the ratio

$$\frac{t_{n+1,k}}{t_{n,k}} = \frac{\binom{n+1}{2k}}{\binom{n}{2k}} = \frac{(n+1)!\,(2k)!\,(n-2k)!}{(2k)!\,(n+1-2k)!\,n!} = \frac{n+1}{n+1-2k}.$$

This leads to

$$p(n,k) = (n+1-2k)\beta_0(n) + (n+1)\beta_1(n)$$

and

$$\bar{t}_{n,k} = \frac{\binom{n}{2k}}{n+1-2k}.$$

For Step 3, we need to analyze the ratio

$$\frac{\bar{t}_{n,k+1}}{\bar{t}_{n,k}} = \frac{\binom{n}{2k+2}(n+1-2k)}{\binom{n}{2k}(n-1-2k)} = \frac{n!\,(2k)!\,(n-2k)!\,(n+1-2k)}{(2k+2)!\,(n-2k-2)!\,n!\,(n-1-2k)}$$

$$= \frac{(n-2k)(n-1-2k)(n+1-2k)}{(2k+2)(2k+1)(n-1-2k)} = \frac{(n-2k)(n-2k+1)}{2(k+1)(2k+1)}.$$

Then we can set $\bar{p}(n,k) = 1$, $q(n,k) = (n-2k)(n-2k+1)$, and $r(n,k) = 2k(2k-1)$. This satisfies the condition in Step 3 of Gosper-Zeilberger, as any combination of choices for a (these are $-n/2$ and $-n/2-1/2$) and b (these are 0 and $-1/2$) does not give $a-b$ to be a positive integer constant independent of n.

According to Step 4, we now want to find a nonzero polynomial $s(n,k)$ satisfying

$$(n+1-2k)\beta_0(n)+(n+1)\beta_1(n) = (n-2k)(n-2k+1)s(n,k+1)-2k(2k-1)s(n,k).$$
$$(10.8)$$

Assuming $s(n,k) = \alpha_d(n)k^d + \alpha_{d-1}(n)k^{d-1} + O(k^{d-2})$, we have $s(n,k+1) = \alpha_d(n)k^d + (d\alpha_d(n) + \alpha_{d-1}(n))k^{d-1} + O(k^{d-2})$. Then the right side of Equation (10.8) becomes

$$(4k^2 + (-4n-2)k + n^2 + n)(\alpha_d(n)k^d + (d\alpha_d(n) + \alpha_{d-1}(n))k^{d-1} + O(k^{d-2}))$$
$$+ (-4k^2 + 2k)(\alpha_d(n)k^d + \alpha_{d-1}(n)k^{d-1} + O(k^{d-2}))$$
$$=(4-4)\alpha_d(n)k^{d+2} + (4d\alpha_d(n) + (4-4)\alpha_{d-1}(n) + (-4n-2+2)\alpha_d(n))k^{d+1}$$
$$+ O(k^d)$$
$$=4(d-n)\alpha_d(n)k^{d+1} + O(k^d).$$

Since this must equal $(n+1-2k)\beta_0(n) + (n+1)\beta_1(n)$, we have $d = 0$. Thus Equation (10.8) simplifies to

$$(n+1-2k)\beta_0(n) + (n+1)\beta_1(n) = (-4nk + n^2 + n)\alpha_0(n).$$

Equating coefficients of k yields the system

$$-2\beta_0(n) = -4n\alpha_0(n),$$
$$(n+1)\beta_0(n) + (n+1)\beta_1(n) = n(n+1)\alpha_0(n).$$

With two equations and three unknowns, we have a nonzero solution with a free parameter. As before, let's set $\alpha_0(n) = 1$. This leads to $\beta_0(n) = 2n$ and $\beta_1(n) = -n$. (Since we're assuming n is a nonnegative integer, we can cancel the $n+1$ factors in the second equation.)

For Step 6, we have

$$T_{n,k} = \frac{2k(2k-1)\binom{n}{2k}}{n-2k+1} = \frac{2k(2k-1)n!}{(2k)!\,(n-2k)!\,(n-2k+1)}$$

$$= \frac{n(n-1)!}{(2k-2)!\,(n-1-(2k-2))!} = n\binom{n-1}{2k-2}.$$

Since $T_{n,k}$ is nonzero for only finitely many values of k, Step 7 gives us the recurrence

$$0 = 2nS(n) - nS(n+1),$$

which means $S(n+1) = 2S(n)$ or $n = 0$.

This solution means that the relationship $S(n+1) = 2S(n)$ is only guaranteed to hold for $n \geq 1$. Direct calculation gives $S(0) = \sum_{k\geq 0}\binom{n}{2k} = \binom{0}{0} = 1$ and $S(1) = \sum_{k\geq 0}\binom{1}{2k} = \binom{1}{0} = 1$. Then the recurrence yields $S(2) = 2S(1) = 2$, $S(3) = 2S(2) = 4$, $S(4) = 2S(3) = 8$, and, in general, $S(n) = 2^{n-1}$ for $n \geq 1$. Putting this all together, we have

$$\sum_{k\geq 0}\binom{n}{2k} = 2^{n-1} + \frac{1}{2}[n = 0].$$

So, Gosper-Zeilberger was smart enough even to handle the piecewise aspect of this identity! $\qquad\square$

Let's take a moment now to review an important theoretical result about the Gosper-Zeilberger algorithm. It is guaranteed to work through Step 6 (i.e., find an antidifference $T_{n,k}$ for a recurrence of the form $\hat{t}_{n,k}$ in Step 1) whenever $t_{n,k}$ is what's known as a "proper hypergeometric term." A *proper hypergeometric term* $t_{n,k}$ is one that can be written in the form

$$t_{n,k} = p(n,k)\frac{\prod_{i=1}^{m}(a_i n + b_i k + c_i)!}{\prod_{i=1}^{m'}(u_i n + v_i k + w_i)!}x^n y^k,$$

where $p(n,k)$ is a polynomial in n and k, the a_i, b_i, c_i, u_i, v_i, and w_i values are specific integer constants, m and m' are nonnegative integer constants, and x and y are nonzero. For a proof of this claim, see, for example, the text *A=B* [52, p. 105].

Why not Steps 7 and 8? Well, sometimes the sum $\sum_k t_{n,k}$ can't actually be taken over all integer values of k. However, in some situations the right modification to Steps 7 and 8 can still be used to produce a closed-form expression for the sum. For our final example, let's take a look at an instance of this.

Identity 23.

$$\sum_{k=0}^{n}\binom{n}{k}\frac{1}{k+1} = \frac{2^{n+1}-1}{n+1}.$$

Proof. The first part of this proof is actually quite similar to that of our proof of Identity 4 in this section, so we'll go quickly over the beginning and then more carefully over the remainder of the proof, where we need to modify what happens in Steps 7 and 8.

We have $t_{n,k} = \binom{n}{k}\frac{1}{k+1}$. The $l = 0$ iteration fails (see Exercise 17). For the $l = 1$ iteration, we get $p(n,k) = (n+1-k)\beta_0(n) + (n+1)\beta_1(n)$, $\bar{t}_{n,k} = t_{n,k}/(n+1-k)$, $\bar{p}(n,k) = 1$, $q(n,k) = n+1-k$, and $r(n,k) = k+1$. Then we need to solve

$$(n+1-k)\beta_0(n) + (n+1)\beta_1(n) = (n+1-k)s(n,k+1) - (k+1)s(n,k).$$

One solution is $s(n,k) = \alpha_0(n) = 1$, $\beta_0(n) = 2$, and $\beta_1(n) = -(n+2)/(n+1)$. This yields

$$T_{n,k} = \frac{(k+1)\binom{n}{k}}{(k+1)(n+1-k)} = \frac{n!}{k!\,(n-k)!\,(n+1-k)} = \frac{(n+1)!}{k!\,(n+1-k)!\,(n+1)}$$
$$= \frac{1}{n+1}\binom{n+1}{k}.$$

Now comes the new part. We cannot sum the equation

$$T_{n,k+1} - T_{n,k} = \hat{t}_{n,k} = 2\binom{n}{k}\frac{1}{k+1} - \binom{n+1}{k}\frac{n+2}{(n+1)(k+1)} \qquad (10.9)$$

over all integer values of k as Step 7 says, since the neither of the terms on the right side is defined when $k = -1$.

However, the sum we are after is $S(n) = \sum_{k=0}^{n}\binom{n}{k}\frac{1}{k+1}$, which means that we won't encounter this problem with $k = -1$ if we simply sum Equation (10.9) over all *nonnegative* integer values of k. In addition, $T_{n,k}$ is nonzero for only finitely many values of k, so this infinite sum will converge. We get

$$\sum_{k=0}^{\infty}(T_{n,k+1} - T_{n,k}) = 2S(n) - \frac{n+2}{n+1}S(n+1).$$

The left side of this equation telescopes (as usual), and since $T_{n,k} = 0$ for $k > n+1$, we end up with the recurrence

$$-T_{n,0} = -\frac{1}{n+1}\binom{n+1}{0} = 2S(n) - \frac{n+2}{n+1}S(n+1)$$
$$\implies (n+2)S(n+1) - 2(n+1)S(n) = 1.$$

If we let $R(n) = (n+1)S(n)$, this recurrence simplifies greatly to

$$R(n+1) - 2R(n) = 1.$$

This is the kind of recurrence we saw in Chapter 7. In fact, Exercise 5 in Chapter 7 asks you to solve several recurrences that look very much like this one. We'll just hit the highlights here.

Rewrite the recurrence we want to solve as $R(n) - 2R(n-1) = 1$, multiply both sides by x^n, and sum as n goes from one to infinity. Standard generating function manipulation of the kind we saw in Chapter 7, plus the fact that $R(0) = S(0) = 1$, yields

$$f_R(x) = \sum_{n=0}^{\infty} R(n)x^n = \frac{1}{(1-x)(1-2x)}.$$

Partial fractions decomposition turns this into

$$f_R(x) = \sum_{n=0}^{\infty} R(n)x^n = \frac{2}{1-2x} - \frac{1}{1-x} = 2\sum_{n=0}^{\infty} 2^n x^n - \sum_{n=0}^{\infty} x^n = \sum_{n=0}^{\infty} (2^{n+1}-1)x^n.$$

Thus $R(n) = 2^{n+1} - 1$, and

$$S(n) = \sum_{k=0}^{n} \binom{n}{k} \frac{1}{k+1} = \frac{2^{n+1}-1}{n+1}.$$

□

What we see in this proof is that even if we can't sum $T_{n+1,k} - T_{n,k} = \hat{t}_{n,k}$ over all values of k, we might still be able to sum this equation over enough values of k to give us a recurrence for $S(n)$—a recurrence that we can then solve.

10.4 Exercises

1. If t_k and s_k are both hypergeometric terms, determine whether the following functions of k are hypergeometric terms as well. Explain your answers.

   (a) c, where c is constant
   (b) a^k, where a is constant
   (c) k^n, where n is a positive integer
   (d) $\cos(k\pi)$
   (e) $k! + (k+1)!$
   (f) $\binom{n}{k} + \binom{n}{k+1}$
   (g) $t_k s_k$
   (h) $t_k + s_k$

2. Show that the $_pF_q$ notation for a hypergeometric series is not unique by finding another way to represent the series

   $$_pF_q \left[\begin{matrix} a_1, & \ldots, & a_p \\ b_1, & \ldots, & b_q \end{matrix} ; x \right] = \sum_{k=0}^{\infty} \frac{(a_1)^{\overline{k}} \cdots (a_p)^{\overline{k}}}{(b_1)^{\overline{k}} \cdots (b_q)^{\overline{k}}} \frac{x^k}{k!}.$$

   (Hint: You'll need different values for p and q.)

3. Sometimes instead of a list of upper or lower parameters in the $_pF_q$ notation you'll see an em-dash. This simply means that the list of parameters is empty. Show that each of the following expressions is equal to a common function, and find a way to represent that function as a hypergeometric series that does not use an em-dash.

(a) $_0F_0[\overline{}; x]$

(b) $_1F_0[\frac{1}{}; x]$ (series converges only for $|x| < 1$, though)

4. Show that

$$\frac{d}{dx} \, _pF_q\left[\begin{matrix} a_1, & \ldots, & a_p \\ b_1, & \ldots, & b_q \end{matrix}; x\right] = \frac{a_1 a_2 \cdots a_p}{b_1 b_2 \cdots b_q} \, _pF_q\left[\begin{matrix} a_1 + 1, & \ldots, & a_p + 1 \\ b_1 + 1, & \ldots, & b_q + 1 \end{matrix}; x\right].$$

5. For each of the functions below, use its given Maclaurin series to prove an associated hypergeometric series representation. (You do not need to prove that the ranges of convergence hold.)

(a) $e^x = \displaystyle\sum_{k=0}^{\infty} \frac{x^k}{k!} = \, _1F_1\left[\begin{matrix} 1 \\ 1 \end{matrix}; x\right]$, valid for all x.

(b) $\sin x = \displaystyle\sum_{k=0}^{\infty} \frac{(-1)^k}{(2k+1)!} x^{2k+1} = x \, _1F_2\left[\begin{matrix} 1 \\ 1, \ 3/2 \end{matrix}; \frac{-x^2}{4}\right]$, valid for all x.

(c) $\cos x = \displaystyle\sum_{k=0}^{\infty} \frac{(-1)^k}{(2k)!} x^{2k} = \, _1F_2\left[\begin{matrix} 1 \\ 1, \ 1/2 \end{matrix}; \frac{-x^2}{4}\right]$, valid for all x.

(d) $\arcsin x = \displaystyle\sum_{k=0}^{\infty} \binom{2k}{k} \frac{x^{2k+1}}{(2k+1)4^k} = x \, _2F_1\left[\begin{matrix} 1/2, \ 1/2 \\ 3/2 \end{matrix}; x^2\right]$, valid for $|x| \le 1$. (The first equation is Identity 161.)

(e) $\arctan x = \displaystyle\sum_{k=0}^{\infty} \frac{(-1)^k}{2k+1} x^{2k+1} = x \, _2F_1\left[\begin{matrix} 1/2, \ 1 \\ 3/2 \end{matrix}; -x^2\right]$, valid for $|x| \le 1$.

(f) $\ln(1 + x) = \displaystyle\sum_{k=1}^{\infty} \frac{(-1)^{k+1}}{k} x^k = x \, _2F_1\left[\begin{matrix} 1, \ 1 \\ 2 \end{matrix}; -x\right]$, valid for $|x| < 1$.

6. Prove

Identity 331.

$$\sum_{k=0}^{n} \binom{-2}{k} = (-1)^n \left(\frac{n}{2} + \frac{3}{4}\right) + \frac{1}{4}.$$

7. In Step 2 of Gosper's algorithm for finding an antidifference of $t_k = \binom{n}{k}$ when n is a negative integer, you're looking for a polynomial $s(k)$ satisfying $(k - n - 1)\frac{-n-1}{} = -s(k+1) - s(k)$. Show that $s(k)$ does exist and is a polynomial of degree $-n - 1$.

8. For n a negative integer, use Gosper's algorithm to show that there is an antidifference of $t_k = \binom{n}{k}$ given by

$$T_k = \frac{(-1)^{k-1}S(k)}{(-n-1)!},$$

where $S(k)$ is a polynomial of degree $-n-1$ satisfying

$$S(k) = \frac{1}{2}k^{-n-1} + O(k^{-n-2}).$$

9. (An extension of Exercise 8.) For n a negative integer and $n \leq -2$, use Gosper's algorithm to show that there is an antidifference of $t_k = \binom{n}{k}$ given by

$$T_k = \frac{(-1)^{k-1}S(k)}{(-n-1)!},$$

where $S(k)$ is a polynomial of degree $-n-1$ satisfying

$$S(k) = \frac{1}{2}k^{-n-1} + \frac{(n+1)^2}{4}k^{-n-2} + O(k^{-n-3}).$$

10. Use Gosper's algorithm to prove Identity 20 in the case where n is not zero or a negative integer.

Identity 20.

$$\sum_{k=0}^{m} \binom{n}{k}(-1)^k = (-1)^m \binom{n-1}{m}.$$

11. Use Gosper's algorithm to prove Identity 20 in the case where n is a negative integer.

Identity 20.

$$\sum_{k=0}^{m} \binom{n}{k}(-1)^k = (-1)^m \binom{n-1}{m}.$$

12. Use Gosper's algorithm to prove Identity 15 in the case where n is not zero, negative one, or a positive integer.

Identity 15.

$$\sum_{k=0}^{m} \binom{n+k}{k} = \binom{n+m+1}{m}.$$

13. Use Gosper's algorithm to prove Identity 15 in the case where n is a positive integer.

Identity 15.

$$\sum_{k=0}^{m} \binom{n+k}{k} = \binom{n+m+1}{m}.$$

14. Use Gosper's algorithm to prove the following, a generalization of Identity 22.

Identity 332.

$$\sum_{k=0}^{m} \binom{n}{k} k(-1)^k = n\binom{n-2}{m-1}(-1)^m.$$

15. Use Gosper's algorithm to prove Identity 107.

Identity 107. *For* $m \geq 2$,

$$\sum_{k=m}^{\infty} \frac{1}{\binom{k}{m}} = \frac{m}{m-1}.$$

16. Show that the $l = 0$ iteration of the Gosper-Zeilberger algorithm fails when applied to the sum $\sum_{k\geq 0} \binom{n}{2k}$.

17. Show that the $l = 0$ iteration of the Gosper-Zeilberger algorithm fails when applied to the sum $\sum_{k\geq 0} \binom{n}{k}\frac{1}{k+1}$.

18. In our proof of Identity 13 in this chapter, we found $T_{n,k} = n\binom{n-1}{2k-2}$ as an antidifference in k for our expression for $\hat{t}_{n,k}$. Prove that this is actually the case; i.e., prove

Identity 333.

$$n\binom{n-1}{2k} - n\binom{n-1}{2k-2} = 2n\binom{n}{2k} - n\binom{n+1}{2k}.$$

19. Use the Gosper-Zeilberger algorithm to prove

Identity 21.

$$\sum_{k=0}^{n} \binom{n}{k} k = n2^{n-1}.$$

20. Use the Gosper-Zeilberger algorithm to prove

Identity 71.

$$\sum_{k=0}^{n} \binom{n}{k} k^2 = n(n+1)2^{n-2}.$$

21. Use the Gosper-Zeilberger algorithm to prove

Identity 49.

$$\sum_{k=0}^{n} \binom{n}{k}\binom{k}{m} = 2^{n-m}\binom{n}{m}.$$

22. Use the Gosper-Zeilberger algorithm to prove the following, a generalization of Identity 24. (Hint: Iteration $l = 0$ succeeds. Also, after reaching Step 6, only sum the antidifference $T_{n,k}$ from 0 to m.)

Identity 44.

$$\sum_{k=0}^{m} \binom{n}{k}\frac{(-1)^k}{k+1} = \frac{1}{n+1}\binom{n}{m+1}(-1)^m + \frac{1}{n+1}.$$

23. Use the Gosper-Zeilberger algorithm to prove

Identity 148.

$$\sum_{k=0}^{n} \binom{n}{k}\frac{(-1)^k}{2k+1} = \frac{(2^n n!)^2}{(2n+1)!}.$$

10.5   Notes

The algorithm for indefinite summation of hypergeometric terms is by Gosper [29]. Zeilberger's extension to Gosper's algorithm appears in Zeilberger [82].

Another technique closely related to the mechanical summation procedures we discuss here is the *WZ method* developed by Wilf and Zeilberger [80]. This method is concerned with *verifying* identities: Starting with an identity (rather than a hypergeometric sum), you can use the WZ method to show that the identity is true by showing that a particular associated rational function exists. The method explicitly uses Gosper's algorithm to help construct the rational function. For details, see the text $A = B$ [52] by Petkovšek, Wilf, and Zeilberger.

In general, $A = B$ [52] is an excellent resource for more details and background on the algorithms presented here. *Concrete Mathematics* [32] works through several examples with Gosper's algorithm and Zeilberger's extension, too. Rosen's *Handbook of Discrete and Combinatorial Mathematics* [58] gives an overview of mechanical summation procedures and a few examples as well.

Appendix A

Hints and Solutions to Exercises

Chapter 1

1. These are all fairly straightforward when using the factorial definition of the binomial coefficients.

2. Set $x = -1, y = 1$ in Identity 3. The special case is $n = 0$, which, as we have defined $0^0 = 1$, yields 1 for the sum.

3. Add Identities 4 and 12 together. The terms with odd values of k cancel out, and we get two copies of each term with an even value of k. Thus we have

$$2 \sum_k \binom{n}{2k} = 2^n + [n = 0].$$

Divide both sides by 2 to obtain Identity 13. Identity 14 can be obtained by subtracting Identity 12 from Identity 4.

4. Use an argument like the one we used to prove $(2) \implies (3)$.

5. Write out the sum as

$$\binom{n}{0} + \binom{n+1}{1} + \binom{n+2}{2} + \cdots + \binom{n+m}{m}.$$

Combine the first two terms to get $\binom{n}{0} + \binom{n+1}{1} = \binom{n+1}{0} + \binom{n+1}{1} = \binom{n+2}{1}$. This term can be combined with the next term, $\binom{n+2}{2}$, to get $\binom{n+3}{2}$, which can then be combined with $\binom{n+3}{3}$, and so forth. Continue in this fashion, repeatedly applying Identity 1, until you obtain $\binom{n+m+1}{m}$ at the end.

6. Substitute $-x$ for x in Identity 3 and add the result to Identity 3. As in the proof of Identity 13, the terms with odd values of k cancel out, and we get two copies of each term with an even value of k. Then divide both sides by 2.

Chapter 2

1. According to the recurrence, $\binom{n}{n} = \binom{n-1}{n} + \binom{n-1}{n-1}$. With the boundary conditions, we have $\binom{n}{n} = \binom{n-1}{n-1} = 1$, and so $\binom{n-1}{n} = 0$ for $n \geq 1$. With the base case $j = 1$ in the claim $\binom{n-j}{n} = 0$ proved, apply induction to show that $\binom{n-j}{n} = 0$ for $1 \leq j \leq n$.

2. Exercise 1 shows that the first recurrence implies the second. To show that the second implies the first, use induction on n to show that the second yields $\binom{n}{k} = 0$ when $0 \leq n < k$. Then show that this implies $\binom{n}{n} = 1$ when n is a nonnegative integer.

   To show that the second recurrence implies $\binom{n}{k} = 0$ when $k < 0$, use induction on j with the claim $\binom{n}{-j} = 0$.

3. If $k < 0$, then the recurrence obviously holds. If $k = 0$ the recurrence yields $1 = 1 + 0$, and $\frac{0^{\underline{0}}}{0!} = 1$. (Remember that an empty product evaluates to 1.) If $k > 0$, write out $\binom{n-1}{k} + \binom{n-1}{k-1}$ as $\frac{(n-1)^{\underline{k}}}{k!} + \frac{(n-1)^{\underline{k-1}}}{(k-1)!}$, factor out like terms, and simplify what remains. The boundary conditions are easy to check: Again, $\frac{0^{\underline{0}}}{0!} = 1$, and $n^{\underline{n}} = n!$.

4. We have $(-1)^{\underline{m}} = (-1)(-2) \cdots (-k)$. Divide this by $k!$ to obtain $(-1)^k$.

5. First show that $(-1/2)^{\underline{n}} = (-1)^n (1)(3)(5) \cdots (2n - 1)/2^n$. Multiply this by $2^n n!/(2^n n!)$ and combine factors to get $(2n)!$ in the numerator.

6. (a) Write out $n^{\overline{k}}$, and note that there are k factors, the largest of which is $n + k - 1$.

   (b) Use part (a); divide by $k!$; apply Identities 17, 19, and 17 successively; and finally multiply by $k!$.

7. (a) Write $\binom{n}{k}$ as $\frac{n!}{k!(n-k)!}$ and convert the falling powers in x and y to binomial coefficients via Identity 17. Apply Identity 57 (the general version for real m and n is proved in Chapter 6), and convert the resulting binomial coefficient back to factorial powers via Identity 17.

   (b) This is similar to part (a), but you need to convert the rising powers in x and y to falling powers with Identity 32 before converting to binomial coefficients. Convert back to rising powers at the end.

8. Use Identity 6 to absorb the factor of k, yielding $n\binom{n-1}{k-1}(k - 1)$ as the summand. Then apply Identity 6 again with $k - 1$ instead of k, followed by Identity 4.

9. This is similar to Exercise 8; apply Identity 6 m times.

10. Use Identity 6 twice, as in Exercise 8, followed by Identity 12. This identity can also be proved by taking $m = 2$ in Identity 26 and multiplying both sides by 2.

11. This is similar to Exercise 10; apply Identity 6 m times. Alternatively, multiply both sides of Identity 26 by $m!$.

12. This is similar to the proof of Identity 23 presented in the chapter. You need to use Identity 6 twice.

13. This is similar to Exercise 12; apply Identity 6 m times.

14. This is similar to the proof of Identity 24 presented in the chapter. Use Identity 6 twice. At the end, express the result using a common denominator.

15. This is similar to Exercise 14. Use Identity 6 m times. After applying Identity 20, express the binomial coefficient in factorial form and cancel factors.

16. Use finite differences, the binomial recursion $\binom{n+1}{k} = \binom{n}{k} + \binom{n}{k-1}$, and the absorption identity in a manner similar to that of the proof of Identity 25 given in the chapter.

17. Mimic the proof of Identity 24 in the text, but use Identity 20 instead of Identity 12.

18. Apply the absorption identity to remove one of the factors of $1/(k+1)$. Then reindex the sum and apply Identity 25.

19. This is similar to Exercise 18; use Identity 43 instead of Identity 25 at the end, though.

20. Use the absorption identity, followed by Identity 14.

21. Multiply the sum by $2/2$ and rewrite the denominator as $1/(2k+2)$. Then use the absorption identity, followed by Identity 13. Remember to subtract off $\binom{n}{0}$ because the resulting sum starts with $\binom{n}{2}$.

22. Swap k and m in Identity 7, and apply the result to the left side of the sum. Switch indices on the sum, and apply Identity 4. (Alternatively, divide both sides of Identity 36 by $m!$.)

23. If $x = 0$ or $y = 0$ then the identity is easy to verify. Otherwise, the proof is the same as that in Exercise 22 until the last step. Instead of applying Identity 4, apply the binomial theorem. Make sure you are careful with the powers on x and y when switching indices.

24. Solve for $\binom{m}{k}$ in Identity 7, and then replace $\binom{m}{k}$ with the result in the sum. Reverse the indexing, and apply parallel summation (Identity 15). Finally, simplify.

25. As in Exercise 24, use trinomial revision and reverse the indexing. Then apply Identity 77, as suggested in the hint. Watch the indexing.

26. From the binomial inversion point of view Identity 20 implicitly defines

$$g(k) = [k \leq m].$$

Applying binomial inversion then yields

Identity 80.

$$\sum_{k=0}^{n} \binom{n}{k}\binom{k-1}{m}(-1)^{k+m} = [n \leq m].$$

27. For the forward direction, substituting $(-1)^k g(k)$ for $g(k)$ in Theorem 1 yields

$$(-1)^n g(n) = \sum_{k=0}^{n} \binom{n}{k} f(k)(-1)^k,$$

which is equivalent to what you need to show. The backward direction is similar.

28. Apply binomial inversion to Identity 41.

29. Apply binomial inversion to Identity 46.

30. Apply binomial inversion to Identity 36.

31. There are two ways to use binomial inversion to prove this identity. One is to apply binomial inversion to Identity 23, move $\sum_{k=0}^{n} \binom{n}{k}(-1)^k/(k+1)$ to the right side, simplify it with Identity 24, and combine terms on the right. Another is to apply binomial inversion to Identity 47, taking $g(k) = (-1)^k/(k+1)$ if k is even and $g(k) = 0$ if k is odd.

A third method starts by using the absorption identity to remove the factor of $1/(k+1)$. Then reindex and apply the binomial theorem.

32. (a) Use the ratio test and the generalized (falling) factorial definition of the binomial coefficient.

(b) Differentiate term-by-term. Then substitute into the differential equation. Identity 11 is quite helpful.

(c) Use the quotient rule and part (b), plus the fact that if a function has a zero derivative it must be a constant function.

Chapter 3

1. Use the fact that if $A \subseteq B$ and $B \subseteq A$ then $A = B$. To show the left side is a subset of the right side use the following "element-chasing argument." Let $x \in \bigcap_{i=1}^{n} A_i$. Argue that this means that there is at least one A_i that does not contain x. Thus x must be in the union of the complements of the A_i's. Since this must be true for all $x \in \bigcap_{i=1}^{n} A_i$, the left side must be a subset of the right side. The other direction is similar, as is the second of DeMorgan's laws.

2. A stars-and-bars argument works nicely here. Imagine a line of n stars. Placing $k - 1$ bars among those n stars defines a solution in nonnegative integers to $x_1 + x_2 + \cdots + x_k = n$, as x_1 is the number of stars before the first bar, x_2 is the number of stars between the first and second bars, and so forth. Placing n stars and $k - 1$ bars in a row, we choose which of the $n+k-1$ positions the $k-1$ bars will go in $\binom{n+k-1}{k-1}$ ways. Alternatively, we're choosing n items from a group of k elements (the k variables) to construct a multiset. (For example, choosing element 2 three times is equivalent to setting $x_2 = 3$.) The number of ways to do this is $\left(\binom{k}{n}\right) = \binom{k+n-1}{k-1}$.

3. The argument is similar to that for Exercise 2, placing $k - 1$ bars among n stars, but now we cannot place two bars next to each other. Instead, we have $n - 1$ choices for where to place the bars; i.e., between any two stars. This gives $\binom{n-1}{k-1}$ solutions in positive integers to $x_1 + x_2 + \cdots + x_k = n$. Alternatively, we choose each variable exactly once, in one way. Then we form a multiset by choosing $n - k$ additional items additional from a group of k elements. The number of ways to do this is $\left(\binom{k}{n-k}\right) = \binom{n-1}{k-1}$.

4. By writing a set of size k chosen from $\{1, 2, \ldots, n\}$ in numerical order you have an increasing sequence of the kind required. There are $\binom{n}{k}$ ways to choose such a set.

5. By writing a multiset of size k chosen from $\{1, 2, \ldots, n\}$ in numerical order you have a nondecreasing sequence of the kind required. Thus this is almost the definition of the multichoose coefficients.

6. Suppose duplicates are allowed in positions $D_1, D_2, \ldots, D_m$. Then the number of allowable sequences is the number of subsets of size k that can be formed from $\{1, 2, \ldots, n, D_1, D_2, \ldots, D_m\}$: $\binom{n+m}{k}$. To see this, choose k values from this set. Then construct a sequence according to the following rules.

   (a) Place all chosen numbers in numerical order.
   (b) Place each D_j value in the jth position in the sequence.
   (c) Finally, replace each D_j with a duplicate of the number in front of it.

This process produces exactly one allowable sequence, and each allowable sequence can be constructed from this process.

7. Both sides count the number of lattice paths from $(0,0)$ to $(n+1, m)$. For the left side, condition on the number of steps taken before crossing the line $y = n + 1/2$.

8. Use Identity 62 to convert $\left(\binom{n}{k}\right)$ to a binomial coefficient. Then write out the binomial coefficient in factorial form.

9. Both sides count the number of ways to distribute m identical candies to $n + 2$ people. The left side conditions on the number of ways to distribute candies to the first $n + 1$ people; the remaining candies must all go to person $n + 2$.

10. For $n, k > 0$, break the multisets counted by $\left(\binom{n}{k}\right)$ into two groups: those that have at least one instance of element n and those that do not. There are $\left(\binom{n}{k-1}\right)$ in the first group and $\left(\binom{n-1}{k}\right)$ in the second. For the boundary conditions, there is only one multiset of size 0, the empty set, and if there are no elements in the source set then there can be no larger multisets formed from it.

11. Both sides count the number of ways to choose, from n people, a committee of size k and a president not on the committee.

12. Both sides count the number of ways to choose, from n people, a committee of size m and a committee of size p such that the two committees are disjoint.

13. Both sides count the number of ways to choose, from n people, a committee of size k with a chair and a secretary who are not the same person.

14. Both sides count the number of ways to choose, from n people, a president and a committee of a size k. The president could either be on the committee or not.

15. Both sides count the number of ways to choose a committee of any size with a chair and a secretary from n people; the same person cannot be both chair and secretary.

16. Both sides count the number of ways to choose a committee of any size with a chair and a secretary from n people; the same person can be both chair and secretary. For the right side, you'll need to add to add together the results from the two cases (same person is both chair and secretary, different people for chair and secretary).

17. This generalizes Exercise 15 to m special positions on the committee held by distinct people.

18. Both sides count the number of ways to partition a group of n people into two chaired committees.

19. Suppose you're playing a game that involves flipping a coin $n + 1$ times and guessing where the last head occurs in the sequence of flips. The guess must be made before the sequence of flips occurs. The two sides count the number of ways that $m + 1$ heads occur and your guess comes before the last head. (The right side does this by counting the number of ways to guess the wrong position for the head minus the number of ways to guess later than the last head actually appears.)

20. Both sides count the number of functions from $\{0, 1, 2\}$ to the set $\{0, 1, \ldots, n\}$ such that $f(0) > f(1)$ and $f(0) > f(2)$. For the left side, condition on the value of $f(0)$. For the right side, consider the cases $f(1) = f(2)$ and $f(1) \neq f(2)$ separately.

21. Both sides count the number of ways to choose a committee of any size containing a subcommittee of size m from a total of n people. For the left side, condition on the size of the committee. For the right side, choose the subcommittee first, and then decide, one-by-one, whether to place the remaining $n - m$ people on the committee.

22. Both sides count the number of ways that a group of n people can be divided into a group of $n - m$ people, each of whom works on one of $x + y$ tasks, plus a committee of m people. For the right side, choose the committee of size m, and then choose tasks for the remaining $n - m$ people. For the left side, condition on the number of people k who are not working on one of the x tasks. There are $\binom{n}{k}$ ways to choose these people. Once that choice is made, there are $\binom{k}{m}$ ways to choose which of these k people will be on the second committee, x^{n-k} ways to assign $n - k$ people to an x task, and y^{k-m} ways to assign the remaining $k - m$ people to a y task.

23. Both sides count the number of ways to choose a committee of size r from a group of people consisting of m men and n women such that the committee chair is a woman.

24. Both sides count the number of ways to choose a committee of size r from a group of people consisting of m men and n women such that the committee chair and secretary are both women. The same woman could be both chair and secretary.

25. The proof is almost the same as that for Identity 61. Suppose we have a total of $m + n + 1$ balls, of which r are red, s are blue, 1 is black, and the rest are green. The two sides count the number of ways to place the balls such that the black ball occurs after all of the red balls but before any of the blue balls. The green balls have no restriction on their position. To understand the index k on the left side, think of the positions of the

balls as occurring on the number line from $-m$ to n. (As the combinatorial proof indicates, this really is just Identity 61 with some index switching.)

For the lattice path proof, this is the same as the lattice path proof of Identity 61, except for variable switching and that the lattice path starts at $(-m, 0)$ instead of $(0, 0)$. Then $m + k$ is the number of steps taken before the path crosses the line $y = r + 1/2$.

26. Using the same ϕ as in the involution proof of Identity 12, the leftover subsets are those of size m that do not contain n.

27. This is similar to the involution proof of Identity 22, modified to reflect the fact that $\binom{n}{k}k(k-1)$ counts the number of committees of size k with a distinct chair and a secretary that can be formed from a committee of size n. The involution ϕ should move around the highest-numbered person who is not chair or secretary. The only committees for $n \geq 2$ for which this ϕ is not defined are those that consist only of a chair and a secretary.

28. This is similar to Exercise 27; however, the chair and the secretary can be the same person.

29. This is similar to Exercise 27; you need m special positions on the subcommittee.

30. The quantity $\binom{n}{k}\binom{k}{m}$ counts the number of ways to choose, from a group of n people, a committee of size k with a subcommittee of size m. When $n < m$ there are clearly no committees satisfying this interpretation, and so the sum is 0. Otherwise, we need a sign-reversing involution on the set (A, B), where A is a committee formed from n people and B is a subcommittee of A that has exactly m people. Let x be the largest-numbered person who is not in B. The involution should move x into or out of A while leaving B untouched. The only committees for $n \geq m$ for which ϕ is undefined are those in which B consists of all n people. (Also, this is just the identity in Exercise 29 multiplied by $m!$ on both sides.)

31. Suppose you have Group 1, consisting of n people, and Group 2, consisting of r people. Suppose we wish to form Committees A and B. Committee A can be any size and consists of people from Group 1. Committee B must be of size m and consists of people from Group 2 and Committee A. The quantity $\binom{n}{k}\binom{r+k}{m}$ counts the number of ways to form these two committees from Groups 1 and 2.

To prove the identity, use the same sign-reversing involution as in Exercise 30, except that x is the largest-numbered person from Group 1 not in B. The only pairs of committees for which ϕ is undefined are those in which B consists of everyone from Group 1. The number of these are the number of ways to choose the remaining $m - n$ people for Committee B from Group 2.

32. This is similar to Exercise 30, but there are a couple of twists. The combinatorial interpretation is that $\binom{n}{k}\binom{k-1}{m}$ counts the number of ways to choose, from a group of n people, a committee A of size k, assign the largest-numbered person out of those k to be committee chair, and then choose a subcommittee B of size m from the remaining $k-1$ people. When $k = 0$ the summand is $\binom{-1}{m} = (-1)^m$, but this does not fit the combinatorial interpretation. Given this, what remains to be proved combinatorially is

$$\sum_{k=1}^{n} \binom{n}{k}\binom{k-1}{m}(-1)^k = (-1)^{m+1}[n > m].$$

Let x be the chair. Let y be the person with largest number who is not in B and who has a smaller number than x. The involution ϕ should move y into or out of A while leaving B alone and retaining x as chair. (The reason for the $y < x$ restriction is to preserve the requirement that x has the largest number in A.) When $n \leq m$ there are no (A, B) pairs that satisfy this interpretation. When $n > m$ the only pair (A, B) for which ϕ is not defined is the pair in which B consists of $\{1, 2, \ldots, m\}$ and person $m + 1$ is the chair.

33. Use an argument like that for Identity 65 but with $|B| = m - k$.

34. While this is the multichoose version of Identity 81, the argument is a bit more involved. Similar to the proof of Identity 65, $\left(\!\binom{n}{k}\!\right)\left(\!\binom{n}{m-k}\!\right)$ can be thought of as counting ordered pairs (A, B), each of which is a multi-subset of $\{1, 2, \ldots, n\}$, such that $|A| = k$ and $|B| = m - k$. However, the corresponding function ϕ based on the symmetric difference of A and B turns out not to be an involution.

Instead, let x be the smallest number in A that has odd multiplicity, and let y be the smallest number in B (with any multiplicity). If x does not exist then let x be denoted ∞; similarly with y. If $x < \infty$ and $x \leq y$, then have $\phi(A, B)$ move one copy of x to B. If $x > y$, then have $\phi(A, B)$ move one copy of y to A. Since ϕ changes $|A|$ by one it is sign-reversing, and some case-checking shows that ϕ is also an involution.

The pairs (A, B) for which ϕ is not defined are those for which $x = y = \infty$. In other words, $B = \emptyset$ and A contains only numbers with even multiplicity. In this case $|A|$ is even and $k = m$. If m is odd, there are no such pairs (A, B), and if m is even the number of such pairs (A, B) is the number of ways to multichoose $m/2$ pairs of numbers from $\{1, 2, \ldots, n\}$, as the numbers must be chosen two at a time in order to guarantee that we have an even number of them.

(This proof is from Lutgen [45].)

35. The summand $\binom{n}{k}k2^{k-1}$ is the number of ways to form, from n people, a chaired committee of size k plus a subcommittee (of any size) from that committee. Modify the ϕ in the involution proof of Identity 22 so that

y is the largest-numbered person who is not the chair and is not on the subcommittee. The committees for which ϕ is not defined are those for which each of the n people are either the chair or on the subcommittee.

36. A combination of the involutions for Identities 22 and 65 works here. Suppose we have a group made up of n men and n women. The number of committees of size n consisting of k men and $n - k$ women that have a male chair is $\binom{n}{k}^2 k$. Number the men from 1 to n, and do the same for the women. For a given committee, let x be the person on the committee with the highest number who is not the chair and whose opposite-gender counterpart is not on the committee. Swap out x for the same-numbered person of opposite gender. This mapping is sign-reversing, as it increases or decreases the number of men on the committee by 1, and it is its own inverse.

There are two cases for which the mapping is not defined: 1) The committee has men and women with exactly the same numbers, excluding the chair, or 2) the committee has men and women with exactly the same numbers, including the chair. In the first case, there must be one more man than woman, so n must be odd. There are n ways to choose the chair, and then $\binom{n-1}{(n-1)/2}$ ways to choose the numbers of the remaining men. The numbers of the women on the committee are determined by these choices. The parity is given by the number of men on the committee, which is $(n+1)/2$. In the second case, there must be as many women as men, so n must be even. The rest of the proof is similar to the odd case.

37. Both sides count the number of onto functions from $\{1, 2, \ldots, m\}$ to $\{1, 2, \ldots, n\}$.

38. Both sides count the number of onto functions from the set $\{1, 2, \ldots, n+1\}$ to the set $\{1, 2, \ldots, n\}$.

39. Both sides count the number of ways of flipping a coin $2n$ times and getting n heads followed by n tails. For the sum, let A_i denote the event that a head is obtained on the ith flip. Also, choose A to be the set of flips where as many heads occur as tails rather than the set of all flips. Then apply Corollary 1.

40. The key to this identity is to recognize the right side as the number of solutions in positive integers to the equation $x_1 + x_2 + \cdots + x_n = m$. For the left side, let A be the set of solutions to this equation in nonnegative integers, and let A_k be the set of solutions to this equation with $x_k \geq 1$. Then apply Corollary 1.

41. Both sides of Identity 88 count the number of permutations on n elements. The right side is straightforward. For the left side, condition on the number of permutations that have exactly k fixed points. Reverse indices on the

sum in Identity 88 to obtain

$$\sum_{k=0}^{n} \binom{n}{k} D_k = n!.$$

Apply Theorem 2, and then reverse indices to produce the result.

Chapter 4

1. Take $x = -2, y = 1$ in Identity 89 and then multiply both sides by -1.

2. Take $x = -1, y = 1$ in Identity 90, and multiply by -1. Remember the convention that n is evaluated before x and y. This means that in the $n = 1$ case, the factor of $nx + y$ becomes $x + y$, and the resulting right side of Identity 90 is just $(x + y)^0$.

3. Multiply Identity 92 by x, differentiate with respect to x, and then substitute $x = y = 1$.

4. Multiply Identity 89 by x^2, differentiate, and substitute $x = y = 1$.

5. Multiply Identity 89 by x^{m-1}, differentiate, and substitute $x = y = 1$.

6. $n! = \binom{n}{j} j! (n-j)!$.

7. After expanding $(e^x - e^{-x})^n$, differentiate both sides n times and let $x = 0$. Alternatively, $(e^x - e^{-x})^n = 2^n \sinh^n x$, and the latter is a bit easier to differentiate successively.

8. This is similar to our calculus proof of Identity 57. You will need to show and use the fact that the rth derivative of $x^{-(n+1)}$, where $n \geq 0$, is $(-1)^r (n+r)^{\underline{r}}/r!$.

9. This is similar to Exercise 8. You'll also want to reverse the index of summation near the end.

10. (a) Starting with Identity 114, let $m = s$. Swap n and r. Replace k with $k+m$. Then replace m with $m-r$. Finally, replace n with $n+m-r-s$. The upper index can be changed to n because the additional terms are all 0.

    (b) Starting with Identity 115, replace k with $k + s$. Then replace r with $r + s$, and finally s with $s - m$. The lower index can be changed to $-s$ because the additional terms are all 0.

11.
$$\sum_{k=0}^{n}\binom{n}{k}\frac{x^{k+1}y^{n-k}}{k+1}=\frac{(x+y)^{n+1}}{n+1}-\frac{y^{n+1}}{n+1}.$$

The extra term on the right comes from the constant of integration, which can be found by setting x to 0. The result is the special case $x_2 = x$, $x_1 = 0$ of Identity 95.

12. It's probably easiest to multiply $z - y$ and $\sum_{k=1}^{n} z^{k-1}y^{n-k}$. Most of the terms in the resulting sum cancel, leaving only $z^n - y^n$.

13. Let $x_1 = x_2 = 0$ and $y = 1$ in Identity 96, and then integrate the resulting function in x_3 from 0 to x. Part (a) is obtained by setting $x = 1$, and part (b) is obtained by setting $x = -1$, getting a common denominator for the resulting fractions, and simplifying. (This calculation could be done in more generality, of course.)

14. Multiply the binomial theorem by x and then integrate from 0 to 1. The substitution $u = x + 1$ makes the integration easier.

15. Let $x_2 = x$, $x_1 = 0$ in Identity 95. Divide both sides by x, and then integrate from x_1 to x_2 with respect to x. For the right side, let $z = x + y$ as in the proof of Identity 100 and switch to integrating in z.

16. Use integration by parts with $u = \left(\frac{x}{1-x}\right)^k$ and $dv = (1 - x)^n dx$. The factor u can be expressed more simply for differentiation purposes as $= \left(\frac{1}{1-x} - 1\right)^k$. When unrolling the recurrence, remember to write $\int_0^1 x^{k-j}(1-x)^{n-k}dx$ as $\int_0^1 x^{k-j}(1-x)^{n-j-(k-j)}dx$.

17. (a) Find coefficients A, B, C, D such that
$$\frac{1+x-x^2}{(1-x+x^2)^2}=\frac{Ax+B}{1-x+x^2}+\frac{Cx+D}{(1-x+x^2)^2}.$$

(b) Complete the square on the expression $1 - x + x^2$. Then use the substitution $x - \frac{1}{2} = \frac{\sqrt{3}}{2}\tan\theta$.

18. Use Identity 106 and proceed as in the proof of Identity 107, although the summation is over k instead of n. You will also need the formula for the partial sum of a geometric series.

19. Use Identity 106 and proceed as in the proof of Identity 108. The calculation is somewhat easier, though, as we do not need to introduce the differentiation operator. The substitution $u = 1/x$ is helpful once the infinite sum is evaluated and only the integral remains.

20. Mimic the proof of Identity 107, although this one does not require differentiation.

21. Use the substitution $x = \sin^2 \theta$ in the definition of the beta integral.

22. Mimic the proof of Identity 107 given in the chapter.

23. Use the substitution $x = 1 - t^2$. The resulting integral can be evaluated geometrically, as the area of a quarter-circle.

24. First method: $B(1/2, 1/2) = (\Gamma(1/2))^2$, by Identity 105. Then use trigonometric substitution to show that the integral form of $B(1/2, 1/2)$ evaluates to π.

    Second method: After applying the substitution $u = t^{1/2}$, denote the resulting integral $2 \int_0^\infty e^{-u^2} \, du$ by A. Then, write A^2 as $4 \left(\int_0^\infty e^{-u^2} \, du \right) \left(\int_0^\infty e^{-v^2} \, dv \right)$, combine the two integrals, switch to polar coordinates, and evaluate.

25. We have, for $a \in (0, 1)$, $\int_a^1 e^{-t} t^{-1} \, dt > e^{-1} \int_a^1 t^{-1} \, dt = -e^{-1} \ln a$. This last expression grows without bound as a approaches zero from the right, so $\int_0^\infty e^{-t} t^{-1} \, dt$ must diverge as well.

26. Use Identity 102 to unroll the numerator and denominator of the left side. Clear fractions by multiplying by $2^n/2^n$. Then use Exercise 24, plus the fact that $(2n)(2n - 2) \cdots (2) = 2^n n!$.

27. (a) With the formula $\int u \, dv = uv - \int v \, du$, let $u = \sin^{n-1} x$. After applying integration by parts, use the identity $\cos^2 x = 1 - \sin^2 x$ and solve for $\int \sin^n x \, dx$.

    (b) Since $\sin(0) = 0$ and $\cos(\pi/2) = 0$, $\int_0^{\pi/2} \sin^n x \, dx = \frac{n-1}{n} \sin^{n-2} x \, dx$. Unroll this recurrence separately for even and odd values. The base cases are $\int_0^{\pi/2} \sin^0 x \, dx = \pi/2$ and $\int_0^{\pi/2} \sin^1 x \, dx = 1$, resulting in

$$I_{2n} = \frac{2n - 1}{2n} \frac{2n - 3}{2n - 2} \cdots \frac{1}{2} \frac{\pi}{2}, \text{ and}$$

$$I_{2n+1} = \frac{2n}{2n + 1} \frac{2n - 2}{2n - 1} \cdots \frac{2}{3}.$$

    (c) This follows directly from part (b).

    (d) We have $\sin^{2n+1} x < \sin^{2n} x < \sin^{2n-1} x$ for $x \in (0, \pi/2)$, which implies $I_{2n+1} < I_{2n} < I_{2n-1}$ and thus $(2n + 1)I_{2n+1} = 2nI_{2n-1} > 2nI_{2n}$.

    (e) The squeeze theorem and (d) imply $I_{2n+1}/I_{2n} \to 1$ as $n \to \infty$. By the relationship in (c), then, the Wallis product converges and equals $\pi/2$.

28. Starting with the partial Wallis product (see Exercise 27)

$$\prod_{i=1}^n \left(\frac{2i}{2i - 1} \cdot \frac{2i}{2i + 1} \right),$$

rewrite the numerator as $4^n (n!)^2$. Then multiply numerator and denominator by $(2^n n!)^2$. This will turn the denominator into $(2n + 1)((2n)!)^2$. Finally, express the partial Wallis product in terms of $\binom{2n}{n}$ and apply the inequality in part (d) of Exercise 27.

Chapter 5

1. Use Identity 2 and Theorem 5 to rewrite $P(X = k)/P(X = k-1)$. Simplify to get $P(X = k) \geq P(X = k - 1)$ if and only if $(n - k + 1)p \geq k(1 - p)$, which simplifies to $k \leq (n + 1)p$. This means that $P(X = k)$ increases until $k = \lfloor (n + 1)p \rfloor$, after which $P(X = k)$ decreases.

2. This is $\sum_{k=0}^{n} \binom{n}{2k} p^{2k} (1 - p)^{n-2k}$. To evaluate this sum, let $x = p$ and $y = 1 - p$ in Identity 16.

3. At first glance this may look like the hypergeometric scenario, but the fact that we're replacing balls changes the probability calculation. In fact, this situation is actually equivalent to flipping a coin until we achieve the first head, where the probability of tossing heads on a single flip is the probability of drawing a red ball from the jar on a single draw. This means that X is negative binomial, with $p = m/(m+n)$ and $r = 1$. Thus, by Theorem 6,

$$P(X = k) = \frac{m}{m + n} \left(\frac{n}{m + n} \right)^{k-1}.$$

4. Suppose he reaches into his left pocket and discovers that its matchbox is empty. If there are k matches remaining in the other matchbox, he must have reached into the left pocket $n + 1$ times and the right pocket $n - k$ times. In other words, he achieved success $n + 1$ on trial $n + 1 + n - k = 2n - k + 1$, where "success" is interpreted to mean reaching into his left pocket. Since the "success" probability is $1/2$, the probability of there being k matches in the right pocket's matchbox when he discovers his left pocket's matchbox is empty is $P(X = 2n - k + 1)$, where X is negative binomial with $r = n + 1$ and $p = 1/2$. Since the analysis is identical for him reaching into his right pocket and discovering that that matchbox is empty, the answer is $2P(X = 2n-k+1)$, which simplifies to the expression in the problem statement.

5. The argument is similar to that in Exercise 4; the answer is

$$\binom{2n - k}{n} p^{n-k} (1 - p)^{n-k} \left(p^{k+1} + (1 - p)^{k+1} \right).$$

6. From a probabilistic standpoint, the left side is $P(X > n)$, where X is

negative binomial with parameters r and p. In terms of a coin-flipping experiment with probability p of tossing heads on a single flip, this is the probability that you have not obtained the rth head by the nth flip. In other words, you've flipped a coin n times and obtained fewer than r heads.

The right side is $P(Y < r)$, where Y is binomial with parameters n and p. In terms of the same coin-flipping experiment, this is the probability that you've tossed a coin n times and obtained fewer than r heads. Since the two sides have the same probabilistic interpretation they must be equal.

It is possible to prove Identity 135 using techniques from Chapter 2, but successfully doing so would be an algebraic *tour de force*.

7. Starting with Identity 128, substitute $n + 1$ for k and then $k + 1$ for r. Finally, substitute x for $1 - p$ and move factors that do not include the index of summation to the right side.

8. Take $p = \dfrac{x}{x + y}$ in Identity 130 and rearrange.

9. Take $p = \dfrac{x}{x + y}$ in Identity 131 and rearrange.

10. Let X_k be 1 if student k chooses her own name and 0 otherwise. We have $E(X_k) = 1/n$, and so the expected total number of students who choose their own names is 1. (Note that this is true regardless of the value of n!)

11. Let X_k be 1 if the people in seats k and $k+1$ are of different genders. After conditioning on the gender of the person in seat k, we find that $E(X_k) = \dfrac{2mn}{(m + n)(m + n - 1)}$. This leads to $\dfrac{2mn}{m + n}$ as the expected number we are after.

12. This is similar to Exercise 11. However, now there are $m + n$ indicator variables to sum over rather than $m+n-1$, and so the answer is $\dfrac{2mn}{m + n - 1}$.

13. This is similar to Exercise 11. However, now we have $m+n$ indicator variables, one for each position, and $E(X_k)$ is different depending on whether $k \in \{1, n\}$ or $k \in \{2, \ldots, n - 1\}$. The former case has $E(X_k)$ the same as in Exercise 11. For the latter, $E(X_k)$ must include the probability that a person of a different gender sits to the left plus the probability that a person of a different gender sits to the right minus the probability that different-gender people sit to both the left and the right (to account for double-counting). The final answer is $\dfrac{mn(3m + 3n - 2)}{(m + n)(m + n - 1)}$.

14. This is similar to Exercise 13 but is simpler because we do not need to break $E(X_k)$ into cases. The answer is $\dfrac{3mn}{m + n - 1}$.

15. With the hint we have $B = B_1 + B_2 + \cdots + B_n$. To find the expected value of B_k, note that there are $k - 1$ cards already obtained and thus $n - k + 1$ left. This means that B_k is a geometric random variable with success probability $\dfrac{n - k + 1}{n}$. This means that $E(B) = nH_n$, where H_n is the nth harmonic number.

16. Use the indicator variable approach for finding $E(X^2)$ in Identity 130 and apply it to the hypergeometric distribution, as in Identity 132. The draws j and k are not independent, however, unlike in Identity 130. The success probability for draw j is $\dfrac{m}{n}$. Successfully drawing a red ball on draw j, though, drops the number of red balls to $m - 1$ and the total number of balls to $m + n - 1$. Thus the success probability for draw k is $\dfrac{m - 1}{m + n - 1}$.

17. The resulting sum is almost a geometric series in the variable $x = 1 - p$. Use the technique of differentiating both sides of the formula for a geometric series, as in the proofs of Identities 107 and 108.

18. As with Exercise 17, the resulting sum is almost a geometric series in the variable $x = 1 - p$. Differentiate both sides of the formula for a geometric series, multiply both sides of the result by x, and differentiate again.

19. Starting with Identity 133, substitute $n + 1$ for k and then $k + 1$ for r. Then substitute x for $1 - p$ and move factors that do not include the index of summation to the right side. You'll also need to use Identity 129. Of course, you could also differentiate Identity 129 with respect to x and then multiply again by x.

20. The hard way to do this is to start with Identity 134, do the appropriate substitutions, and rearrange. It's easier simply to differentiate Identity 137 with respect to x and then multiply again by x.

21. If you proceed as in Proof 2 of Identity 133 you eventually obtain the recurrence $E(X_r^2) = E(X_{r-1}^2) + \dfrac{2r}{p} - \dfrac{1}{p}$. (You'll need to use $E(X_r) = r/p$ from Identity 133.) This recurrence is a bit more difficult to solve because of the presence of the $\dfrac{2r}{p}$ term. If you unroll it, though, you can evaluate the result with help from the formula $\displaystyle\sum_{k=1}^{n} k = \dfrac{n(n + 1)}{2}$.

22. (a) If X is the number of successes, then $\binom{X}{2}$ is the number of pairs of successes, by the combinatorial definition of the binomial coefficients. Since $X_j X_k = 1$ only when both events j and k are successes, the right side counts the number of pairs of successes as well.

(b) Since X_j and X_k are independent, $E(X_j X_k) = p^2$. There are $\binom{n}{2}$ terms on the right side of the equation in (a).

(c) Argue that the expected number of subsets of size m that can be formed from the successes is given by $E\binom{X}{m}$ and by $\binom{n}{m}p^m$.

(d) Write out x^3 in falling factorial powers, and then apply the result from (c).

23. (a) This is similar to Exercise 22. However, now X_j and X_k are not independent. We have $E(X_jX_k) = P(X_jX_k = 1) = \dfrac{m}{m+n}\dfrac{m-1}{m+n-1}$, since the act of selecting a red ball on draw j means that there are only $m-1$ out of $m+n-1$ red balls available for draw k. There are $\binom{r}{2}$ terms of the form $E(X_jX_k)$ over which to sum.

(b) Generalize part (a) to the expected number of subsets of size p that can be formed from the red balls chosen.

24. Use the absorption identity m times, reindex the sum, and use the fact that the negative binomial probabilities sum to 1 (Identity 128).

25. Write out the definition of $E[X^q]$ using the binomial distribution. Use the absorption identity (Identity 6), and then do a variable switch on the sum.

26. (a) Let $q = 1$ in Theorem 10. Since $(Y+1)^0 = 1$, its expected value is 1.

(b) Let $q = 2$ in Theorem 10, and use Identity 130.

(c) Let $q = 3$ in Theorem 10, and use Identity 131. It may help to write $n^2p^2 + np(1-p)$ as $np + n(n-1)p^2$.

27. Write out the definition of $E[X^q]$ using the negative binomial distribution. Use the absorption identity (Identity 6), and then do a variable switch on the sum. (In other words, the procedure is the same as with Exercise 25.)

28. The approach here is the same as that in Exercise 26.

(a) Let $q = 1$ in Theorem 11. Since $(Y-1)^0 = 1$, its expected value is 1.

(b) Let $q = 2$ in Theorem 11, and use Identity 133.

(c) Let $q = 3$ in Theorem 11, and use Identity 134.

29. This is similar to Exercises 25 and 27. However, you need to use the absorption identity twice.

30. The approach here is the same as that in Exercises 26 and 28.

(a) Let $q = 1$ in Theorem 12. Since $(Y+1)^0 = 1$, its expected value is 1.

(b) Let $q = 2$ in Theorem 12, and use Identity 132.

(c) Let $q = 3$ in Theorem 12, and use Identity 136.

31. First, prove the result for $n = 2$. Use the fact that, as sets, $A_2 = (A_1 \cap A_2) \cup (\bar{A}_1 \cap A_2)$ so that $P(A_2) = P(A_1 \cap A_2) + P(\bar{A}_1 \cap A_2)$. Similarly, $A_1 \cup A_2 = A_1 \cup (\bar{A}_1 \cap A_2)$, and so $P(A_1 \cup A_2) = P(A_1) + P(\bar{A}_1 \cap A_2)$. Putting these together yields $P(A_1 \cup A_2) = P(A_1) + P(A_2) - P(A_1 \cap A_2)$.

Now assume the result is true up to a fixed value of n. Then $P\left(\bigcup_{i=1}^{n+1} A_i\right) = P\left(\bigcup_{i=1}^{n} A_i\right) + P(A_{n+1}) - P\left(\left(\bigcup_{i=1}^{n} A_i\right) \cap A_{n+1}\right)$. Distribute the intersection over the union on the third term on the right. Finally, apply the induction hypothesis to the first and third terms on the right.

32. Start with $P\left(B \cap \left(\bigcap_{i=1}^{n} A_i\right)\right) = P(B) P\left(\bigcap_{i=1}^{n} A_i | B\right)$, and then apply inclusion-exclusion to $P\left(\bigcap_{i=1}^{n} A_i | B\right)$.

33. Let X be the number of people who get their own names back. To find $P(X = k)$, first choose the k people who get their own name in $\binom{n}{k}$ ways. The number of ways the other $n - k$ people don't get their own names back is, according to Identity 68, $D_{n-k} = (n-k)! \sum_{j=0}^{n-k} \frac{(-1)^j}{j!}$. Multiplying these expressions together and multiplying by k, summing over all possible values of k, and finally dividing by $n!$ (the total number of ways the names can be distributed) yields, with the answer of 1 from Exercise 10, the identity

Identity 334.

$$\sum_{k=0}^{n} \sum_{j=0}^{n-k} \binom{n}{k} \frac{k(n-k)!(-1)^j}{n!\,j!} = 1.$$

This is more compactly expressed (albeit without the binomial coefficient) as

$$\sum_{k=0}^{n-1} \sum_{j=0}^{n-1-k} \frac{(-1)^j}{j!\,k!} = 1,$$

or, since we might as well replace $n - 1$ with n,

$$\sum_{k=0}^{n} \sum_{j=0}^{n-k} \frac{(-1)^j}{j!\,k!} = 1.$$

(Incidentally, our argument here for $P(X = k)$ contains the essential elements of the argument to prove Identity 88 in Exercise 41 of Chapter 3.)

34. Use the same interpretation for B and the A_i's as in the probabilistic proof of Identity 41 except that A_i is now the event that ball $n + 1$ is drawn before ball i.

35. Suppose you have a jar containing $m - 1$ red balls, n blue balls, and a black ball. Use Corollary 2 to show that the two sides each give the probability that, if you select them one-by-one from the jar, you get all the red balls,

followed by the black ball, followed by the blue balls. Let B be the event that the black ball appears after all the red balls, and let A_i be the event that the black ball appears before the ith blue ball.

36. Suppose you have a jar containing $m-1$ red balls, n blue balls, and a black ball. Use Corollary 2 to show that the two sides each give the probability that, if you select them one-by-one from the jar, you get the black ball first. Let B be the event that the black ball is drawn before all of the red balls, and let A_i be the event that the black ball is drawn before the ith blue ball.

37. Suppose you have a jar containing n numbered red balls, n numbered blue balls, and a black ball. Both sides represent the probability that, for each number i, the black ball is drawn after red ball i or blue ball i. Let A_i be the event that the black ball is drawn after red ball i or blue ball i. For the left side, use Theorem 9. The intersection of any k of the $\bar{A}_i$'s is the event that the black ball is drawn as the first of $2k+1$ specific balls. For the right side, the order of the first appearance of the numbers can be done in $n!$ ways, and their colors can be chosen in 2^n ways. Denote the ordering of numbers as $x_1, x_2, \ldots, x_n$. The black ball goes immediately after all of these. Then choose the place for the second appearance of the number x_n: It can go after the black ball or after the first appearance of x_n, for two choices. Then choose the place for the second appearance of x_{n-1}: It can go after the black ball, after either of the appearances of x_n, or after the first appearance of x_{n-1}, for four choices. In general, there will be $2k$ choices for the second appearance of number x_{n-k+1}. Since $(2)(4)(6)\cdots(2n) = 2^n n!$, multiplying all of this together gives us $(2^n n!)^2$ ways to choose an acceptable ordering of the balls, out of $(2n+1)!$ total ways to order $2n+1$ balls.

Chapter 6

1. Both parts follow directly from addition and scalar multiplication properties of summations and functions.

2. The sequence is $1, 1, 1, \ldots$ (the infinite sequence of 1's).

3. Differentiate both sides of the geometric series formula and then multiply by x.

4. The proofs of these properties are quite similar to those for Theorem 15. They are even slightly easier, as you don't have to deal with the extra factor of $k!$ in the denominator of the series defining the generating functions.

5. (a) Apply Theorem 17 and the results of Exercises 2 and 3.

(b) Differentiate both sides of the result in Exercise 3 and multiply by x.

(c) Apply part (a) of Theorem 18.

6. Multiply both sides of Identity 149 by x^n, shift indices on the sum, and remember that $\binom{k}{n} = 0$ when $0 \le k < n$. (This is of course similar to manipulations we use in this chapter to answer the question in Exercise 3 in Chapter 3.)

7. First, the infinite series in Identity 149 converges for $|x| < 1$ because the geometric series converges for $|x| < 1$, and differentiation of power series does not change the interior of the interval of convergence. Then use the substitution $k = n + r$ and apply Identity 149.

8. The argument is similar to that for our solution in this chapter to Exercise 3 from Chapter 3. The answer is $\binom{n-m-1}{m-1}$.

9. The argument is similar to that for Exercise 8. The answer is $\binom{n-m(r-1)-1}{m-1}$.

10. The value of a_n is the number of solutions to the equation $x_1 + 5x_2 + 10x_3 + 25x_4 + 50x_5 + 100x_6 = n$, where x_i is a nonnegative integer. Using the ideas in the first section of this chapter, we find that the generating function for (a_n) is

$$\sum_{n=0}^{\infty} a_n x^n = \left(x^0 + x^1 + x^2 \cdots\right)\left(x^0 + x^5 + x^{10} \cdots\right)\left(x^0 + x^{10} + x^{20} \cdots\right)$$
$$\times \left(x^0 + x^{25} + x^{50} \cdots\right)\left(x^0 + x^{50} + x^{100} \cdots\right)$$
$$\times \left(x^0 + x^{100} + x^{200} \cdots\right)$$

Applying the geometric series formula, we get the answer

$$\left(\frac{1}{1-x}\right)\left(\frac{1}{1-5x}\right)\left(\frac{1}{1-10x}\right)\left(\frac{1}{1-25x}\right)\left(\frac{1}{1-50x}\right)\left(\frac{1}{1-100x}\right).$$

This doesn't yield a closed formula for a_n, but a computer algebra system can easily produce a_n for a specific value of n.

11. Replace x with t in Identity 150, and then integrate both sides from 0 to x. Finally, divide both sides by x.

12. Show that

$$\binom{2n+1}{n} = \frac{2n+1}{n+1}\binom{2n}{n} = 2\binom{2n}{n} - \frac{1}{n+1}\binom{2n}{n}.$$

Then use Identities 150 and 159.

13. Replace x with $t^2/4$ in Identity 150 and then integrate both sides from 0 to x.

14. Replace x with t in Identity 150, subtract 1 from both sides, and then divide by t. Integrate both sides from 0 to x. Do a substitution with $u = 2\sqrt{t}$ followed by trigonometric substitution. You may need to use logarithm properties and/or multiplying by the conjugate to obtain the form of the generating function in Identity 162.

15. Replace x with $\sqrt{x}/2$ in Identity 153.

16. Since $\binom{n}{k} = \binom{n}{n-k}$, the left side of this identity is the convolution of $\binom{n}{k}$ with itself. Thus the right side is the coefficient of x^n in $(1+x)^{2n}$. (Alternatively, this is the special case of Vandermonde's identity, Identity 57, when $m = n = r$.)

17. Multiply the generating functions $(1+x)^m$ and $(1+x)^{-m-1}$ together. The result generates the sequence $(-1)^n$. By the convolution property, this must be equal to the convolution of $\binom{m}{k}$ and $\binom{m+k}{k}(-1)^k$. Multiply both sides by $(-1)^n$ to obtain the identity.

18. The sum is the convolution of the sequences $\binom{k}{m}$ and $\binom{k}{r}$. Use the convolution property and Identity 129 to obtain the generating function

$$\frac{x^{m+r}}{(1-x)^{m+r+2}} = \frac{1}{x}\frac{x^{m+r+1}}{(1-x)^{m+r+2}}$$

for the right side. The factor $1/x$ shifts the index on n by 1.

19. Replace x with $x^2/4$ in Identity 150 to obtain the series for $1/\sqrt{1-x^2}$. Divide Identity 161 by x to obtain the series for $\arcsin x/x$. Finally, apply the convolution property and Identity 152.

20. After swapping the order of summation, use Identity 129.

21. Write out $\binom{n}{m}$ in factorials, cancel the $n!$ expressions, reindex the sum, and apply the Maclaurin series for e^x.

22. Use Newton's generalized binomial series (Identity 18), followed by the Maclaurin series for e^x.

23. For the even numbers, add the Maclaurin series for e^x and e^{-x} together and divide by 2. The argument is similar for the odd numbers.

24. In either case, write out the generating functions $f_a(x)$ and $f_a(-x)$ in terms of the sequence (a_n). Then cancel terms that add to zero.

25. Since (thanks to the binomial theorem) the ordinary generating function of row n of the binomial coefficients is $(1+x)^n$, Exercise 24 implies

$$\sum_k \binom{n}{2k} x^{2k} = \frac{(1+x)^n + (1-x)^n}{2}.$$

Replace x with x/y and then multiply both sides by y^n to complete the proof. (If $y = 0$ then both sides equal x^n [n is even], so we may consider that case separately.)

26. Applying Property 4 of Theorem 15 iteratively, we see that $x^2 f_a(x)$ forward shifts the sequence (na_{n-1}) and then multiplies it by n to produce the sequence $(n(n-1)a_{n-2})$. By the same argument, $x^3 f_a(x)$ generates the sequence $(n(n-1)(n-2)a_{n-3})$. Generalizing this, we have that $x^m f_a(x)$ generates $(n^{\underline{m}} a_{n-m})$. Thus $x^m e^x$ generates $(n^{\underline{m}})_{n=0}^\infty$. By Theorem 14, the exponential generating function for the left side of Identity 36 is $x^m e^{2x}$, which generates the sequence $(n^{\underline{m}} 2^{n-m})_{n=0}^\infty$.

27. (a) This is the special case $m = 1$ of Identity 167. Alternatively, differentiate the Maclaurin series for e^x and then multiply both sides by x.

   (b) Apply Theorem 17, part (a) of this exercise, and the Maclaurin series for e^x.

   (c) Differentiate the result in part (a) and multiply by x.

   (d) Apply Property 1 of Theorem 15.

28. (a) $\displaystyle\sum_{n=0}^\infty n! \frac{x^n}{n!} = \sum_{n=0}^\infty x^n = \frac{1}{1-x}$.

   (b) Apply Property 2 of Theorem 15 to the result in part (a).

   (c) Apply Property 3 of Theorem 15 to the result in part (a).

29. The exponential generating function $1/(1-x)$ generates the sequence given by $n!$. The binomial convolution of this sequence with itself is

$$\sum_{k=0}^n \binom{n}{k} k!(n-k)! = \sum_{k=0}^n n! = n!(n+1) = (n+1)!.$$

30. Let $f_a(x)$ be the exponential generating function of (a_n). Since e^x is the exponential generating function for the sequence of 1's, Theorem 15 says that the exponential generating function for $(m^n)_{n=0}^\infty$ is e^{mx}. By Theorem 14, then, the exponential generating function for the binomial convolution of (a_n) with $(m^n)_{n=0}^\infty$ is $e^{mx} f_a(x)$. This is also the exponential generating function of $B_m(a_n)$, by Theorem 16.

31. Apply Identity 23 and Theorem 16 to obtain the exponential generating function for the left side. Then apply the fifth property of Theorem 15 to see that this generates the sequence given by $a_n = 1/(n+1)$.

32. In the generating function proof of Identity 23 we see that the exponential generating function of $(1/(n+1))_{n=0}^\infty$ is $(e^x - 1)/x$. Differentiate this (accomplishing the backward shift, thanks to Theorem 15) and

replace x with $-x$ to obtain the exponential generating function for $((-1)^n/(n+2))_{n=0}^\infty$. Multiply by e^x, producing the exponential generating function $(e^x - x - 1)/x^2$ for the left side. To obtain the exponential generating function for the right side, start with $(e^x - 1)/x$ generating $(1/(n+1))_{n=0}^\infty$ and apply Property 5 of Theorem 15, yielding $(e^x - x - 1)/x^2$ as the generator for $\left(\frac{1}{(n+1)(n+2)}\right)_{n=0}^\infty$.

Chapter 7

1. Successively apply the recurrence to get $a_n = a_0 + 1 + 2 + \cdots + n$. Then apply Identity 59.

2. (a) The characteristic equation is $r^2 = 7r - 12$, with roots $r_1 = 3$ and $r_2 = 4$. Solving for the initial conditions yields $a_n = 4^n - 3^n$ as the solution.

   (b) The procedure is the same as in part (a). The roots are $r_1 = 8$ and $r_2 = -2$, and the solution is $a_n = 2 \cdot 8^n + 3 \cdot (-2)^n$.

3. The repeated root is $r = 3$, and the solution is $a_n = 2 \cdot 3^n - n3^n$.

4. The outline given in the problem statement pretty much tells you what to do.

5. Generally speaking, these are in increasing order of difficulty. For each, use the same approach as the solution of the recurrence $a_n = a_{n-1} + n$ given in the chapter.

   (a) $f_a(x) = \dfrac{5}{1 - 3x}$, $a_n = 5 \cdot 3^n$.

   (b) $f_a(x) = \dfrac{3}{(1-x)^2} - \dfrac{2}{1-x}$, $a_n = -2 + 3\binom{n+1}{1} = 3n + 1$.

   (c) This one requires partial fractions decomposition, after which you have $f_a(x) = \dfrac{2}{1-4x} - \dfrac{1}{1-x}$, Then $a_n = 2 \cdot 4^n - 1$.

   (d) This one also requires partial fractions decomposition, after which you have $f_a(x) = \dfrac{5/2}{1-x} + \dfrac{3/2}{1-3x}$, Then $a_n = \dfrac{5}{2} + \dfrac{3}{2} \cdot 3^n$.

   (e) $f_a(x) = \dfrac{2x}{(1-x)^3}$, $a_n = 2\binom{n+1}{2} = n(n+1)$.

   (f) $f_a(x) = \dfrac{2}{1-2x} + \dfrac{2x}{(1-2x)^3}$, $a_n = (n^2 + n + 4)2^{n-1}$.

6. Parts (a) and (b) are immediate. Parts (c) and (d) are very similar to part (e) of Exercise 5 and the solution to the problem involving straight lines on a piece of paper in this chapter.

7. Follow the approach of Exercise 6 and use the result of part (b) of Exercise 5 in Chapter 6 to obtain $f_S(x) = \dfrac{x(1+x)}{(1-x)^4} = \dfrac{x}{(1-x)^4} + \dfrac{x^2}{(1-x)^4}$.
   Then apply Identity 149 and part (d) of Theorem 18.

8. To find the exponential generating function, mimic the proof of Identity 158 in this chapter. Start by multiplying both sides of the recurrence by $x^n/n!$ and summing as n goes from 1 to infinity. Manipulate the infinite series expressions to find the exponential generating function for a_n to be

$$f_a(x) = \frac{1}{(1-x)^2} - \frac{1}{1-x}.$$

   Using this and the results of Exercise 28 from Chapter 6, show that the answer is $a_n = n(n!)$.

9. Use the fact that $\Delta a_n = a_{n+1} - a_n$ to rewrite $\Delta(c\,a_n + d\,b_n)$, and then regroup terms.

10. Remember that $\binom{n}{k} = n^{\underline{k}}/k!$ (Identity 17). Multiply both sides of Identity 58 by $m!$.

11. Write out

$$\sum_{k=0}^{n} a_k \Delta b_k + \sum_{k=0}^{n} \Delta a_k b_{k+1}.$$

   Two of the four terms in the resulting summand will cancel, and the other two will telescope when you sum them.

12. For the base case the identity reduces to $a_m = a_m$. For the inductive step, replace k with $k-1$ on the sum that includes the a_{m+1+k} term. Then combine the two terms using Identity 1.

13. Apply the binomial theorem (Identity 3), with $y = -1$ and $x = m$, and Identity 174.

14. By Identity 21, $\sum_{k=0}^{n} \binom{n}{k} k = n2^{n-1}$. Applying the variant of binomial inversion in Theorem 2 and Identity 174, the answer is $(n2^{n-1})_{n=0}^{\infty}$.

15. Identity 84 says that, for $n > m$, $\sum_{k=0}^{n}(n-k)^m(-1)^k = 0$ or, equivalently, $\sum_{k=0}^{n} k^m(-1)^{n-k} = 0$. This means that, via Identity 174, for $p \le m < n$ the right diagonal of the difference triangle for the sequence given by $a_k = k^p$ consists of zeros from row n on. Running the construction of the difference triangle backwards, this means that the subtriangle with entry n of row n as the apex of the triangle consists entirely of zeros. Since the left diagonal of this subtriangle is precisely the sequence

$(\Delta^n a_0, \Delta^n a_1, \Delta^n a_2, \ldots)$, the result follows for power functions of the form $f(x) = x^p$ (again, $p \leq m$). The result for polynomials of degree m then follows by linearity of summation.

16. Rewrite $x^{\underline{n}}$ as $n!\binom{x}{n}$ (Remember the definition of the generalized binomial coefficient in Identity 17.) Take the finite difference, and apply Pascal's recurrence.

17. Take the nth finite difference of the sequence $a_m = m^{\underline{r}}$. Rewrite the falling powers as regular factorials via $n^{\underline{k}} = n!/(n-k)!$ and then combine into binomial coefficients.

18. Starting with Identity 176, replace k with $k + s$. Then replace m with $m - s$.

19. This is similar to the finite difference proof of Identity 109. Use $a_m = (m-1)^{\underline{-r-1}}$.

20. This is a generalization of the proof of Identity 53 given in the chapter. Use $m - 1$ finite differences and Identity 178.

21. Starting with Identity 49,

$$\sum_{k=0}^{n} \binom{n}{k}\binom{k}{m} = 2^{n-m}\binom{n}{m},$$

apply Identity 184. Use Pascal's recurrence (Identity 1) to simplify to the form on the right side of the identity.

22. Starting with Identity 68,

$$\sum_{k=0}^{n} \binom{n}{k}(n-k)!(-1)^k = D_n,$$

reverse the order of summation and move the $(-1)^n$ to the right side. Then apply Identity 184. Factor out the $(-1)^{n+1}$ on the right, and then move it to the left. Finally, reverse the order of summation again.

23. Use induction to prove that

$$\Delta^m \left((-1)^n d_n\right) = (-1)^{n+m} \sum_{k=0}^{m} \binom{m}{k} d_{k+m}.$$

Then apply Identity 185.

24. Starting with the case $m = 1$, $r = 1$ of Identity 207 (or just with Identity 41), multiply by -1, and then apply Identity 186.

25. Let $m = 1$, $r = m$ in Identity 207, multiply by -1, and then apply Identity 186.

26. The finite difference of $-1/(k+1)$ is $\frac{1}{(k+1)(k+2)}$. Then apply Identity 188 to Identity 23.

27. We have $\Delta H_k = H_{k+1} - H_k = 1/(k+1)$. Then use Identity 188 with Identity 23.

28. This is similar to Exercise 27, but use Identities 189 and 24 instead.

29. The proof is almost identical to that for Identity 190.

30. The proof is almost identical to that for Identity 192. (It is slightly easier, though.)

31. Take $a_k = k + m$ and $\Delta b_k = k^{\underline{m-2}}$. Thanks to Identity 177, this gives us $b_k = k^{\underline{m-1}}/(m - 1)$ (up to an arbitrary constant, which we can take to be 0). Then apply summation by parts; use Identity 204 to evaluate the resulting sum. After this, collect common factors, remembering that $(n - m)^{\underline{m}} = (n - m)(n - m - 1)^{\underline{m-1}}$, and crunch through the remaining algebra.

32. The proof is similar to that for Identity 194, although it does not require the additional manipulations in order to apply Identity 204.

33. This is similar to the proof of Identity 23 in Chapter 2. First, multiply the summand by $2/2$ and $(2k + 1)/(2k + 1)$. Then apply the absorption identity (Identity 6) twice. Shift indices, and use Identities 13 and 192.

34. The proof here is almost identical to that of Identity 196. The differences are that you need Identity 191, and the initial conditions are $f_0 = 0$, $f_1 = 1$. An alternate approach that avoids the use of Identity 191 is to add $\sum_k \binom{n}{2k} x^k$ and $\sum_k \binom{n}{2k+1} x^{k+1/2}$ to obtain $\sum_{k=0}^n \binom{n}{k} \sqrt{x}^k$. Then apply the binomial theorem and Identity 196. Finally, solve for $\sum_k \binom{n}{2k+1} x^k$.

35. Starting with the summation by parts formula (Identity 205), let $a_k = \binom{k}{m}$ and $b_k = a_k$. Use Pascal's recurrence (Identity 1) to evaluate Δa_k.

36. Divide both sides of Identity 200 by $r!$. Rewrite the two rising factorial power expressions as falling factorial powers, and apply Identity 17.

37. Let $a_k = (-1)^k$. Then $\Delta a_k = -2(-1)^k$. Apply Identity 197. With $g_m = \sum_{k=0}^n \binom{k}{m}(-1)^k$, this yields a recurrence in g_m and g_{m+1} (where n is fixed). Then use the generating function technique for solving recurrences we saw in the proof of Identity 158 in Section 7.1. You'll need to show that $g_0 = [n \text{ is even}]$. The resulting generating function for g_m is

$$G(x) = \frac{\frac{(-1)^n}{2} \sum_{m=1}^{\infty} \binom{n+1}{m} x^m + [n \text{ is even}]}{1 + x/2}.$$

Thus $(g_m)_{m=0}^{\infty}$ is the convolution of $((-1/2)^k)_{k=0}^{\infty}$ and the sequence (b_k), with $b_0 = [n \text{ is even}]$ and $b_k = \frac{(-1)^n}{2}\binom{n+1}{k}$ for $k \geq 1$. The identity follows.

38. Use Theorem 21 and Identity 2; the solution is $n!/k!$, for $n \geq k$.

39. Since $(n - k)(n + k) = n^2 - k^2$, show by induction on n that $\left|\begin{smallmatrix}n\\k\end{smallmatrix}\right| = (n!)^2$, for $n \geq k$. (Don't forget the boundary conditions.)

40. Start from Exercise 39 and apply Theorem 21. The solution is

$$\frac{(n!)^2}{(n-k)!(k!)^2} = \binom{n}{k}\frac{n!}{k!} = \binom{n}{k}^2 (n-k)!,$$

provided $n \geq k$.

41. (a) In order to draw exactly k red balls in n trials, we must have drawn k red balls in $n - 1$ trials and then draw a blue ball on the nth trial, or we must have drawn $k - 1$ red balls in $n - 1$ trials and then draw a red ball on the nth trial. In the former case, there are k blue balls out of m total in the jar; thus the probability is $\frac{k}{m}\left|\begin{smallmatrix}n-1\\k\end{smallmatrix}\right|$. In the latter case, there are $m - k + 1$ red balls out of m total in the jar; thus the probability is $\frac{m-k+1}{m}\left|\begin{smallmatrix}n-1\\k-1\end{smallmatrix}\right|$.

   (b) This follows from Theorem 21, with $h(n) = \frac{1}{m}$ and $g_2(k) = m - k + 1$.

   (c) In order to draw the kth red ball on the nth trial, we must have drawn $k - 1$ red balls through $n - 1$ trials and then draw a red ball on the nth trial. By the argument in part (a) and the answer in (b), this is $\frac{m-k+1}{m}\left\{\begin{smallmatrix}n-1\\k-1\end{smallmatrix}\right\}\frac{m^{k-1}}{m^{n-1}}$, which combines to yield the given solution.

42. Mimic the proof of Identity 201, being careful with the powers of x and y when reindexing the sum in the second line.

Chapter 8

1. With the Fibonacci recurrence $F_n = F_{n-1} + F_{n-2}$ (Identity 170), this is a straightforward induction proof.

2. Since $F_0 = 0$, we can start the sum at 1. Replace n with m. Then, using the Fibonacci recurrence $F_n = F_{n-1} + F_{n-2}$ (Identity 170), replace F_{2k} in Identity 187 with $F_{2k-1} + F_{2k-2}$. The resulting sum actually adds the Fibonacci numbers from 1 to $2m - 1$. This takes care of the case with n odd. To cover the case with n even, just add F_{2m} to both sides and use the Fibonacci recurrence again.

3. A male bee has just one parent, a female, which agrees with $F_2 = 1$. This mother has two parents, a female and a male, so the original male bee has two grandparents. This agrees with $F_3 = 2$. In general, the number of bees n generations back will be the number of males $n - 1$ generations back plus twice the number of females $n - 1$ generations back. But the number of females $n - 1$ generations back is the same as the total number of bees $n - 2$ generations back, since each bee has exactly one mother. Thus the number of bees n generations back is the number of bees $n - 1$ generations back plus the number of bees $n - 2$ generations back, satisfying the Fibonacci recurrence.

4. Let S_n be the number of subsets of $\{1, 2, \ldots, n\}$ that do not contain consecutive elements. These subsets can be partitioned in two groups: (1) Those that contain n, and (2) Those that do not. Argue that there are S_{n-2} of the first and S_{n-1} of the second. Finally, check that S_n satisfies the same initial conditions as F_{n+2}.

5. The argument is the same as that for the combinatorial proof of Identity 183 except that you are tiling a $1 \times (2n + m - 1)$ board.

6. Both sides count the number of ways to tile a $1 \times 2n$ board using at least one square. For the left side, condition on the position of the last square. Since there are an even number of spaces on the board, and dominoes take up two spaces, the last square can only appear in an even-numbered space.

7. (a) You want to place n elements into two nonempty sets. First, create two sets: a set with element 1 and an empty set. Then, for each of the $n - 1$ elements numbered 2 through n, either (1) place that element in the set containing element 1 or (2) place that element in the set not containing element 1. Finally, subtract off the case in which all the elements end up in the set containing element 1.

   (b) Placing n objects into $n - 1$ nonempty subsets means that there are $n - 2$ singleton sets and one set with two elements. There are $\binom{n}{2}$ ways to choose which two elements go in the same set. (Compare this with Exercise 38 in Chapter 3.)

   (c) Partitioning n objects into $n - 2$ nonempty subsets requires either (1) $n - 4$ singleton subsets and two 2-subsets, or (2) $n - 3$ singleton sets and one 3-subset. In the first case, there are $\binom{n}{4}$ ways to choose which elements will go in the 2-subsets and three ways to place those elements into 2-subsets. In the second case, there are $\binom{n}{3}$ ways to choose the elements that go in the 3-subset.

8. In the notation of Identity 201, the recurrence for the Stirling numbers of the first kind has $\alpha' = \beta = \beta' = \gamma = 0$ and $\alpha = \gamma' = 1$. Identity 201 then gives $n!$ for the row sum.

9. Apply Identities 222 and 140, plus linearity of expectation (Theorem 8). Alternatively, instead of linearity of expectation, swap the order of summation.

10. Use the same approach as Exercise 9 but with Identity 142 instead of Identity 140.

11. Use the same approach as Exercises 9 and 10 but with Identity 143 instead of Identity 140 or 142.

12. Use induction. For the inductive step, use Identity 225 on $\sum_{k=0}^{n} \left\{ {n \atop k} \right\} x^{\underline{k}}$. Then separate into two sums, factor out an x and reindex the second sum, recombine the sums, and apply the inductive hypothesis.

13. Relabel n as m, x as n, and use Identity 17 to rewrite $n^{\underline{k}}$ in terms of binomial coefficients. Summing to m and summing to n are equivalent because you are effectively summing to the minimum of m and n. Then apply Theorem 2 and swap the roles of n and m.

14. Both sides count the number of functions from an m-element set to the *subsets* of an n-element set. For the left side, condition on the number of elements in the codomain. For the right side, condition on the number of elements in the image. (Some of the ideas in the proof of Identity 227 may be helpful.)

15. Apply Identity 222 to k^m, swap the order of summation, and apply Identity 38.

16. Use the same interpretation as in Exercise 14. The positive and negative-signed functions are those with an even and odd number, respectively, of elements in the codomain. For a sign-reversing involution, find the largest-numbered element that is not in the image of a particular function. If it is in the codomain, take it out. If it is not in the codomain, put it in. The leftover functions are those in which every element in the n-element set is in the image of the function; i.e., the onto functions from an m-element set to an n-element set.

17. Both sides count the number of ways to partition $n+1$ elements into $m+1$ nonempty subsets. For the left side, condition on the number of elements k that do not appear in the same subset as element 1.

18. For the non-combinatorial proof, apply the variation of binomial inversion (Theorem 2) to Identity 255.

   For the combinatorial proof, $\binom{n}{k} \left\{ {k+1 \atop m+1} \right\}$ counts the number of ways to choose k balls from a jar containing balls numbered $1, 2, \ldots, n$ and then form $m + 1$ nonempty subsets from these k balls plus a ball numbered $n + 1$. The parity of one of these collections of subsets is the number of balls left in the jar. To construct a sign-reversing involution, let y be the smallest element that is either in the jar or in the subset containing $n+1$. If y is in the jar, move it to the subset containing $n+1$. If y is in the subset containing $n+1$, move it to the jar. The mapping is an involution, and it is sign-reversing because it changes the number of balls left in the jar by one. The collections of subsets for which y does not exist are those in which all n balls from the jar are placed in subsets not containing $n+1$. There are $\left\{{n \atop m}\right\}$ ways to do this, and all ways have even parity.

19. Both sides count the number of ways to partition n numbers into two ordered lists, a first list and a second list. For the left side, $\left[{n \atop i+j}\right]$ partitions the n numbers into $i+j$ disjoint cycles. Then $\binom{i+j}{i}$ chooses i of these cycles to form the first ordered list. (A permutation can be uniquely expressed as a set of disjoint cycles; see the second proof of Identity 230.) The leftover j cycles form the second ordered list in a similar fashion.

    For the right side, add a new symbol the the n numbers, like $\star$. Then create a permutation on these $n+1$ symbols. This can be done in $(n+1)!$ ways. The $\star$ symbol separates the first ordered list from the second ordered list.

20. Both sides count the number of ways to partition $n+1$ elements into $m+1$ nonempty cycles. For the left side, condition on the number of elements k that do not appear in the cycle with element 1. The $n-k$ elements that do appear in the cycle with 1 can be placed in that cycle in $(n-k)!$ ways.

21. Use Identity 32 to convert $x^{\underline{n}}$ to a rising power, then apply Identity 228.

22. The easiest proof is probably to use Identity 218, with (thanks to Identity 229) $\alpha = \gamma' = 1$, $\alpha' = \beta = \beta' = \gamma = 0$.

23. Start with $x^{\overline{n}}$, convert to ordinary powers and then back to rising powers. Since the two sides are polynomials, their coefficients must be equal.

24. Apply the Stirling inversion formula (Theorem 26) to the binomial power sum (Identity 223).

25. For the double sum with the Stirling numbers of the first kind, replace x with y in Identity 228 and multiply by $x^n/n!$. Use Identity 32, sum for $n \geq 0$ and apply Newton's binomial series (Identity 18).

    For the double sum with the Stirling numbers of the second kind, replace x with y in Identity 222 and multiply by $x^n/n!$. Sum for $n \geq 0$ and use the Maclaurin series for e^x.

26. Starting with the left side, replace ϖ_k using Identity 234, swap the order of summation, and apply Identity 255. Remember that $\left\{{n+1 \atop 0}\right\} = 0$ for $n \geq 0$.

27. Both sides count the number of ways to partition a set of $m+n$ objects. For the left side, partition the m objects into j subsets, choose k of the n objects to be partitioned into new subsets, and distribute the remaining

$n - k$ objects among the j subsets formed from the set of size m. Sum over all possible values of j and k.

28. Both sides count the number of permutations on $n + m$ elements. For the left side, place the m elements into exactly j cycles. Then choose k of the n elements and form a permutation from them. Finally, distribute the remaining $n - k$ elements among the j cycles by placing the first remaining element after one of the m elements in the j cycles, the next remaining element after the now $m + 1$ elements in the j cycles, and so forth. Then sum over all possible values of j and k.

29. The first equality follows upon applying Identity 184 to Identity 235, with $m = 1$. The second equality can be shown by replacing $\left\{ {n+1 \atop k} \right\}$ by $\left\{ {n+2 \atop k} \right\} - \left\{ {n+1 \atop k-1} \right\}$ (Identity 225) and summing.

Alternatively, all three expressions count the number of partitions of $n + 2$ elements in which elements $n + 1$ and $n + 2$ are not in the same subset. For the first expression, condition on the number of elements $k + 1$ not in the set containing $n + 2$. For the second expression, the number of partitions in which $n + 1$ and $n + 2$ are in the same subset is ϖ_{n+1}. For the third expression, condition on the number of subsets in a partition of $n + 1$ elements, and then multiply by the number of choices for where element $n + 2$ could go (in a subset by itself or in one of the subsets not containing element $n + 1$).

30. Apply the variation of binomial inversion in Theorem 2 to Identity 235.

31. Apply Identity 184 to Identity 235.

32. Multiply Identity 267 by $(-1)^n$, apply Identity 186, multiply by -1, and add ϖ_0.

33. If $G(x)$ is the exponential generating function for the Bell numbers, then Identity 235 implies, by Theorems 15 and 16, $e^x G(x) = G'(x)$. This is a differential equation in G and x and can be be solved via separation of variables. Use $G(0) = \varpi_0 = 1$ to find the constant of integration.

34. First, show that $E[X^{\underline{m}}] = 1$ for all $m \geq 0$ when X is Poisson(1). (Use the definition of $E[X^{\underline{m}}]$, then do a variable switch and apply the Maclaurin series for e^x.) Then, with $E[X^n]$, convert to falling powers (Identity 222), apply linearity of expectation (Theorem 8), and use Identity 234.

35. To create a ranking of n people in which ties are allowed, condition on the number k of distinct scores (where two people with the same score are tied). There are $\left\{ {n \atop k} \right\}$ ways to partition the n people into k score groups. Then order those score groups in $k!$ ways and sum up.

36. Starting with $x^{\underline{n}}$, convert to rising factorial powers via Identity 32, apply Identity 237, and then use Identity 32 again.

37. Mimic the proof of Theorem 29, using Identity 26 in place of Identity 240.

38. Each side is the coefficient of $x^{\underline{m}}$ when converting $x^{\overline{n}}$ to falling powers. The left side uses Identity 237. For the right side, convert from rising powers to ordinary powers (Identity 228) and then from ordinary powers to falling powers (Identity 222).

39. Swap the order of summation and use Identity 230. Then apply Identity 272.

40. As with Exercise 35, condition on the number k of distinct scores, where two people with the same score are tied. There are $\left\{ {n \atop k} \right\}$ ways to partition n people into k score groups. Then create a permutation (ordering) of the k score groups by partitioning them into m nonempty cycles, in $\left[{k \atop m} \right]$ ways. (See the second proof of Identity 230.) Summing over k gives the number of rankings with ties allowed that contain m cycles, and summing this result over m gives the total number of rankings of n people in which ties are allowed.

41. The easiest approach is probably by induction. To find the $n+1$ derivative, differentiate the nth derivative, switch variables, and apply the Lah recurrence in Identity 239. (Daboul et al. [19] give four other proofs.)

42. Argue that $P_0 = P_1 = 1$. Then show that P_n satisfies the Catalan recurrence in Identity 242. To do this, condition on the number k of pairs of parentheses between the first open parenthesis and its corresponding closing parenthesis in a list of $n+1$ pairs of parentheses. Alternatively, show a bijection between n valid pairs of parentheses and lattice paths from $(0,0)$ to (n, n) that never go above the diagonal.

43. Show that $L_0 = L_1 = 1$. Then show that L_n satisfies the Catalan recurrence in Identity 242. To do this for $n+1$ "+" and $n+1$ "−" symbols, condition on the number of "+" symbols that occur before the second-to-last zero partial sum. (The last zero partial sum is the full, entire sum.) Alternatively, show a bijection between valid sequences of n "+" symbols and n "−" symbols and either (A) lattice paths from $(0,0)$ to (n, n) that never go above the diagonal, or (B) n valid pairs of parentheses (see Exercise 42).

44. Suppose we have $2(n+1)$ people. Number them in order around the table from 1 to $2n+2$. Condition on the number j of the person who is shaking hands with person 1. If j is odd then there are an odd number of people from 2 to $j-1$. They must all shake hands with each other, as otherwise one of them would cross arms with person 1 or person j. But an odd number of people can't shake hands with each other. Thus j must be even, and so $j = 2k+2$ for some k, $0 \le k \le n$.

Continuing this reasoning, person 1 and person $2k+2$ divide the table into two groups: one consisting of $2k$ people and one consisting of $2(n - k)$ people. The people in each of these groups must shake hands with another person in their group. Thus X_n satisfies the recurrence $X_n = \sum_{k=0}^{n} X_k X_{n-k}$, which is the Catalan recurrence.

Alternatively, find a bijection between valid sets of handshakes among $2n$ people and lattice paths from $(0,0)$ to (n,n) that never go above the the the diagonal. (Hint: An up step corresponds to person i shaking hands with a higher-numbered person, and a right step corresponds to person i shaking hands with a lower-numbered person.)

45. Use the explicit formula for the Catalan numbers (Identity 244), the result of Exercise 26 of Chapter 4, and Identity 102, the recurrence for the gamma function.

46. Any lattice path from $(0,0)$ to (n,m) that never goes above the diagonal is, by inserting a right step at the beginning of the path, equivalent to a lattice path from $(0,0)$ to $(n+1,m)$ that never touches the diagonal after $(0,0)$. The ballot theorem says that there are $\frac{n+1-m}{n+1+m}\binom{n+1+m}{n+1}$ of the latter. Then apply the absorption identity (Identity 6).

47. The left side is the convolution of $\binom{2n}{n}$ and C_n. Hence the generating function for the right side is, by Theorem 13, the product of the generating functions for $\binom{2n}{n}$ and C_n. You'll need Identities 150, 243, and 160.

Combinatorially, the right side counts the number of lattice paths from $(0,0)$ to $(n+1,n)$. To show that the left side does as well, condition on the last point (k,k) that a path touches the line $y = x$.

48. Both sides count the number of lattice paths from $(0,0)$ to $(n+r,n)$. For the left side, condition on the first time a lattice path intersects a point $(k+r,k)$ on the line $y = x-r$. (Use the ballot theorem to count the number of lattice paths from $(0,0)$ to $(k+r,k)$ that touch the line $y = x - r$ for the first time at $(k+r,r)$ by rotating and translating the 2D plane.)

49. Both sides count the number of lattice paths from $(0,0)$ to $(n+1,m)$ that never touch the line $y = x$ after the start. The right side follows from the ballot theorem (Theorem 31). For the left side, condition on the coordinate $(n+1,k)$ where the path first touches the line $x = n+1$. Argue that the number of such paths is the number of lattice paths from $(0,0)$ to (n,k) that never touch $y = x$. These are also counted by the ballot theorem.

50. The argument is the same as that in Exercise 49 except that lattice paths can touch (but not go above) $y = x$ and you need the weak ballot theorem (Theorem 36).

51. For a path from $(0,0)$ to (n,m), condition on the last step. If it is a right step, it is counted by $D(n-1,m)$. If it is an up step, it is counted by $D(n,m-1)$. If it is an up-right step, it is counted by $D(n-1,m-1)$.

52. This is an easy induction proof. Alternatively, use Theorem 16 to show that the exponential generating function of (a_n) is the zero function.

53. Use Exercise 24 of Chapter 6 with Identity 247 to obtain an expression for the exponential generating function of the sequence $(0, B_1, 0, B_3, 0, B_5, \ldots)$. Then find a common denominator, factor, and cancel. You'll be left with $-x/2$, which says that $B_1 = -1/2$ and all other odd Bernoulli numbers are zero.

54. Use Exercise 24 in Chapter 6 and Identity 247 to find an expression for the exponential generating function of the sequence $(B_0, 0, B_2, 0, B_4, 0, \ldots)$. Then get a common denominator of $e^{x/2} - e^{-x/2}$ and rewrite the expression until it looks like $x \cosh(x/2)/(2 \sinh(x/2))$, which is the right side.

55. Show that the exponential generating function of the sequence $(1/(n + 1))_{n=0}^{\infty}$ is $(e^x - 1)/x$. The result then follows from Identity 247 and the binomial convolution formula in Theorem 14.

56. The exponential generating function for the left side is $e^x B(x)$; that for the right side is $B(-x)$. With Identity 247, show that the two are equal.

Chapter 9

1. By the product rule, $f'(x) = e^{-ix}(-\sin x + i \cos x) - ie^{-ix}(\cos x + i \sin x)$, which simplifies to 0. A function with a zero derivative must be constant, and $f(0) = 1$, so $f(x) = 1$ and Euler's formula follows.

2. Use Euler's formula on the right side and then rewrite using the triangle definitions of sine and cosine.

3. On the right side of both identities, replace e^{ix} and e^{-ix} with $\cos x + i \sin x$ and $\cos(-x) + i \sin(-x)$, respectively, and simplify. Remember that cosine is even and sine is odd.

4. Write $e^{i(A+B)}$ as $e^{iA}e^{iB}$, expand both of these expressions using Euler's formula, and then equate real parts and imaginary parts.

5. Let $x = 3$ in Identities 288 and 289.

6. Replace x in Identity 196 with $-x$. Then write $1 + i\sqrt{x}$ in complex exponential form, as $\sqrt{1 + x}\, e^{i \arctan \sqrt{x}}$. Similarly, write $1 - i\sqrt{x}$ as $\sqrt{1 + x}\, e^{i \arctan(-\sqrt{x})}$. Since $\arctan x$ is an odd function, the latter is equal to $\sqrt{1 + x}\, e^{-i \arctan(\sqrt{x})}$. Finally, use the identity $\cos x = (e^{ix} + e^{-ix})/2$ from Exercise 3.

7. The proof here is almost identical to that in Exercise 6, although you need the identity for sine from Exercise 3.

8. Use the difference identity for sine on Identity 293. (Replace B with $-B$ in the $\sin(A + B)$ identity in Exercise 4 to obtain the difference identity for sine. Remember that sine is an odd function.)

9. Use the difference identity for cosine on Identity 294. Then factor a $\sqrt{2}$ out of the result. (Replace B with $-B$ in the $\cos(A+B)$ identity in Exercise 4 to obtain the difference identity for cosine. Remember that cosine is an even function.)

10. This is almost identical to the proof of Identity 290.

11. Let $x = 3$ in Identities 296 and 297.

12. Proving both of these identities is similar to proving Identity 304. The extra factors of ζ^{-1}, ζ^{-2}, and ζ^{-4} should be converted to $e^{-2\pi i/3}$, $e^{-4\pi i/3} = e^{2\pi i/3}$, and $e^{-8\pi/3} = e^{-2\pi i/3}$, respectively.

13. For Identity 324, take $r = 3$, $m = 0$, and $x = -1$ in Identity 303. Note that $(-1)^{-3k} = (-1)^k$. Move the factor of $(-1)^n$ from the left to the right. Convert $1 - e^{2\pi i/3}$ and $1 - e^{4\pi i/3}$ to the complex exponentials $\sqrt{3}e^{-i\pi/6}$ and $\sqrt{3}e^{i\pi/6}$, respectively. Finally, use the identity $\cos x = (e^{ix} + e^{-ix})/2$ from Exercise 3.

   The proofs of Identities 325 and 326 are similar. One difference is that the extra factors of $(\zeta^j)^{-m}$, when converted to complex exponentials, produce the cosine angle shifts of 4π and -4π, respectively. There is an additional factor of -1 in Identity 325 that comes from the factor $(-1)^{n-3k-1}$ on the left side of Identity 303.

14. This is similar to Exercises 12 and 13. Let $r = 4$ and $x = 1$ in Identity 303. The primitive root ζ is equal to $e^{2\pi i/4} = e^{\pi i/2} = \cos(\pi/2) + i\sin(\pi/2) = i$, which makes the calculations easier. The case $j = 2$ on the sum has to be handled carefully, though. Since $i^2 = -1$, $i^j + 1 = 0$ when $j = 2$. Thus the $j = 2$ term contributes nothing to the sum, except for the case $n = 0$, in which we have the $0^0 = 1$ situation. Thus when $n = 0$, the $j = 2$ term contributes $0^0(i^2)^{-m} = (-1)^m$ to the sum. When multiplied by the 1/4 outside the sigma, we have the extra $\frac{(-1)^m}{4}[n = 0]$ term on the right side of the identity.

15. (a) This is pretty much straightforward substitution into the sine formula from Exercise 3.

    (b) Let $\sinh x = y$, so that $x = \sinh^{-1} y$. Using $\sin(ix) = iy$, solve for x.

    (c) Let $x = \sinh y = (e^y - e^{-y})/2$, and solve for y. To do so, you will need to multiply through by e^y. The result will be a quadratic equation in e^y. Use the quadratic formula, and argue that only the positive root leads to real values for y.

(d) Let $x = -1$ in Identity 163, and use parts (c) and (d) to make sense of $\arcsin(i/2)$. The series converges for $|x| < 4$, by the generating function proof of Identity 108.

16. (a) $AB = \begin{bmatrix} 19 & 22 \\ 43 & 50 \end{bmatrix}$

(b) $BA = \begin{bmatrix} 23 & 34 \\ 31 & 46 \end{bmatrix}$

(c) $A^{-1} = \begin{bmatrix} -2 & 1 \\ 3/2 & -1/2 \end{bmatrix}$

(d) $\det A = -2$

17. If $V\mathbf{c} = \mathbf{0}$ then, for $k \in \{0, 1, \ldots, m\}$, $c_0 + c_1 x_k + \cdots + c_m x_k^m = 0$. This means that, for each k, x_k is a root of the same degree m polynomial. The only way that a polynomial of degree m can have $m + 1$ roots is for it to be the zero polynomial. This means that the column vectors of V are linearly independent and so V is invertible. Thus V's row vectors are also linearly independent.

18. Use induction on m. For the first step on the matrix V, multiply each column i by x_0 and subtract the result from column $i + 1$. This does not change the value of the determinant. The first row of the resulting matrix has a 1 in the first entry and 0's elsewhere. Now, expand by the first row. There is only one minor to consider. Factor $x_j - x_0$ from the jth row of the matrix corresponding to this minor. The determinant calculation at this point is

$$\prod_{j=1}^{m}(x_j - x_0)\det \begin{bmatrix} 1 & x_1 & \cdots & x_1^{m-1} \\ \vdots & & \ddots & \vdots \\ 1 & x_m & \cdots & x_m^{m-1} \end{bmatrix}.$$

Continue this process to complete the argument.

19. Let $\mathbf{g} = (g(1), g(2), \ldots, g(n))$. Let $\mathbf{f} = P_n\mathbf{g}$. Then $\mathbf{g} = P_n^{-1}\mathbf{f}$. The forward direction for Theorem 2 is just row n of this matrix argument. The backward direction is similar.

20. In both cases this follows directly from Identity 26.

21. Let $f(n) = b_n$ and $g(k) = a_k$ in Theorem 2. Then, for $1 \le n \le m$, the left equation in Theorem 2 can be expressed in matrix form as $\mathbf{b} = P_m\mathbf{a}$, where $\mathbf{a}$ and $\mathbf{b}$ are vectors in $\mathbb{R}^{m+1}$. Since P_m is invertible, we have $\mathbf{a} = P_m^{-1}\mathbf{b}$. The right side of Theorem 2 gives the coefficients needed to transform $\mathbf{b}$ into $\mathbf{a}$, which must therefore be the entries in P_m^{-1} (as it is unique).

22. Let $r = m$ in Vandermonde's identity. Then rewrite $\binom{m}{m-k}$ as $\binom{m}{m}$ and $\binom{n+m}{m}$ as $\binom{n+m}{n}$.

23. By Identity 308 and the fact that P_m and R_m are lower and upper-triangular matrices, respectively, $Q_m = P_m R_m$. Thus $\det Q_m = \det(P_m R_m) = (\det P_m)(\det R_m)$. Since P_m and R_m are both triangular matrices, $\det P_m = \det R_m = 1$.

24. Since $a(i,j) = b(i,j) = 0$ when $j > i$, entry (i,j) in the product $A_m B_m$ when $j > i$ is necessarily 0. The fact that $a(i,j) = b(i,j) = 0$ when $j > i$ plus $\sum_{k=0}^{i} a(i,k)b(k,j) = [i = j]$ means that (1) for $i < j$, entry (i,j) in the product $A_m B_m$ is 0, and (2) entry (i,i) is 1.

25. The (i,j) entry in $S_m(x)T_m(x)$ is given by

$$\sum_{k=0}^{m} \begin{bmatrix} i \\ k \end{bmatrix} x^{i-k} \begin{Bmatrix} k \\ j \end{Bmatrix} x^{k-j}.$$

Simplify and apply Identity 241.

26. Starting with the left side of Identity 313, rewrite the Lah numbers in terms of binomial coefficients. Cancel $k!$, shift indices on the sum, and apply Identity 50.

27. First, show that $\mathcal{F}^{-1} = \begin{bmatrix} 0 & 1 \\ 1 & -1 \end{bmatrix}$, which is easy to do using standard techniques for inverting 2×2 matrices. Then use an induction argument like the one in the proof of Identity 315 for nonnegative n.

28. The argument for the value of A_n is a straight induction proof using the Fibonacci recurrence (Identity 170). The two determinant properties stated in the exercise mean that $\det A_n = -\det(A_{n-1})$. With $\det A_0 = 1$, we get $\det A_n = (-1)^n$.

29. Reverse the process of constructing A_n from A_{n-1} to show how to construct A_{-1} from A_0, A_{-2} from A_{-1}, and, in general, A_{n-1} from A_n. In other words, to construct A_{n-1} from A_n you need to swap the two rows of A_n and then subtract the second row of A_n from its first row. Using this process, we still have $A_n = \begin{bmatrix} F_{n-1} & F_n \\ F_n & F_{n+1} \end{bmatrix}$ and $\det A_n = (-1)^n$ for negative n.

Chapter 10

1. (a) Yes.

   (b) Yes.

   (c) Yes.

(d) Yes, $\cos(k\pi) = (-1)^k$, which is hypergeometric.

(e) Yes, $k! + (k+1)! = k!(k+2)$, which is hypergeometric.

(f) Yes, $\binom{n}{k} + \binom{n}{k+1} = \binom{n+1}{k+1}$ (Identity 1), which is hypergeometric.

(g) Yes.

(h) Not necessarily. For example, ϕ^k and ψ^k are both hypergeometric, but $\phi^k + \psi^k$ is not. (See the Fibonacci example in the text.)

2. The hypergeometric series

$$_{p+1}F_{q+1}\left[\begin{matrix} a_1, & \ldots, & a_p, & 1 \\ b_1, & \ldots, & b_q, & 1 \end{matrix}; x\right] = \sum_{k=0}^{\infty} \frac{(a_1)^{\overline{k}} \cdots (a_p)^{\overline{k}} 1^{\overline{k}}}{(b_1)^{\overline{k}} \cdots (b_q)^{\overline{k}} 1^{\overline{k}}} \frac{x^k}{k!}$$

is the same as the one in the problem statement.

3. The alternative hypergeometric representations for these functions are in the text, actually.

   (a) This is $\sum_{k=0}^{\infty} \frac{x^k}{k!} = e^x = {}_1F_1\left[\begin{matrix} 1 \\ 1 \end{matrix}; x\right]$.

   (b) This is $\sum_{k=0}^{\infty} x^k = \frac{1}{1-x} = {}_2F_1\left[\begin{matrix} 1, & 1 \\ 1 \end{matrix}; x\right]$.

4. Rewrite the hypergeometric series using Equation (10.1), differentiate, reindex the series, and use the fact that $c^{\overline{k+1}} = c(c+1)^{\overline{k}}$.

5. For all of these, construct the ratio t_{k+1}/t_k to find the a_i and b_j parameters. Remember that if there is no factor of $k+1$ in the denominator of t_{k+1}/t_k, you'll need to multiply by $(k+1)/(k+1)$ to ensure that this factor is present.

   (a) This one is done in the text.

   (b) This is fairly straightforward, but remember to rewrite $(2k+3)(2k+2)$ as $4(k+3/2)(k+1)$.

   (c) This is similar to part (b).

   (d) The $\binom{2k}{k}$ term is handled like the Catalan example in the text.

   (e) This is fairly straightforward.

   (f) Since there is no term with $k = 0$, reindex the infinite series as $\sum_{k=0}^{\infty}(-1)^k x^{k+1}/(k+1)$ before working with the ratio t_{k+1}/t_k .

6. This is very similar to the proof of Identity 330 given in the text. For a check, $s(k) = -k/2 - 1/4$.

7. Suppose $s(k)$ is of the form $s(k) = \sum_{j=0}^{d} \alpha_j k^j$. We need to show that these α_j's exist and that $d = -n - 1$. By expansion of $s(k+1)$, the leading coefficient of $-s(k+1) - s(k)$ involves just α_d. By setting this equal to the leading coefficient of $(k-n-1)^{\underline{-n-1}}$ we see that d must equal $-n-1$.

We can also find α_{-n-1}. Then, by expanding $s(k+1)$, we can see that the coefficient of k^{-n-2} in $-s(k+1) - s(k)$ involves α_{-n-1} and α_{-n-2}. Since we have already found the former, we can set this equation equal to the coefficient of k^{-n-2} in $(k-n-1)^{\underline{-n-1}}$ and solve for α_{-n-2}. Similarly, the coefficient of k^{-n-3} in $-s(k+1) - s(k)$ involves α_{-n-1}, α_{-n-2}, and α_{-n-3}, allowing us to solve for α_{-n-3}. Continuing in this vein, we can solve for all the α_j's.

From another point of view, if we collected in matrix form all the equations needed to find the α_j's, we would have a $-n \times -n$ triangular matrix. Thus there must be unique solutions for the $-n$ values of the α_j's.

8. Use Gosper's algorithm and follow the argument in the text leading up to the statement of Identity 330. At this point you're looking for a polynomial $s(k)$ such that $(k-n-1)^{\underline{-n-1}} = -s(k+1) - s(k)$. The existence of this polynomial and the fact that it has degree $-n-1$ are given by Exercise 7. Let the leading coefficient of $s(k)$ be α_{-n-1}. Since the leading coefficient of $(k-n-1)^{\underline{-n-1}}$ is 1, you get $\alpha_{-n-1} = -1/2$.

Step 4 of Gosper's algorithm says

$$T_k = \frac{s(k)\binom{n}{k}}{(k-n-1)^{\underline{-n-1}}}.$$

Use Identity 19 to rewrite $\binom{n}{k}$ and cancel with $(k-n-1)^{\underline{-n-1}}$, leaving $(-1)^k/(-n-1)!$. Let $S(k) = -s(k)$, and you're done.

9. This is similar to Exercise 8, but there's additional work needed to find the coefficient of k^{-n-2}. With $s(k) = -k^{\underline{-n-1}}/2 + \alpha_{-n-2}k^{-n-2}$, the coefficient of k^{-n-2} in $-s(k+1) - s(k)$ is $-2\alpha_{-n-2} + (-n-1)/2$.

To find the coefficient of k^{-n-2} in $(k-n-1)^{\underline{-n-1}}$, write out the factors as $(k-n-1)(k-n-2)\cdots(k+1)$. Since the second coefficient in a polynomial $\prod_{j=1}^{m}(x+a_j)$ is the sum of the a_j's (multiply out the polynomial to check this), we have that the coefficient of k^{-n-2} in $(k-n-1)^{\underline{-n-1}}$ is (by Identity 59) $1 + 2 + \cdots + (-n-1) = -n(-n-1)/2$. Setting this to $-2\alpha_{-n-2} + (-n-1)/2$ and solving yields $\alpha_{-n-2} = -(n+1)^2/4$. Finally, let $S(k) = -s(k)$.

10. This proceeds like the argument in the text for $t_k = \binom{n}{k}$, except that we get $q(k) = k - n$. We still have $r(k) = k$. With n not a negative integer, the condition in Step 1 of Gosper's algorithm is satisfied, and so we move to Step 2. We need $s(k)$ satisfying

$$1 = (k-n)s(k+1) - ks(k).$$

If the leading term in $s(k)$ is $\alpha_d k^d$, the leading term of $(k-n)s(k+1)$ is $\alpha_d k^{d+1}$, and the leading term of $-ks(k)$ is $-\alpha_d k^{d+1}$. Since these add to 0, the leading term of $(k-n)s(k+1) - ks(k)$ must actually be $(\alpha_{d-1} + (d -$

$n)\alpha_d - \alpha_{d-1})k^d = (d - n)\alpha_d$. Since this must equal 1, we have $d = 0$ and $s(k) = \alpha_0 = -1/n$.

With Step 4, then, we have

$$T_k = \frac{k\binom{n}{k}(-1)^k}{-n} = \binom{n-1}{k-1}(-1)^{k-1},$$

by the absorption identity (Identity 6). With this antidifference,

$$\sum_{k=0}^{m} \binom{n}{k}(-1)^k = T_{m+1} - T_0 = \binom{n-1}{m}(-1)^m.$$

11. This proceeds much like the argument in the text for $t_k = \binom{n}{k}$ in the case where n is a negative integer. The difference is that $q(k) = 1$; we still have $r(k) = 1$ and $p(k) = (k - n - 1)\frac{-n-1}{}$. Then we need to find $s(k)$ satisfying

$$(k - n - 1)\frac{-n-1}{} = s(k+1) - s(k) = \Delta s(k).$$

As the text indicates, Identity 177 is helpful here, with $x = k - n - 1$, and we get $s(k) = (k - n - 1)\frac{-n}{}/(-n)$. Then

$$T_k = \frac{(k - n - 1)\frac{-n}{}\binom{n}{k}(-1)^k}{(-n)(k - n - 1)\frac{-n-1}{}} = \frac{k\binom{n}{k}(-1)^k}{-n},$$

and the rest of the argument proceeds as in Exercise 10.

12. With $t_k = \binom{n+k}{k}$, we get $t_{k+1}/t_k = (n + k + 1)/(k + 1)$. Letting $r(k) = k$ and $q(k) = k + n + 1$, the condition in Step 1 of Gosper's algorithm is satisfied as long as n is not 0 or a positive integer. From here the argument proceeds much like that in the proof of Identity 58 in this chapter. We get $s(k) = 1/(n+1)$ and the antidifference T_k satisfying $T_k = \binom{n+k}{k-1}$. Finally, $\sum_{k=0}^{m} \binom{n+k}{k} = T_{m+1} - T_0 = \binom{n+m+1}{m}$, which is what we want.

13. As in Exercise 12, with $t_k = \binom{n+k}{k}$ we get $t_{k+1}/t_k = (n + k + 1)/(k + 1)$. However, the assignment $r(k) = k$ and $q(k) = k + n + 1$ does not work, as the condition in Step 1 of Gosper's algorithm is not satisfied when n is a positive integer. Follow the three-step procedure described in the text for handling this situation to get $r(k) = q(k) = 1$ and $p(k) = (k + n)\underline{n}$. To solve $(k + n)\underline{n} = s(k + 1) - s(k)$, use Identity 177; the solution is $s(k) = (k + n)\underline{n+1}/(n + 1)$. Then

$$T_k = \frac{(k + n)\underline{n+1}\binom{n+k}{k}}{(n + 1)(k + n)\underline{n}},$$

which simplifies to $T_k = \binom{n+k}{k-1}$, just as in Exercise 12.

15. This is similar to the proof of Identity 58 given in the text. You'll get $p(k) = 1$, $q(k) = k - m + 1$, and $r(k) = k$, leading to $s(k) = -1/(m-1)$. Then

$$T_k = \frac{-k}{(m-1)\binom{k}{m}}.$$

Thus

$$\sum_{k=m}^{n} \frac{1}{\binom{k}{m}} = T_{n+1} - T_m,$$

and taking the limit as n goes to ∞ yields $m/(m-1)$.

16. Following the steps of Gosper-Zeilberger, we get $p(n, k) = \beta_0(n)$ and $\bar{t}_{n,k} = \binom{n}{2k}$. In Step 3, $\bar{p}(n, k) = 1$, $q(n, k) = (n - 2k)(n - 2k - 1)$, and $r(n, k) = (2k)(2k - 1)$. Solving the equation in Step 4 leads to the same situation as in the $l = 0$ iteration for the proof of Identity 154: Either $d = -1$ or $s(n, k) = \beta_0(n) = 0$, neither of which is allowed.

17. Following the steps of Gosper-Zeilberger, we get $p(n, k) = \beta_0(n)$ and $\bar{t}_{n,k} = \binom{n}{k}\frac{1}{k+1}$. In Step 3, $\bar{p}(n, k) = 1$, $q(n, k) = n - k$, and $r(n, k) = k+1$. Solving the equation in Step 4 leads to same problem as in Exercise 16: Either $d = -1$ or $s(n, k) = \beta_0(n) = 0$, neither of which is allowed.

18. This can be proved with repeated application of Identity 1.

19. Iteration $l = 0$ fails. With iteration $l = 1$, $p(n, k) = (n+1-k)\beta_0(n)+(n+1)\beta_1(n)$ and $\bar{t}_{n,k} = \binom{n}{k}k/(n+1-k)$. Then $\bar{p}(n, k) = 1$, $q(n, k) = n+1-k$, and $r(k) = k - 1$. Then $s(n, k) = 1$, $\beta_0(n) = 2$, and $\beta_1(n) = -n/(n+1)$. This leads to the recurrence $2S(n) = nS(n+1)/(n+1)$. Thus $S(0) = 0$, and $S(n+1) = 2(n+1)S(n)/n$ for $n \geq 1$. With $S(1) = 1$, one can unroll the recurrence to solve it.

20. Iteration $l = 0$ fails. With iteration $l = 1$, $p(n, k) = (n+1-k)\beta_0(n)+(n+1)\beta_1(n)$ and $\bar{t}_{n,k} = \binom{n}{k}k^2/(n+1-k)$. Then $\bar{p}(n, k) = k$ (note that this is different from all the examples in the text!), $q(n, k) = n+1-k$, and $r(n, k) = k-1$. One solution to the resulting equation is $s(n, k) = k - (n+1)/(n+2)$, $\beta_0(n) = 2$, $\beta_1(n) = -1+2/(n+2)$. This leads to the recurrence $nS(n+1) = 2(n+2)S(n)$. So $S(0) = 0$, and $S(n+1) = 2(n+2)S(n)/n$ for $n \geq 1$. With $S(1) = 1$, one can unroll the recurrence to solve it.

21. Iteration $l = 0$ fails. With iteration $l = 1$, $p(n, k) = (n+1-k)\beta_0(n)+(n+1)\beta_1(n)$ and $\bar{t}_{n,k} = \binom{n}{k}\binom{k}{m}/(n+1-k)$. Then $\bar{p}(n, k) = 1$, $q(n, k) = n+1-k$, and $r(n, k) = k - m$. A solution of the resulting equation is $s(n, k) = 1$, $\beta_0(n) = 2$, $\beta_1(n) = -1+m/(n+1)$. This leads to the recurrence $2(n+1)S(n) = (n+1-m)S(n+1)$. Since $S(0) = 0$, we get $S(n) = 0$ for $1 \leq n \leq m - 1$ as well. But $S(m) = 1$, and unrolling the recurrence $S(n+1) = 2(n+1)S(n)/(n+1-m)$ gives $S(n) = 2^{n-m}n^{\underline{n-m}}/(n-m)!$, which is what we want.

22. We have $p(n,k) = \beta_0(n)$ and $\bar{t}_{n,k} = \binom{n}{k}(-1)^k/(k+1)$. Then $\bar{p}(n,k) = 1$, $q(n,k) = k-n$, and $r(n,k) = k+1$. A solution of the resulting equation is $s(n,k) = 1$, $\beta_0(n) = -(n+1)$. Then $T(n,k)$ simplifies to $\binom{n}{k}(-1)^k$, and the rest follows from some algebra.

23. Here, $t_{n,k} = \binom{n}{k}(-1)^k/(2k+1)$. Iteration $l=0$ fails. With iteration $l=1$, $p(n,k) = (n+1-k)\beta_0(n) + (n+1)\beta_1(n)$ and $\bar{t}_{n,k} = t_{n,k}/(n+1-k)$. Then $\bar{p}(n,k) = 1$, $q(n,k) = (k-n-1)(2k+1)$, and $r(n,k) = k(2k+1)$. This has $a-b=0$ for $a = 1/2$, $b = 1/2$, which is fine. A solution of the resulting equation is $s(n,k) = 1$, $\beta_0(n) = -2(n+1)$, $\beta_1(n) = 1 + 2(n+1)$. This leads to the recurrence $S(n+1) = 2(n+1)S(n)/(2n+3)$, with $S(0) = 1$. Unrolling the recurrence, we have

$$S(n) = \frac{2^n n!}{3(5)\cdots(2n+1)},$$

which can be seen to equal the right side of Identity 148 after multiplying the numerator and denominator by $2^n n!$.

Bibliography

[1] Martin Aigner. *A Course in Enumeration*. Springer, 2010.

[2] Désiré André. Solution directe du problème résolu par M. Bertrand. *Comptes Rendus de l'Académie des Sciences, Paris*, 105:436–437, 1887.

[3] Valerio De Angelis. Pairings and signed permutations. *American Mathematical Monthly*, 113(7):642–644, 2006.

[4] Arthur T. Benjamin and Naiomi T. Cameron. Counting on determinants. *American Mathematical Monthly*, 112(6):481–492, 2005.

[5] Arthur T. Benjamin, Naiomi T. Cameron, Jennifer J. Quinn, and Carl R. Yerger. Catalan determinants—a combinatorial approach. *Congressus Numerantum*, 200:27–34, 2010.

[6] Arthur T. Benjamin, Bob Chen, and Kimberly Kindred. Sums of evenly spaced binomial coefficients. *Mathematics Magazine*, 83(5):370–373, 2010.

[7] Arthur T. Benjamin and Jennifer J. Quinn. *Proofs that Really Count*. MAA, 2003.

[8] Arthur T. Benjamin and Jacob N. Scott. Third and fourth binomial coefficients. *Fibonacci Quarterly*, 49(2):99–101, 2011.

[9] Jacob Bernoulli. *Ars Conjectandi*. Basel, 1713.

[10] J. Bertrand. Solution d'un problème. *Comptes Rendus de l'Académie des Sciences, Paris*, 105:369, 1998.

[11] Paul Blanchard, Robert L. Devaney, and Glen R. Hall. *Differential Equations*. Brooks/Cole, Boston, MA, fourth edition, 2012.

[12] Khristo N. Boyadzhiev. Binomial transform and the backward difference. *Advances and Applications in Discrete Mathematics*, 13:43–63, 2014.

[13] Robert Brawer and Magnus Pirovino. The linear algebra of the Pascal matrix. *Linear Algebra and Its Applications*, 174:13–23, 1992.

[14] Gregory S. Call and Daniel J. Velleman. Pascal's matrices. *American Mathematical Monthly*, 100(4):372–376, 1993.

[15] Charalambos A. Charalambides. *Enumerative Combinatorics*. CRC Press, Boca Raton, Florida, 2002.

[16] Young-Ming Chen. The Chung-Feller theorem revisited. *Discrete Mathematics*, 308(7):1328–1329, 2008.

[17] Gi-Sang Cheon and Jin-Soo Kim. Stirling matrix via Pascal matrix. *Linear Algebra and Its Applications*, 329:49–59, 2001.

[18] Kai Lai Chung and W. Feller. On fluctuations in coin tossing. *Proceedings of the National Academy of Sciences of the USA*, 35:605–608, 1949.

[19] Siad Daboul, Jan Mangaldan, Michael Z. Spivey, and Peter Taylor. The Lah numbers and the nth derivative of $e^{1/x}$. *Mathematics Magazine*, 86(1):39–47, 2013.

[20] Yingpu Deng and Yanbin Pan. The sum of binomial coefficients and integer factorization. *Integers*, 16, 2106. Article A42.

[21] Keith Devlin. Cracking the DaVinci code. *Discover*, pages 65–69, June 2004.

[22] John D. Dixon and Brian Mortimer. *Permutation Groups*. Springer, 1996.

[23] S. Douady and Y. Couder. Phyllotaxis as a dynamical self organizing process part I: The spiral modes resulting from time-periodic iterations. *Journal of Theoretical Biology*, 178:255–274, 1996.

[24] Peter Doubilet, Gian-Carlo Rota, and Richard Stanley. On the foundations of combinatorial theory VI: The idea of generating function. In *Proc. Sixth Berkeley Symp. on Math. Statist. and Prob.*, volume 2, pages 267–318. University of California Press, 1972.

[25] Alan Edelman and Gilbert Strang. Pascal matrices. *American Mathematical Monthly*, 111(3):189–197, 2004.

[26] William Feller. *An Introduction to Probability Theory and Its Applications*, volume I. John Wiley & Sons, New York, second edition, 1957.

[27] J. Fernando Barbero G., Jesús Salas, and Eduardo J. S. Villaseñor. Bivariate generating functions for a class of linear recurrences: General structure. *Journal of Combinatorial Theory, Series A*, 125:146–165, 2014.

[28] Frank R. Giordano, William P. Fox, and Steven B. Horton. *A First Course in Mathematical Modeling*. Cengage, fifth edition, 2014.

[29] R. William Gosper. Decision procedure for indefinite hypergeometric summation. *Proceedings of the National Academy of Sciences of the United States of America*, 75:40–42, 1978.

[30] Henry W. Gould. *Combinatorial Identities*. Morgantown Printing, Morgantown, WV, 1972.

[31] H.W. Gould. Series transformations for finding recurrences for sequences. *Fibonacci Quarterly*, 28(2):166–171, 1990.

[32] Ronald L. Graham, Donald E. Knuth, and Oren Patashnik. *Concrete Mathematics*. Addison-Wesley, Reading, MA, second edition, 1994.

[33] Ralph P. Grimaldi. *Fibonacci and Catalan Numbers: An Introduction*. Wiley, 2012.

[34] Geoffrey Grimmett and David Stirzaker. *Probability and Random Processes*. Oxford University Press, third edition, 2001.

[35] David R. Guichard. Sums of selected binomial coefficients. *College Mathematics Journal*, 26(3):209–213, 1995.

[36] Pentti Haukkanen. Formal power series for binomial sums of sequences of numbers. *Fibonacci Quarterly*, 31(1):28–31, 1993.

[37] Katherine Humphreys. A history and survey of lattice path enumeration. *Journal of Statistical Planning and Inference*, 140:2237–2254, 2010.

[38] Luke A.D. Hutchison, Natalie M. Myres, and Scott R. Woodward. Growing the family tree: The power of DNA in reconstructing family relationships. In *Proceedings of the First Symposium on Bioinformatics and Biotechnology (BIOT-04)*, pages 42–49, September 2004.

[39] Charles Jordan. *Calculus of Finite Differences*. Chelsea, New York, third edition, 1965.

[40] Thomas Koshy. *Fibonacci and Lucas Numbers with Applications*. John Wiley & Sons, New York, 2001.

[41] Thomas Koshy. *Catalan Numbers with Applications*. Oxford University Press, 2009.

[42] D. H. Lehmer. Interesting series involving the central binomial coefficient. *American Mathematical Monthly*, 92(7):449–457, 1985.

[43] Mario Livio. *The Golden Ratio: The Story of PHI, the World's Most Astonishing Number*. Broadway Books, New York, 2003.

[44] Nicholas A. Loehr and T. S. Michael. The combinatorics of evenly spaced binomial coefficients. *Integers*, 18, 2018. Article A89.

[45] Jeffrey Lutgen. A combinatorial proof of an alternating convolution identity for multichoose numbers. *American Mathematical Monthly*, 123(2):192–196, 2016.

[46] George Markowsky. Misconceptions about the golden ratio. *The College Mathematics Journal*, 23(1):2–19, January 1992.

[47] István Mező. The dual of Spivey's Bell number formula. *Journal of Integer Sequences*, 15(2), 2012. Article 12.2.4.

[48] Ronald E. Mickens. *Difference Equations: Theory, Applications and Advanced Topics*. CRC Press, Boca Raton, FL, third edition, 2015.

[49] Sri Gopal Mohanty. *Lattice Path Counting and Applications*. Academic Press, New York, 1979.

[50] Erich Neuwirth. Recursively defined combinatorial functions: extending Galton's board. *Discrete Mathematics*, 239(1–3):33–51, 2001.

[51] Jonathon Peterson. A probabilistic proof of a binomial identity. *American Mathematical Monthly*, 120:558–562, 2013.

[52] Marko Petkovšek, Herbert S. Wilf, and Doron Zeilberger. $A = B$. A. K. Peters, Wellesley, MA, 1996.

[53] Helmut Prodinger. Some information about the binomial transform. *Fibonacci Quarterly*, 32(5):412–415, 1994.

[54] James Propp. Exponentiation and Euler measure. *Algebra Universalis*, 29(4):459–471, 2003.

[55] Jocelyn Quaintance and H. W. Gould. *Combinatorial Identities for Stirling Numbers*. World Scientific, 2016.

[56] Marc Renault. Lost and found in translation: André's *actual* method and its application to the generalized ballot problem. *American Mathematical Monthly*, 115(4):358–363, 2008.

[57] John Riordan. *Combinatorial Identities*. John Wiley & Sons, New York, 1968.

[58] Kenneth H. Rosen, editor. *Handbook of Discrete and Combinatorial Mathematics*. CRC Press, Boca Raton, Florida, 2000.

[59] Sheldon M. Ross. *A First Course in Probability*. Pearson, ninth edition, 2012.

[60] Neil J. A. Sloane. *A Handbook of Integer Sequences*. Academic Press, New York, 1973.

[61] Neil J. A. Sloane and Simon Plouffe. *The Encyclopedia of Integer Sequences*. Academic Press, San Diego, CA, 1995.

[62] Michael Z. Spivey. Fibonacci identities via the determinant sum property. *College Mathematics Journal*, 37(4):286–289, 2006.

[63] Michael Z. Spivey. Combinatorial sums and finite differences. *Discrete Mathematics*, 307(24):3130–3146, 2007.

[64] Michael Z. Spivey. A generalized recurrence for Bell numbers. *Journal of Integer Sequences*, 11(2), 2008. Article 08.2.5.

[65] Michael Z. Spivey. On solutions to a general combinatorial recurrence. *Journal of Integer Sequences*, 14(9), 2011. Article 11.9.7.

[66] Michael Z. Spivey. Enumerating lattice paths touching or crossing the diagonal at a given number of lattice points. *Electronic Journal of Combinatorics*, 19(3), 2012. Article P24.

[67] Michael Z. Spivey. A combinatorial proof for the alternating convolution of the central binomial coefficients. *American Mathematical Monthly*, 121(6):537–540, 2014.

[68] Michael Z. Spivey. The binomial recurrence. *Mathematics Magazine*, 89(3):192–195, 2016.

[69] Michael Z. Spivey. The Chu-Vandermonde identity via Leibniz's identity for derivatives. *College Mathematics Journal*, 47(3):219–220, 2016.

[70] Michael Z. Spivey. Probabilistic proofs of a binomial identity, its inverse, and generalizations. *American Mathematical Monthly*, 123(2):175–180, 2016.

[71] Michael Z. Spivey and Laura L. Steil. The k-binomial transforms and the Hankel transform. *Journal of Integer Sequences*, 9(1), 2006. Article 06.1.1.

[72] Michael Z. Spivey and Andrew M. Zimmer. Symmetric polynomials, Pascal matrices, and Stirling matrices. *Linear Algebra and Its Applications*, 428(4):1127–1134, 2008.

[73] Renzo Sprugnoli. Riordan arrays and combinatorial sums. *Discrete Mathematics*, 132:267–290, 1994.

[74] Renzo Sprugnoli. Sums of reciprocals of the central binomial coefficients. *Integers*, 6, 2006. Article A27.

[75] Richard P. Stanley. *Enumerative Combinatorics*, volume I. Cambridge University Press, second edition, 2012.

[76] Richard P. Stanley. *Catalan Numbers*. Cambridge University Press, 2015.

[77] Marta Sved. Counting and recounting: The aftermath. *Mathematical Intelligencer*, 6(4):44–46, 1984.

[78] M. Werman and D. Zeilberger. A bijective proof of Cassini's Fibonacci identity. *Discrete Mathematics*, 58:109, 1986.

[79] Herbert S. Wilf. *Generatingfunctionology*. Academic Press, Boston, second edition, 1994.

[80] Herbert S. Wilf and Doron Zeilberger. Rational functions certify combinatorial identities. *Journal of the American Mathematical Society*, 3:147–158, 1990.

[81] Sheng-liang Yang and Zhan-ke Qiao. Stirling matrix and its property. *International Journal of Applied Mathematics*, 14(2):145–157, 2003.

[82] Doron Zeilberger. The method of creative telescoping. *Journal of Symbolic Computation*, 11:195–204, 1991.

[83] Zhizheng Zhang. The linear algebra of the generalized Pascal matrix. *Linear Algebra and Its Applications*, 250:51–60, 1997.

Index of Identities and Theorems

343

Identity 294. $\sum_{k\geq 0} \binom{n}{2k+1} k(-1)^k = 2^{(n-3)/2} \left(n\cos\frac{(n-1)\pi}{4} - \sqrt{2}\sin\frac{n\pi}{4}\right)$, 247, 261, 329

Identity 295. For real $y > 0$,
$$\sum_{k=0}^n \binom{n}{k}\frac{(i^{k+1}-x^{k+1})y^{n-k}}{k+1} = \frac{(y^2+1)^{(n+1)/2}e^{i(n+1)\arctan(1/y)}-(x+y)^{n+1}}{n+1},$$
247

Identity 296. For real $x > 0$,
$$\sum_{k\geq 0} \binom{n}{2k}\frac{x^k(-1)^k}{2k+1} = \frac{(x+1)^{(n+1)/2}\sin((n+1)\arctan\sqrt{x})}{\sqrt{x}(n+1)},\ 247\text{--}248,\ 329$$

Identity 297. For real $x > 0$, $\sum_{k\geq 0} \binom{n}{2k+1}\frac{x^k(-1)^k}{k+1} =$
$$\frac{2}{x(n+1)}\left(1 - (x+1)^{(n+1)/2}\cos\left((n+1)\arctan\sqrt{x}\right)\right),\ 247\text{--}248,\ 329$$

Identity 298. $\sum_{k\geq 0} \binom{n}{2k}\frac{(-1)^k}{2k+1} = \frac{2^{(n+1)/2}}{n+1}\sin\frac{(n+1)\pi}{4}$, 248

Identity 299. $\sum_{k\geq 0} \binom{n}{2k+1}\frac{(-1)^k}{k+1} = \frac{2}{n+1}\left(1 - 2^{(n+1)/2}\cos\frac{(n+1)\pi}{4}\right)$, 248

Identity 300. $\cos nx = \sum_{k\geq 0} \binom{n}{2k}\sin^{2k}x\cos^{n-2k}x\,(-1)^k$, 248

Identity 301. $\sin nx = \sum_{k\geq 0} \binom{n}{2k+1}\sin^{2k+1}x\cos^{n-2k-1}x\,(-1)^k$, 248

Identity 302. $\sum_{k=0}^{r-1} \zeta^{sk} = r[r|s]$, 249–250

Identity 303. $\sum_{k\geq 0} \binom{n}{rk+m}x^{n-rk-m} = \frac{1}{r}\sum_{j=0}^{r-1}(\zeta^j + x)^n(\zeta^j)^{-m}$, 249–251, 265, 329

Identity 304. $\sum_{k\geq 0} \binom{n}{3k} = \frac{1}{3}\left(2^n + 2\cos\frac{n\pi}{3}\right)$, 250, 265, 329

Identity 305. $\sum_{k\geq 0} \binom{n}{3k+1} = \frac{1}{3}(2^n + 2\cos\frac{(n-2)\pi}{3})$, 250, 261

Identity 306. $\sum_{k\geq 0} \binom{n}{3k+2} = \frac{1}{3}(2^n + 2\cos\frac{(n+2)\pi}{3})$, 250, 262

Identity 307. For $x \in \mathbb{Z}$, $P_m(x) = P_m^x$, 253–254

Identity 308. $PR = Q$, 256, 331

Identity 309. $\sum_{k=0}^n \binom{n}{k}\binom{m}{k} = \binom{n+m}{n}$, 256, 264

Identity 310. $S_m^{-1} = T_m(-1)$, 256–257

Identity 311. $T_m^{-1} = S_m(-1)$, 257

Identity 312. $L_m^{-1} = L_m(-1)$, 257

Identity 313. $\sum_{k=0}^n L(n,k)L(k,m)x^{n-k}y^{k-m} = (x+y)^{n-m}L(n,m)$, 258, 264, 331

Identity 314. For $x \in \mathbb{Z}$, $L_m^x = L_m(x)$, 258

Identity 315. For integers n, $\mathcal{F}^n = \begin{bmatrix} F_{n+1} & F_n \\ F_n & F_{n-1} \end{bmatrix}$, 259, 264, 331

Identity 316 (Cassini's Identity). For integers n,
$$F_{n+1}F_{n-1} - F_n^2 = (-1)^n,\ 241,\ 259,\ 265,\ 266$$

Identity 317. $\sum_{k\geq 0} \binom{n}{2k}3^k(-1)^k = 2^n\cos\frac{n\pi}{3}$, 260

Identity 318. $\sum_{k\geq 0} \binom{n}{2k+1}3^k(-1)^k = \frac{2^n}{\sqrt{3}}\sin\frac{n\pi}{3}$, 260

Identity 319. $\sum_{k\geq 0} \binom{n}{2k}k(-1)^k = n2^{n/2-2}(\cos\frac{n\pi}{4} - \sin\frac{n\pi}{4})$, 261

Identity 320. $\sum_{k\geq 0} \binom{n}{2k+1}k(-1)^k = 2^{n/2-2}\left(n\cos\frac{n\pi}{4} + (n-2)\sin\frac{n\pi}{4}\right)$, 261

Identity 321. For real $y > 0$,
$$\sum_{k=0}^n \binom{n}{k}k^{\underline{m}}i^k y^{n-k} = i^m n^{\underline{m}}(1 + y^2)^{(n-m)/2}e^{i(n-m)\arctan(i/y)},\ 261$$

Identity 322. $\sum_{k\geq 0} \binom{n}{2k}\frac{3^k(-1)^k}{2k+1} = \frac{2^{n+1}}{\sqrt{3}(n+1)}\sin\frac{(n+1)\pi}{3}$, 261

Theorem 25. The Stirling numbers of the first kind $\left[{n \atop k}\right]$ count the number of ways that a set of size n can be partitioned into k nonempty cycles, 196–199, 207, 324, 326

Theorem 26. For functions f and g,
$$f(n) = \sum_{k=0}^{n} \left\{{n \atop k}\right\} g(k)(-1)^k \iff g(n) = \sum_{k=0}^{n} \left[{n \atop k}\right] f(k)(-1)^k, \; 201,$$
206, 207, 256, 324

Theorem 27. The value of ϖ_n is the number of ways to partition a set of n elements, 202–203, 325

Theorem 28. The Lah numbers $L(n, k)$ count the number of ways that a set of size n can be partitioned into k nonempty tuples, 204–207

Theorem 29. For functions f and g, $f(n) = \sum_{k=0}^{n} L(n, k) g(k)(-1)^k \iff g(n) = \sum_{k=0}^{n} L(n, k) f(k)(-1)^k$, 206–207, 256, 326

Theorem 30. For $n \geq 1$, the number of lattice paths from $(0, 0)$ to (n, n) that do not go above or touch the main diagonal except at the starting and ending points is C_{n-1}, 209–210, 215, 218

Theorem 31 (Ballot Theorem). The number of lattice paths from $(0, 0)$ to (n, m), $n > m$, that touch the main diagonal only at $(0, 0)$ is given by $\frac{n-m}{n+m}\binom{n+m}{n}$, 216–219, 236, 238, 327

Theorem 32. The number of lattice paths from $(0, 0)$ to (n, n) that touch the main diagonal exactly k times after $(0, 0)$ without going above it is $\frac{k}{2n-k}\binom{2n-k}{n}$, 218–220, 238

Theorem 33. The number of lattice paths from $(0, 0)$ to (n, n) that touch the main diagonal exactly k times after $(0, 0)$ is $\frac{k2^k}{2n-k}\binom{2n-k}{n}$, 220–221, 238

Theorem 34. For $n \geq 2$, the number of ways to triangulate a convex n-gon is C_{n-2}, 221–224

Theorem 35. For $n \geq 0$, the number of binary trees on n vertices is C_n, 223–224

Theorem 36 (Weak Ballot Theorem). The number of lattice paths from $(0, 0)$ to (n, m), $n \geq m$, that never go above the main diagonal is given by $\frac{n+1-m}{n+1}\binom{n+m}{n}$, 236, 327

Theorem 37. P_m^{-1} is the $(m + 1) \times (m + 1)$ matrix whose (i, j) entry is $\binom{i}{j}(-1)^{i-j}$, 241, 252–253, 257, 263

Theorem 38. The series $\sum_{k=0}^{\infty} t_k$ is hypergeometric if and only if $t_0 = 1$ and there exist nonzero polynomials $P(k)$ and $Q(k)$ such that, for all k, $Q(k) t_{k+1} = P(k) t_k$, 269–270

Main Index